Workshop Statistics

Second Edition

NOTES

T 1-3   45/45

T 1-5   35/35

T 6-8   50/50

Note to Instructors

For desk copies and information about supplementary materials to accompany this text, details can be found at the following addresses:

Authors' Web Sites:
http://www.rossmanchance.com/ws2gc
 Materials include downloadable lists and programs for use with TI calculators, Instructor's Guide, and more.
Key College Publishing can be contacted at:
 1-888-877-7240 or at http://www.keycollege.com

Workshop Statistics

Discovery with Data and the Graphing Calculator

Second Edition

Allan J. Rossman

DICKINSON COLLEGE

Beth L. Chance

CALIFORNIA POLYTECHNIC STATE UNIVERSITY
SAN LUIS OBISPO

J. Barr von Oehsen

CLEMSON UNIVERSITY

 **Key College Publishing**
Innovators in Higher Education

www.keycollege.com

Allan J. Rossman
Beth L. Chance
California Polytechnic State University
San Luis Obispo, CA 93407

J. Barr von Oehsen
Clemson University
Clemson, SC 29634-0910

Key College Publishing opened its doors in 1999 as a division of Key Curriculum Press in cooperation with Springer-Verlag New York. Key College Publishing publishes texts and courseware in mathematics, statistics, and mathematics and statistics education, concentrating on innovation in content, curriculum, and delivery.

Key College Publishing
1150 65th Street
Emeryvillle, CA 94608
www.keycollege.com
(510) 595-7000

Library of Congress Cataloging-in-Publication Data
Rossman, Allan J.
    Workshop statistics : discovery with data & the graphing calculator.—2nd ed. / Allan J.
  Rossman, Beth L. Chance, J. Barr von Oehsen.
      p.   cm.
    Includes index.
    ISBN 1-930190-05-0 (alk. paper) — ISBN 1-930190-04-2 (pbk : alk. paper)
    1. Mathematical statistics.   I. Chance, Beth L.   II. Title.
  QA276.12.R6728   2001
  001.4´22—dc21   2001029897

The first edition of *Workshop Statistics: Discovery with Data and the Graphing Calculator* was published in © 1997 by Springer-Verlag New York, Inc.

Editorial Director: Richard J. Bonacci
Production Editor: Lesley Poliner
Manufacturing Manager: Jacqui Ashri
Cover Design: Joseph Piliero
Cover Illustration: Roy Weimann
Composition: Beth L. Chance using FrameMaker files.
Printer/Binder: Hamilton Printing Co., Rensselaer, NY

Printed in the United States of America.

9 8 7 6 5 4 3 2   04 03 02

ISBN 1-930190-05-0 Key College Publishing SPIN 10764583 (hardcover)
ISBN 1-930190-04-2 Key College Publishing SPIN 10764575 (softcover)

To Eileen, Frank, and Shari

Contents

Preface to the Second Edition

Before we describe the changes and new features in this second edition, we want to highlight what has not changed: *Workshop Statistics* aims to provide students and instructors with a self-contained, learner-centered resource of activities through which students can discover statistical concepts, explore statistical properties, and apply statistical techniques. The features that distinguished the first edition, detailed in the original preface reprinted below, have been retained: emphases on active learning, conceptual understanding, genuine data, and use of calculators.

WHAT'S NEW

The changes in this second edition are substantial. They have been informed by the reactions and suggestions of instructors and students who have used the first edition, as well as by our own experiences and by our study of the research literature regarding students' learning of statistics.

New Topics

New topics have been added on sampling distributions and the Central Limit Theorem for sample means, chi-square tests for two-way tables, and inference for correlation and regression. Many other topics are given a much fuller treatment, such as probability and experimental design. Concepts and techniques that have been added within topics include mean absolute deviation, relative risk, stratification, and blocking.

New and Revised Activities

Many of the in-class activities have been rewritten and new ones developed to better focus students' attention on the statistical principles being introduced. Many of the new activities address conceptual understanding in particular. Numerous homework activities have been added to allow instructors more choices in activities to assign.

New and Updated Data

Many of the time-dependent data sets in the first edition, most recorded in the early 1990s, have been updated to more current values. These include Senators' and Justices' years of service, Hollywood and Broadway box office revenues, golfers' winnings, Presidential election results, governors' salaries, and baseball results, to name just a few. The use of hypothetical datasets has been decreased.

New Organization

The book continues to be arranged around six units, each consisting of multiple topics. Some of these units and topics have been reorganized. Unit II addresses comparisons as well as relationships, with relationships between categorical variables moved earlier and considered as comparisons. The issue of data collection is now the focus of the third unit. The unit on principles of inference now encompasses inference for population means as well as proportions, eliminating the need for a unit on inference for measurements. The final unit presents inference for both comparisons and relationships.

New Formatting

As in the first edition, expository paragraphs are interspersed among the activities to reinforce the ideas that students are to garner from the activities. The more important of these expository passages have been highlighted within boxes to ensure that students do not miss the most important ideas. The header now helps students and instructors to find activities by number more easily, and page references are now given for the numerous activities that refer to previous ones.

TECHNOLOGY AND DATA

Students are intended to use technology with this book, both as a tool for analyzing data and as a vehicle through which to explore statistical concepts. Roughly half of the activities require the use of technology. This verison of the book assumes that students will use a graphing calculator.

Instructions for using a Texas Instruments calculator, specifically a TI-83 or TI-83 Plus, have been integrated into the activities. By no means do we intend this book to be a user's manual for the calculator. Rather, the focus is on helping students to use the calculator as an aid in discovering statistical concepts and exploring statistical principles.

We intend for this text to be flexible enough to support a variety of implementations by instructors. With few exceptions we recommend that students investigate a concept through hand-drawn displays and calculator-assisted calculations from formulas before proceeding to use the statistical functions of the graphing calculator to check their work and explore larger data sets. In most homework activities, whether or not a student should use a graphing calculator has been left to the instructor's discretion. Our goal has been to provide detailed enough instructions that students can complete activities using a graphing calculator outside of class time with a minimum of instructor support.

All of the data sets in the book are available for downloading from the web in a variety of formats, so there is no need to type in data by hand. Several activities also rely on TI programs that are also available on the web. See the publisher's web site at www.keycollege.com or the authors' web site at www.rossmanchance.com/ws/ for links to these data files and programs.

ACCOMPANYING RESOURCES

Instructors and students who use this book should be aware of a variety of supporting materials that are available. A complete Guide for Instructors is also available, as well as solutions to selected activities, sample syllabi, and sample exams. As mentioned above, downloadable data sets are also available, as are some Java applets to accompany selected activities. Please check the publisher's web site at www.keycollege.com or the authors' web site at www.rossmanchance.com/ws/ for links to these materials.

This second edition of *Workshop Statistics* is available in four versions. For the most part, the topics and activities in the various versions are identical. The differences are the technology-specific instructions that accompany the activities in the Minitab®, Graphing Calculator, and Fathom™ versions, providing detailed instructions appropriate to the specific software package or calculator. Also available is *Workshop Statistics: Discovery with Data, A Bayesian Approach* by James H. Albert and Allan J. Rossman.

Acknowledgments

We gratefully acknowledge the very helpful feedback that we have received on the first edition of *Workshop Statistics* from the following instructors:

Jim Albert	John Zhang	Jean Werner	Charles Bertness
Chuck Biehl	Skip Allis	Patricia Bassett	Gordon Bril
Marilyn Byers	Jim Bohan	Cheri Boyd	Al Coons
Carolyn Cuff	Julie Clark	Benjamin Collins	Carolyn Dobler
Clark Engel	Christine Czapleski	L.J. Davis	Brian Gray
Dorothea Grimm	Christa Fratto	Steve Friedberg	Anne Kaufman
Bruce King	Bill Halteman	Alice Hankla	Todd Lee
Jerry Moreno	Larry Langley	Suzanne Larson	Bill Rinaman
Charlie Scheim	Sue Peters	Gina Reed	Bernie Schroeder
Joanne Schweinsberg	Ned Schillow	Brian Schott	Sue Suran
Sam Tumulo	Sallie Scudder	Mike Seyfried	Don Weimer
Rhonda Weissman	Kathryn Voit	Sheila Young	Thomas Zachariah

We also thank Dickinson College students Jason Herr and Mary Joan LaFrance for their assistance with compiling data for the second edition, Lisa Mannarelli for her help with the web solutions, Matthew Weber for designing and setting up the web site of teacher resources, and Kari Lock for compiling the TI data files for the web site. We especially thank Robin Lock for his careful reading and valuable suggestions on this manuscript, and David Kramer for his copyediting.

ALLAN J. ROSSMAN
BETH L. CHANCE
J. BARR VON OEHSEN
April 2001

Preface to the First Edition

> Shorn of all subtlety and led naked out of the protective fold of educational research literature, there comes a sheepish little fact: lectures don't work nearly as well as many of us would like to think.
>
> —George Cobb (1992)

This book contains activities that guide students to discover statistical concepts, explore statistical principles, and apply statistical techniques. Students work toward these goals through the analysis of genuine data and through interaction with each other, with their instructor, and with technology. Providing a one-semester introduction to fundamental ideas of statistics for college and advanced high school students, *Workshop Statistics* is designed for courses that employ an interactive learning environment by replacing lectures with hands-on activities. The text contains enough expository material to stand alone, but it can also be used to supplement a more traditional textbook.

Some distinguishing features of *Workshop Statistics* are its emphases on active learning, conceptual understanding, genuine data, and the use of technology. The following sections of this preface elaborate on each of these aspects and also describe the unusual organizational structure of this text.

ACTIVE LEARNING

> Statistics teaching can be more effective if teachers determine what it is they really want students to know and to do as a result of their course, and then provide activities designed to develop the performance they desire.
>
> —Joan Garfield (1995)

This text is written for use with the workshop pedagogical approach, which fosters active learning by minimizing lectures and eliminating the conventional distinction between laboratory and lecture sessions. The book's activities require students to collect data, make predictions, read about studies, analyze data, discuss findings, and write explanations. The instructor's responsibilities in this setting are to check students' progress, ask and answer questions, lead class discussions, and deliver "mini-lectures" where appropriate. The essential point is that every student is actively engaged with learning the material through reading, thinking, discussing, computing, interpreting, writing, and reflecting. In this manner students construct their own knowledge of statistical ideas as they work through the activities.

The activities also lend themselves to collaborative learning. Students can work together through the book's activities, helping each other to think through the material. Some activities specifically call for collaborative effort through the pooling of class data.

This text also stresses the importance of students' communication skills. As students work through the activities, they constantly read, write, and talk with each other. Students should be encouraged to write their explanations and conclusions in full, grammatically correct sentences, as if to an educated layperson.

CONCEPTUAL UNDERSTANDING

> Almost any statistics course can be improved by more emphasis on data and on concepts at the expense of less theory and fewer recipes.
>
> —David Moore (1992)

This text focuses on "big ideas" of statistics, paying less attention to details that often divert students' attention from larger issues. Little emphasis is placed on numerical and symbolic manipulations. Rather, the activities lead students to explore the meaning of concepts such as variability, distribution, outlier, tendency, association, randomness, sampling, sampling distribution, confidence, significance, and experimental design. Students investigate these concepts by experimenting with data, often with the help of technology. Many of the activities challenge students to demonstrate their understanding of statistical issues by asking for explanations and interpretations rather than mere calculations.

To deepen students' understandings of fundamental ideas, the text presents these ideas repetitively. For example, students return to techniques of exploratory data analysis when studying properties of randomness and also in conjunction with inference procedures. They also encounter issues of data collection not just when studying randomness but also when investigating statistical inference.

GENUINE DATA

> We believe that data should be at the heart of all statistics
> education and that students should be introduced to statistics
> through data-centered courses.
> —*Thomas Moore and Rosemary Roberts (1989)*

The workshop approach is ideally suited to the study of statistics, the science of reasoning from data, for it forces students to be actively engaged with genuine data. Analyzing genuine data not only exposes students to what the practice of statistics is all about, it also prompts them to consider the wide applicability of statistical methods and often enhances their enjoyment of the material.

Some activities ask students to analyze data that they collect in class about themselves, while most present students with genuine data from a variety of sources. Many questions in the text ask students to make predictions about data before conducting their analyses. This practice motivates students to view data not as naked numbers but as numbers with a context, to identify personally with the data, and to take an interest in the results of their analyses.

The data sets in *Workshop Statistics* do not concentrate in one academic area but come from a variety of fields of application. These fields include law, medicine, economics, psychology, political science, and education. Many examples come not from academic disciplines but from popular culture. Specific examples therefore range from such pressing issues as testing the drug AZT and assessing evidence in sexual discrimination cases to less crucial ones of predicting basketball salaries and ranking *Star Trek* episodes.

USE OF TECHNOLOGY

> Automate calculation and graphics as much as possible.
> —*David Moore (1992)*

The text assumes that students using this text have access to technology for creating visual displays, performing calculations, and conducting simulations. The preferable technology is a statistical software package, although a graphing calculator can do almost as well. Roughly half of the activities ask students to use technology; students typically perform small-scale displays, calculations, and simulations by hand before letting the computer or calculator take over those mechanical chores.

This workshop approach employs technology in three distinct ways. First, technology performs the calculations and presents the visual displays necessary to analyze genuine data sets that are often large and cumbersome. Second, technology conducts simulations that allow students to visualize and explore the long-term behavior of sample statistics under repeated random sampling.

The most distinctive use of technology with the workshop approach is to enable students to explore statistical phenomena. Students make predictions about a particular statistical property and then use the computer to investigate their predictions, revising their predictions and iterating the process as necessary. For example, students use technology to investigate the effects of outliers on various summary statistics and the effects of sample sizes on confidence intervals.

Activities requiring the use of technology are integrated throughout the text, reinforcing the idea that technology is not to be studied for its own sake but as an indispensable tool for analyzing genuine data and a convenient device for exploring statistical phenomena.

Specific needs of the technology are to create visual displays (dotplots, histograms, boxplots, scatterplots), calculate summary statistics (mean, median, quartiles, standard deviation, correlation), conduct simulations (with binary variables), and perform inference procedures (z-tests and z-intervals for binary variables, t-tests and t-intervals for measurement variables).

ORGANIZATION

> Judge a statistics book by its exercises, and you cannot go
> far wrong.
> —George Cobb (1987)

For the most part this text covers traditional subject matter for a first course in statistics. The first two units concern descriptive and exploratory data analysis, the third introduces randomness and probability, and the final three delve into statistical inference. The six units of course material are divided into smaller topics, each topic following the same structure:

- *Overview:* a brief introduction to the topic, particularly emphasizing its connection to earlier topics;
- *Objectives:* a listing of specific goals for students to achieve in the topic;
- *Preliminaries:* a series of questions designed to get students thinking about issues and applications to be studied in the topic and often to collect data on themselves;
- *In-class activities:* the activities that guide students to learn the material for the topic;
- *Homework activities:* the activities that test students' understanding of the material and ability to apply what they have learned in the topic;
- *Wrap-up:* a brief review of the major ideas of the topic emphasizing its connection to future topics.

In keeping with the spirit of the workshop approach, hands-on activities dominate the book. Preliminary questions and in-class activities leave enough space for students to record answers in the text itself. While comments and explanations are interspersed among the activities, these passages of exposition are purposefully less thorough than in traditional textbooks. The text contains very few solved examples, further emphasizing the idea that students construct their own knowledge of statistical ideas as they work through the activities.

While the organization of content is fairly standard, unusual features include:

- Probability is not treated formally but is introduced through simulations. The simulations give students an intuitive sense of random variation and the idea that probability represents the proportion of times that something would happen in the long run. Since students often have trouble connecting the computer simulation with the underlying process that it models, the text first asks students to perform physical simulations involving dice and candies to help them understand the process being modeled.

- The Central Limit Theorem and the reasoning of statistical inference are introduced in the context of a population *proportion* rather than a population *mean*. A population proportion summarizes all of the relevant information about the population of a binary variable, allowing students to concentrate more easily on the concepts of sampling distribution, confidence, and significance. These ideas are introduced through physical and computer simulations that are easier to conduct with binary variables than with measurement variables. Dealing with binary variables also eliminates the need to consider issues such as the underlying shape of the population distribution and the choice of an appropriate parameter.

- Exploratory data analysis and data production issues are emphasized throughout, even in the units covering statistical inference. Most activities that call for the application of inference procedures first ask students to conduct an exploratory analysis of the data; these analyses often reveal much that the inference procedures do not. These activities also guide students to question the design of the study before drawing conclusions from the inference results. Examples used early in the text to illustrate Simpson's paradox and biased sampling reappear in the context of inference, reminding students to be cautious when drawing conclusions.

Acknowledgments

I am privileged to teach at Dickinson College, where I enjoy an ideal atmosphere for experimenting with innovative pedagogical strategies and curriculum development. I thank my many colleagues and students who have helped me in writing this book. Nancy Baxter Hastings has directed the Workshop Mathematics Program, of which *Workshop Statistics* forms a part, with assistance from Ruth Rossow, Joanne Weissman, and Sherrill Goodlive. Barry Tesman, Jack Stodghill, Peter Martin, and Jackie Ford have taught with the book and provided valuable feedback, as have Barr von Oehsen of Piedmont College and Kevin Callahan of California State University at Hayward. Students who have contributed in many ways include Dale Usner, Kathy Reynolds, Christa Fratto, Matthew Parks, and Jennifer Becker. I also thank Dean George Allan for his leadership in establishing the productive teaching/learning environment that I enjoy at Dickinson.

I appreciate the support given to the Workshop Mathematics Program by the Fund for the Improvement of Post-Secondary Education, U.S. Department of Education, and by the National Science Foundation, as well as by Dickinson College. I also thank Springer-Verlag for their support, particularly Jerry Lyons, Liesl Gibson, and Steve Pisano. I thank Sara Buchan for help with proofreading.

Much of what I have learned about statistics education has been shaped by the writings from which I quote above. I especially thank Joan Garfield, George Cobb, Tom Short, and Joel Greenhouse for many enlightening conversations.

Finally, I thank my wonderful wife, Eileen, without whose support and encouragement I would not have completed this work. Thanks also to my feline friends Eponine and Cosette.

ALLAN J. ROSSMAN
December 1995

References

Cobb, George W. (1987), "Introductory Textbooks: A Framework for Evaluation," *Journal of the American Statistical Association* 82, 321–339.

Cobb, George W. (1992), "Teaching Statistics," in *Heeding the Call for Change: Suggestions for Curricular Action*, ed. Lynn Steen, MAA Notes Number 22, 3–43.

Garfield, Joan (1995), "How Students Learn Statistics," *International Statistical Review* 63, 25–34.

Moore, David S. (1992), "Teaching Statistics as a Respectable Subject," in *Statistics for the Twenty-First Century*, eds. Florence and Sheldon Gordon, MAA Notes Number 26, 14–25.

Moore, Thomas L. and Rosemary A. Roberts (1989), "Statistics at Liberal Arts Colleges," *The American Statistician* 43, 80–85.

To the Instructor

We want to emphasize from the outset that there is no one "right" way to teach with this book. We hope that it will prove useful to students and instructors in a wide variety of settings. Naturally, we think that the text will work best in a classroom environment that promotes the features extolled in the preface: active learning, conceptual understanding, genuine data, and use of technology.

The following suggestions are based on our own experiences and on those of many instructors who have taught with the original version of *Workshop Statistics:*

1. Take control of the course.

While this may seem obvious, we feel that the "control" needed in the course differs from the traditional lecture setting but is still quite important. Students need to see that the instructor is monitoring and facilitating the progress of the course and that there is a pedagogical purpose behind all of the classroom activities.

2. Keep the class roughly together.

Part of the control that needs to be taken is to keep the students roughly together with the material, not letting some groups get too far ahead while others lag far behind.

3. Allow students to discover.

We encourage you to resist the temptation to tell students too much. Rather, let them discover the ideas and conduct analyses for themselves, while you point them in the right direction as needed. This principle of self-discovery enables students to construct their own knowledge, ideally leading to a deeper understanding of fundamental ideas and a heightened ability to apply these ideas beyond this course.

4. Promote collaborative learning among students.

This course provides a natural occasion for encouraging students to work in groups, allowing them to collaborate and learn from each other as well from you and the book.

5. Encourage students' guessing and development of intuition.

We believe that much can be gained by asking students to think and make predictions about issues and data before detailed analysis. We urge you to give students time to think about and respond to "Preliminaries" questions in the hope that these questions lead students to care more about the data they will analyze, as well as to gradually develop their own statistical intuition.

6. Lecture when appropriate.

By no means do we propose that you never speak to the class as a whole. In many circumstances interrupting the class for a "mini-lecture" is appropriate and important. As a general rule, though, we advocate lecturing on an idea only after students have begun to grapple with it first themselves.

7. Have students do some work by hand.

While we believe strongly in using technology to explore statistical phenomena as well as to analyze genuine data, we think that students have much to gain by first performing small-scale analyses by hand. We feel particularly strongly about this point in the context of simulations, where students can better comprehend the process of simulation through physical examples before proceeding to computer simulations. We also encourage instructors to assign a mixture of problems to be solved by hand and with the computer.

8. Use technology as a tool.

The counterbalance to the previous suggestion is that students should come to regard technology as an invaluable tool both for analyzing data and for studying statistics. After you have given students the chance to do some small-scale displays and calculations by hand, we suggest that you then encourage them to use technology to alleviate their computational burdens.

9. Be proactive in approaching students.

As your students work through the activities, we strongly suggest that you not wait for them to approach you with questions. Rather, approach them to check their work and provide quick feedback.

10. Give students access to "right" answers.

Some students are fearful of a self-discovery approach because they worry about discovering the "wrong" things. We appreciate this objection, and feel that it makes a strong case for giving students regular and consistent feedback, including access to right answers.

11. *Provide plenty of feedback.*

This suggestion closely follows the two previous ones about being proactive and providing "right" answers. An instructor can supply much more personalized, in-class feedback with this "workshop" approach than in a traditional lecture setting.

12. *Stress good writing.*

We regard writing-to-learn as an important aspect of *Workshop Statistics*, although it is certainly a feature that students resist. Many activities call for students to write interpretations and explanations of their findings, and we urge you to insist that students relate these to the context at hand.

13. *Implore students to read well.*

Students can do themselves a great service by taking their time and reading carefully. By reading directions and questions well, students can better know what is expected in an activity. Moreover, the book's expository passages interspersed among the activities contain a great deal of information that is essential for students to understand.

14. *Have fun!*

We sincerely hope that you and your students will enjoy a dynamic and productive learning environment as you study with *Workshop Statistics*.

15. *Make use of our web resources.*

Our web sites contain a wide variety of resources ranging from downloadable worksheets and programs to an instructor's guide to sample exams to Java applets. They also contain links to sites of other instructors using *Workshop Statistics*. They can be found at:

http://www.keycollege.com/
http://www.rossmanchance.com/ws/

To the Student

We hope that you will find statistics to be both an important and an engaging subject to study. We want you to be aware of three principles that guided our writing of this book; the first two relate to the study of statistics generally, and the third pertains to the distinctive nature of this book:

1. Statistics is not number-crunching.

Contrary to its popular perception, statistics involves much more than numerical computations. In this book you will be asked to concentrate on understanding statistical concepts and on interpreting and communicating the results of statistical analyses. In other words, you will be expected to learn to construct and analyze numerical arguments, using data to support your statements. In contrast to most mathematics courses, you will be using phrases such as "there is strong evidence that..." and "the data suggest that..." rather than "the exact answer is..." and "it is therefore proven that...." In order to allow you to better focus on this understanding and communication, you will use technology to alleviate computational drudgery. Technology can also help present the ideas in a visual, interactive environment, allowing you to more easily understand the concepts and their properties.

2. Statistics involves the analysis of genuine data.

Supporting our contention that statistics is applicable in everyday life and in most fields of academic endeavor, you will analyze genuine data from a wide variety of applications throughout the course. Many of these data sets involve information that you will collect about yourselves and your peers; others will come from sources such as almanacs, journals, magazines, newspapers, and books. We hope that by spanning a wide variety of subject matter, the contexts will be of interest to a general audience.

3. Understanding results from investigation and discovery.

The structure of this text asks you to spend most of your time actively engaged with the material as opposed to passively taking notes. The activities have been carefully designed and tested to lead you to discover fundamental statistical ideas for yourself, in collaboration with your peers, your instructor, and the calculator.

You should try to read very carefully, particularly the expository passages of the text that present the most important pieces of information, to supplement the knowledge that you construct.

Our advice to you for success in this course can be summed up in two words: *think* and *participate.* This course will ask you to think critically and to defend your arguments. Moreover, you will be asked to make guesses and collect data and draw conclusions and write summaries and discuss findings and explore alternatives and investigate scenarios and … . You must have an open and active mind in order to complete these tasks; in other words, you must accept responsibility for your own learning. Our responsibility as authors has been to provide you with a resource that will facilitate this learning process and lead you on the path toward understanding statistics.

Two final words of advice: *Have fun!* We sincerely hope that you will enjoy a dynamic and interactive learning environment as you study statistics!

List of Activities

NOTE: In-Class Activities appear in **boldface**.

Unit One

Exploring Data: Distributions

$$\begin{array}{c}\rule{2cm}{1pt}\blacksquare\rule{2cm}{1pt}\end{array}$$

Topic 1:

DATA AND VARIABLES

OVERVIEW

Statistics is the science of reasoning from **data**, so a natural place to begin your study is by examining what is meant by the term "data." The most fundamental principle in statistics is that of **variability**. Indeed, if the world were perfectly predictable and showed no variability, there would be no reason to study statistics. Thus, you will also discover the notion of a **variable** and consider different classifications of variables. You will also begin to explore the notion of the **distribution** of a set of data measuring a particular variable.

OBJECTIVES

- To begin to appreciate that **data** are numbers with a context that are studied for a purpose.
- To learn to recognize different classifications of **variables**.
- To become familiar with the fundamental concept of **variability**.
- To discover the notion of the **distribution** of a variable.
- To encounter **bar graphs** and **dotplots** as visual displays of a distribution.
- To gain some exposure to the types of questions that statistical reasoning addresses.
- To begin to gather experience describing distributions of data verbally.

PRELIMINARIES

1. Write one sentence (please make it a complete, grammatically correct sentence!) explaining your primary reason for taking this course.

2. Since you will be expected to participate actively in your own learning in this course, please sign your name (first and last name, as you would on a check) below to certify that you accept this responsibility.

3. Since this course is about analyzing data, please gather some data by measuring your signature's length (left to right) and height (at its highest point), and record these values in centimeters below. Also record how many letters are in the names that you wrote.

   length: height: letters:

4. Place a check beside each state that you have visited (or lived in or even just driven through), and count how many states you have visited.

State	Visited?	State	Visited?	State	Visited?	State	Visited?
Alabama		Indiana		Nebraska		South Carolina	
Alaska		Iowa		Nevada		South Dakota	
Arizona		Kansas		New Hampshire		Tennessee	
Arkansas		Kentucky		New Jersey		Texas	
California		Louisiana		New Mexico		Utah	
Colorado		Maine		New York		Vermont	
Connecticut		Maryland		North Carolina		Virginia	
Delaware		Massachusetts		North Dakota		Washington	
Florida		Michigan		Ohio		West Virginia	
Georgia		Minnesota		Oklahoma		Wisconsin	
Hawaii		Mississippi		Oregon		Wyoming	
Idaho		Missouri		Pennsylvania			
Illinois		Montana		Rhode Island			

   number of states visited:

5. Take a wild guess as to the number of different *states* that have been visited (or lived in) by a typical student at your school. Also guess what the fewest and most states visited by the students in your class will be.

fewest: typical: most:

6. Record in a table like the one below the following information concerning each student in this class:
  • gender
  • length, height, and number of letters in signature
  • number of states visited
  • responses to the following
  (1) Which of the following terms best describes your political views: liberal, moderate, or conservative?
  (2) Do you think that the United States should retain or abolish the penny as a coin of currency?
  (3) Rate your opinion of the value of statistics in society on a numerical scale of 1 (completely useless) to 9 (incredibly important).

student	gender	length	height	letters	states	politics	penny	value
1								
2								
3								
4								
:								

IN-CLASS ACTIVITIES

Activity 1-1: Word Lengths

(a) For each word that you wrote in response to question 1, record the number of letters in the word:

The numbers that you have recorded are **data**. Not all numbers are data, however. Data are numbers collected in a particular context. For example, the numbers 3 and 7, 35 and 19 do not constitute data in and of themselves. They are data, however, if they refer to the number of letters in your first two words or the number of states visited by two of the students in this class.

(b) Did every word that you wrote contain the same number of letters?

The answer to this obvious question reveals the most fundamental principle of statistics: **variability**. Data vary, and variability abounds both in everyday life and in academic study. The students in this class vary with respect to gender, length of signature, number of states visited, and lots of other variables.

> A **variable** is any characteristic of a person or thing that can be assigned a number or a category. The person or thing to which the number or category is assigned is called the **observational unit** or case. A **quantitative** or measurement variable is typically one that measures a numerical characteristic, while a **categorical** or qualitative variable is one that simply records a category designation. **Binary** variables are categorical variables for which only two possible categories exist. These designations can be quite important, for one typically employs different statistical tools depending on the type of variable measured.

Activity 1-2: Types of Variables

Consider the students in your class as observational units.

(a) Which of the following are legitimate variables that can be measured on those observational units? [*Hint*: Ask yourself whether the value can change from observational unit to observational unit.] For each that is a variable, indicate whether it is a quantitative or a categorical variable. If it is a categorical variable, indicate whether or not it is a binary variable.

- hair color:

- number of students with red hair:

- height of tallest student in your class:

- whether or not a student has red hair:

- height:

- instructor's age:

- zip code of home town:

(b) If the observational units had been classes at your school, would "number of students in the class with red hair" be a variable? Explain.

As the term "variable" suggests, the values assumed by a variable can differ from observational unit to observational unit. For example, hair color differs from person to person, and whether or not a student has red hair differs as well. However, the number of students with red hair in your class summarizes the variable for all students in your class and so does not vary from student to student.

(c) The observational units for the variables listed below are students in your class. For each variable, indicate whether it is a quantitative or a categorical variable. If it is a categorical variable, indicate whether or not it is a binary variable.

- gender:

- length of signature:

- number of states visited:

- political identification:

- penny question:

(d) What are the observational units for the "letters per word" variable on which you recorded data in Activity 1-1 on page 5?

(e) Suppose that instead of recording the number of letters in each word of your sentence, you had been asked to classify each word according to the following criteria:

1–3 letters:	small word
4–6 letters:	medium word
7–9 letters:	big word
10 or more letters:	very big word

In this case, what type of variable is "size of word"?

(f) Considering the states as the observational units in question 4 of the "Preliminaries" section, what type of variable is "whether or not you have visited the state"?

> As the term "variable" suggests, the values assumed by a variable differ from observational unit to observational unit. In other words, data display **variability**. The *pattern* of this variability is called the **distribution** of the variable. Much of the practice of statistics concerns distributions of variables, from displaying them visually to summarizing them numerically to describing them verbally.

Activity 1-3: Penny Thoughts

(a) How many students responded to the question about whether the United States should retain or abolish the penny? How many of these voted to retain the penny? What proportion of the respondents is this?

(b) How many and what proportion of the respondents voted to abolish the penny?

(c) Create a visual display of this distribution of responses by drawing rectangles whose heights correspond to the proportions voting for each option.

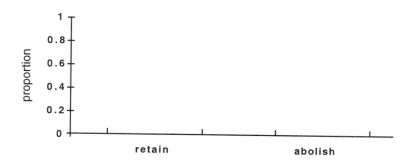

> The visual display that you have constructed above is called a **bar graph**. Bar graphs display the distribution of categorical variables.

(d) Write a sentence or two describing what your analysis reveals about the attitudes of students in this class toward the penny.

Activity 1-4: Value of Statistics

When a quantitative variable can assume a fairly small number of possible values, one can **tally** the data by counting the **frequency** of each possible numerical response.

Consider the question of students ratings of the value of statistics in society on a numerical scale of 1 to 9.

(a) Tally the responses by counting how many students answered 1, how many answered 2, and so on. Record the results in the table below:

rating	1	2	3	4	5	6	7	8	9
tally (count)									

(b) Is there one value that was chosen more than any other? If so, what is it?

(c) How many and what proportion of students gave a response (strictly) above 5? Below 5?

(d) Based on these data, write a sentence or two interpreting how your class generally seems to feel about the value of statistics in society. Specifically comment on the degree to which these students seem to be in agreement. Also address whether students seem to be generally optimistic, pessimistic, or undecided about the value of statistics.

(e) Consider the following frequency tables for hypothetical classes A–E. (All empty cells should be considered counts of zero.) For each of the descriptions below, identify which of the class tables fits the description best. Match each class to one description.

rating	1	2	3	4	5	6	7	8	9
Class A count	1	4	7	1		1	3	5	2
Class B count	2	3	6	5	2	1	2	3	
Class C count						1	19	4	
Class D count		2	3		2	5	6	2	4
Class E count	2	3	1	3	2	4	5	1	3

- The class is in considerable agreement that it is useful. Class: _____
- The class generally feels that statistics is useful but to varying degrees and with a few disagreements.   Class: _____

- •The class displays a wide range of opinions, with a slight preference toward feeling that statistics is useful. Class: _____
- •The class is sharply divided on the issue. Class: _____
- •The class generally feels that statistics is not useful but displays a range of opinions. Class: _____

_____

Activity 1-5: Students' Travels

(a) Create a visual display of the distribution of the numbers of states visited. A horizontal scale has been drawn below; you are to place a dot for each student above the appropriate number of states visited. For repeated values, just stack the dots on top of each other.

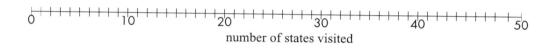

number of states visited

> The visual display that you constructed for the states visited is a ***dotplot***. Dotplots are useful for displaying the distribution of relatively small data sets of quantitative variables.

(b) Circle your own value on the dotplot. Where do you seem to fall in relation to your peers with regard to number of states visited?

(c) Based on this display, comment on the accuracy of your guesses in the "Preliminaries" section.

(d) Write a paragraph of at least four sentences describing various features of the distribution of states visited. Imagine that you are trying to explain what this distribution looks like to someone who cannot see the display and has absolutely no idea about how many states people visit. (*Advice:* Here and throughout the course, please relate your comments to the context. Remember that data are not just numbers but numbers with a context, in this case the numbers of states visited by you and your classmates.)

Activity 1-6: Parents' Ages

The following dotplots display the distributions of the ages at which a sample of 200 mothers had their first child and the ages at which a sample of 200 fathers had their first child.

(a) Trace over the shape of each graph with a smooth curve. Comment on any similarities and differences in the distributions of ages between these first-time mothers and fathers.

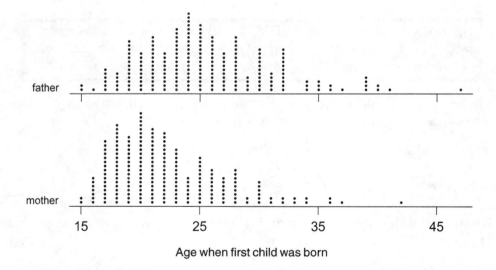

Age when first child was born

Comments:

(b) Identify an age such that roughly half of the mothers were older and half were younger than that age when they had their first child (i.e., find the middle age). Then do the same thing for fathers. Comment on how these ages compare to each other and what that reveals.

mothers: fathers:

WRAP-UP

Since statistics is the science of **data**, this topic has tried to give you a sense of what data are and a glimpse of what data analysis entails. Data are not mere numbers: Data are collected for some purpose and have meaning in some context. The guessing exercises in these activities have not been simply for your amusement; they have tried to instill in you the inclination to *think* about data in their context and to anticipate reasonable values for the data to be collected and analyzed.

You have encountered two very important concepts in this topic that will be central to the entire course: **variability** and **distribution**. You have also learned to distinguish between **quantitative** and **categorical** variables. These activities have also hinted at a fundamental principle of data analysis: One should always begin analyzing data by looking at a visual display (i.e., a "picture") of the data. You have discovered two simple techniques for producing such displays: **bar graphs** for categorical variables and **dotplots** for quantitative variables.

The next topic will continue your study of data and variables by introducing you to the use of the graphing calculator, to the idea of manipulating variables, and to the issue of whether a variable adequately measures what it purports to.

HOMEWORK ACTIVITIES

Activity 1-7: Types of Variables (*cont.*)

Suppose that each of the following is a variable that you are to measure for each student in this class. Indicate whether it is a quantitative variable or a categorical variable; if it is categorical, indicate whether it is also binary.

(a) height
(b) armspan
(c) ratio of height to armspan
(d) time spent sleeping last night
(e) whether or not the individual went to sleep before midnight last night
(f) month of birth
(g) numerical score (out of a possible 100 points) on the first exam in this course
(h) whether or not the individual scores at least 70 points on the first exam in this course
(i) distance from home
(j) whether the individual has a cellular phone
(k) how many e-mail messages the person has sent or received in the last 24 hours
(l) whether the person has sent at least one e-mail message in the last 24 hours
(m) the number of letters in the last name

Activity 1-8: Types of Variables (*cont.*)

For each of the following variables, indicate whether it is a quantitative variable or a categorical (possibly binary) variable. Also identify the observational unit (case) involved. (You will encounter each of these variables later in the book.)

(a) whether a spun penny lands "heads" or "tails"
(b) the color of a Reese's Pieces candy
(c) the number of calories in a fast food sandwich
(d) the life expectancy of a nation
(e) whether an American household owns a cat or does not own a cat
(f) the year in which a college was founded
(g) the comprehensive fee charged by a college
(h) for whom an American voted in the 1996 Presidential election
(i) whether or not a newborn baby tests HIV-positive
(j) the running time of an Alfred Hitchcock movie
(k) the age of an American penny
(l) the weight of an automobile
(m) whether an automobile is foreign or domestic to the United States
(n) the classification of an automobile as small, midsize, or large
(o) whether or not an applicant for graduate school is accepted
(p) the occupational background of a Civil War general
(q) whether or not an American child lives with both parents
(r) whether a college student has abstained from the use of alcohol for the past month
(s) whether or not a participant in a sport suffers an injury in a given year
(t) a sport's injury rate per 1000 participants
(u) a state's rate of automobile thefts per 1000 residents
(v) the airfare to a selected city from Baltimore
(w) the average low temperature in January for a city

(**x**) the age of a bride on her wedding day

(**y**) whether the bride is older, younger, or the same age as the groom in a wedding couple

(**z**) the difference in ages (groom's age minus bride's age) of a wedding couple

Activity 1-9: Types of Variables (*cont.*)

In 1998, the American Film Institute selected the 100 best American films of all time. The following variables use these films as observational units. Identify the type of each of the following variables.

(**a**) year produced

(**b**) number of years since production

(**c**) decade produced

(**d**) whether or not it was produced before 1960

(**e**) whether it won an Academy Award for Best Picture

(**f**) whether or not you have seen it

(**g**) the number of people in your class who have seen it

Activity 1-10: Variables of State

Suppose that the fifty states are the observational units (cases) of interest. Identify which of the following are legitimate variables and which are not. For those that are variables, report or make a guess as to the value of the variable for your home state. (Checking whether you can do this is a good way to determine whether it is a legitimate variable.) For those that are not variables, provide a brief explanation.

(**a**) the number of states that have a female governor

(**b**) the percentage of the state's residents over 65 years of age

(**c**) the highest speed limit in the state

(**d**) whether or not the state's name consists of one word

(**e**) the average income of an adult resident of the state

(**f**) the number of births per 1000 residents of the state in 1999

(**g**) how many states were settled before 1865

(**h**) the average number of Congressional representatives per state

(**i**) the number of Congressional representatives from the state

Activity 1-11: Super Bowls and Oscar Winners

Select *either*

(**a**) National Football League Super Bowls, or

(**b**) Movies that have won the Academy Award for Best Picture

as the *observational units* of interest in a study. List two quantitative variables and two binary categorical variables that one might study about those cases. (Be sure that your variables are really characteristics of the game or movie that can be assigned a number or a category.)

Activity 1-12: Natural Light and Achievement

A recent study by the Heschong Mohone group, based near Sacramento, found that students who took their lessons in classrooms with more natural light scored as much as 25 percent higher on standardized tests than other students in the same school district.
(a) Identify the observational units in this study.
(b) Identify the two primary variables of interest. Be sure to state each of these variables as characteristics that change from observational unit to observational unit.
(c) Indicate whether you believe that the above variables were measured as categorical or quantitative data.

Activity 1-13: Children's Television Viewing

Researchers at Stanford studied whether reducing children's television viewing might help to prevent obesity. Third and fourth grade students at two public elementary schools in San Jose were the subjects. One of the schools incorporated a curriculum designed to reduce watching television and playing video games, while the other school made no changes to its curriculum. At the beginning and end of the study a variety of variables were measured on each child. These included body mass index, triceps skinfold thickness, waist circumference, waist-to-hip ratio, weekly time spent watching television, and weekly time spent playing video games.
(a) Identify the observational units in this study.
(b) Identify some of the variables measured on these units, making sure to phrase them as variables. You should identify at least one quantitative and at least one categorical variable.

Activity 1-14: Students' Political Views

Consider the students' self-descriptions of political inclination as liberal, moderate, or conservative.
(a) Calculate the proportion of students who identified themselves as liberal, the proportion who regard themselves as moderate, and the proportion who lean toward the conservative.
(b) Create a bar graph to display this distribution of political inclinations.
(c) Comment in a sentence or two on what your calculations and bar graph reveal about the distribution of political inclinations among these students.

Activity 1-15: Parents' Ages (*cont.*)

Refer to Activity 1-6 on page 12, where you examined the distribution of ages for having a first child for groups of mothers and fathers.

(a) Consider the variable "was the person over age 30 when his or her first child was born?" Is this a quantitative or a categorical variable? Explain.

(b) Produce a bar graph of this variable for the group of mothers.

(c) Produce a bar graph of this variable for the group of fathers.

(d) Comment on the differences in these bar graphs, being sure to relate your comments to the context.

Activity 1-16: Word Lengths (*cont.*)

(a) Create a dotplot of the lengths (number of letters) of words that you recorded in Activity 1-1 on page 5.

(b) Is there one particular length that occurs more often than any other? If so, what is it?

(c) Try to identify a length such that about half of your words are longer than that length and about half are shorter.

(d) Write a few sentences describing this distribution of word lengths.

Activity 1-17: Signature Measurements

(a) Create separate dotplots of the signature lengths for men and for women students. Be sure to put them adjacent to each other on the same scale (as in Activity 1-6 on page 12).

(b) Write a paragraph comparing and contrasting the distributions.

Activity 1-18: Variables of Personal Interest

Please list three *variables* that you would be interested in studying. These can be related to anything at all and need not be related to topics that are feasible for us to study in class. Be sure, however, that these correspond to the definition of a variable given above. Also indicate in each instance what the *observational unit* is. Please be very specific.

Topic 2:

DATA, VARIABLES, AND CALCULATORS

OVERVIEW

In the first topic you encountered many sets of data and explored the notion of a **variable** and its **distribution**. This topic introduces you to using your TI calculator to analyze data. It also leads you to perform elementary manipulations of variables and to ask whether the data at hand satisfactorily address the question or issue one has in mind.

OBJECTIVES

- To begin to use your graphing calculator as an indispensable tool for analyzing data.
- To understand the need to ask whether a variable actually measures the property that it purports to.
- To recognize that elementary manipulations of variables can often produce a more appropriate variable to measure the desired property.
- To appreciate the use of rates or percentages for making meaningful comparisons in many situations.
- To recognize limitations of a variable for measuring some properties.

PRELIMINARIES

1. Record how many letters are in your last name.

2. Determine and record how many "Scrabble points" are in your last name by assigning to each letter of your name its number of Scrabble points and then adding the points for all of your letters. The Scrabble points of the letters of the alphabet are given here:

A	B	C	D	E	F	G	H	I	J	K	L	M
1	3	3	2	1	4	2	4	1	8	5	1	3
N	O	P	Q	R	S	T	U	V	W	X	Y	Z
1	1	3	10	1	1	1	1	4	4	8	4	10

Total Scrabble points in your last name:

3. Record this information for all of the students in your class in a table like the following:

Student	Letters	Points
1		
2		
3		
...		

4. Take a guess as to how much it cost a family of four to attend a Major League Baseball game in 1999.

5. Identify your favorite Major League Baseball team. (If you do not have a favorite, identify one that you have heard of.)

6. Name an area of medicine in which you suspect women physicians often choose to specialize. Also name an area in which you suspect women physicians seldom choose to specialize.

7. Take a guess as to the percentage of physicians in the United States who are women (as of 1997).

8. For each of the following pairs of sports, identify the one that you consider *more hazardous* to its participants:

bicycle riding or football?

ice hockey or soccer?

swimming or skateboarding?

IN-CLASS ACTIVITIES

Activity 2-1: Scrabble Names

The following table reports the number of letters and Scrabble points in the last names of famous statisticians highlighted in David Moore's *The Basic Practice of Statistics*:

name	letters	points	name	letters	points
Nightingale	11	16	Gosset	6	7
Tukey	5	12	Norwood	7	11
Fisher	6	12	Pearson	7	9
Blackwell	9	20	Deming	6	10
Neyman	6	11	Galton	6	7

(a) Enter these data into your TI-83 graphing calculator. To do this, use the following instructions:
- Turn on your calculator.
- Press the ⎡ 2nd ⎤ button (this accesses the commands written in yellow above the keys) followed by the ⎡ (⎤ button. You now should see a { on your screen. Now type the number of letters in the last names of these famous statisticians, using the comma key to separate the numbers. Once the numbers have been typed, press the ⎡ 2nd ⎤ and ⎡) ⎤ buttons to end the list with a }.
- Now push the ⎡ STO▶ ⎤ button (this enables you to store these data into a named list). To type the name of the list, you will need to use your ⎡ ALPHA ⎤ button to access the green letters above the keys. Using this button before selecting each character, type the name LTTRS (the name can consist of at most five characters). Your home screen should now look like the following:

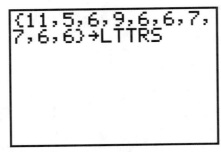

- Now press the ENTER button.

(b) Download the program DOTPLOT.83p into your calculator. To download this file into your calculator you must use the TI-83 TI-Graph Link™ cable. Connect your calculator to a computer via this cable and then download. (You will use this method to download all files and programs from a computer.)

Use the DOTPLOT program to produce a dotplot of the number of letters:
- Open the program menu using the PRGM button, select DOTPLOT, and then press the ENTER button twice to start the program.
- Select "ONE PLOT" and then enter the LTTRS list at the prompt. You can find the LTTRS list by pressing 2nd and then STAT to get into the LIST directory. Use the arrow key to move down the list until you find LTTRS. [*TI Hint*: To jump to the list names beginning with "L," press the ALPHA key and then the) button.]
- Once you have the correct name selected, press the ENTER button. Your screen should look like the following:

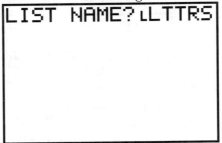

The calculator will display the dotplot after you press the ENTER button. Use your TRACE button to get information about your plot. Comment briefly on the distribution of the number of points.

The DOTPLOT program turns off the coordinate axes. If you want to see the coordinate axes displayed, press ⟨2nd⟩ and then ⟨ZOOM⟩ to view the FORMAT window. [*TI hint*: We will often refer to this key sequence as [FORMAT], without always reminding you to use the 2nd key or to look among the yellow words for the command.] Highlight AxesOn and press ⟨ENTER⟩, then press ⟨GRAPH⟩.

(c) Using a similar process as above, create a list named POINT and produce a dotplot for the number of points. [*TI Hint*: You can use the A-LOCK feature when repeatedly entering letters.] Comment briefly on the distribution of the number of points.

(d) Which of these statisticians' names has the most letters? Which has the most points? Do they belong to the same person?

most letters: most points:

same person?

(e) Who has the fewest letters? Who has the fewest points? Are they the same person?

fewest letters: fewest points:

same person?

(f) Use your calculator to create a new variable: ratio of points to letters. To do this:
- Use the LIST directory (⟨2nd⟩ ⟨STAT⟩) to select the POINT list and press ⟨ENTER⟩.
- Press the ⟨÷⟩ key.
- Select the LTTRS list from the LIST directory.
- Press the ⟨STO▶⟩ button to store this new list into a named list.
- Use the ALPHA-LOCK feature (⟨2nd⟩ ⟨ALPHA⟩) to type in the list name RATIO.

Your home screen should look like the following:

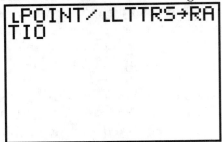

Then press ENTER .

Record the ratio values in the table below (you can use the right arrow to see subsequent values in the list displayed on your home screen). Produce at a dotplot of the distribution of ratios and comment briefly.

name	letters	points	ratio	name	letters	points	ratio
Nightingale	11	16		Gosset	6	7	
Tukey	5	12		Norwood	7	11	
Fisher	6	12		Pearson	7	9	
Blackwell	9	20		Deming	6	10	
Neyman	6	11		Galton	6	7	

Comment:

(g) To view the lists, you can place them into the Stat List Editor. To do this, select the STAT menu and select option SetUpEditor (by using the down arrow or by pressing its corresponding number, such as 5). When SetUpEditor is displayed in your home screen, use the LIST menu repeatedly to select the LTTRS, POINT, and RATIO lists, using the comma key between list names. Your home screen should now look like the following:

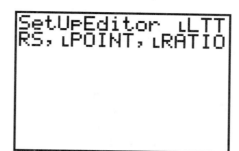

```
SetUpEditor LLTT
RS, LPOINT, LRATIO
```

Press ENTER , and the calculator will tell you when it is done. To view the lists press STAT and select Edit. You should now see the lists within your Stat List Editor. You will be using the Stat List Editor frequently in this class to enter lists and to edit lists.

(h) Identify who has the highest ratio, and explain why that person's ratio is so high.

Activity 2-2: Gender of Physicians

Suppose that you want to study the gender breakdown of physicians by medical specialty in an effort to identify which areas have more and which have less participation by women. For each of 37 medical specialties, the table below lists the numbers of men and women physicians who identified themselves as practicing that specialty as of December 31, 1997, taken from the *1999 World Almanac and Book of Facts*.

	specialty	men	women		specialty	men	women
1	aerospace medicine	548	39	20	nuclear medicine	1,173	261
2	allergy & immunology	3,024	748	21	obstetrics/gynecology	26,725	12,532
3	anesthesiology	27,023	6,707	22	occupational medicine	2,565	484
4	cardiovascular disease	17,939	1,348	23	ophthalmology	15,432	2,390
5	child psychiatry	3,425	2,195	24	orthopedic surgery	22,132	755
6	colon/rectal surgery	974	59	25	otolaryngology	8,348	748
7	dermatology	6,328	2,734	26	pathology–anat./clin.	13,069	5,167
8	diagnostic radiology	16,258	3,859	27	pediatric cardiology	1,033	329
9	emergency medicine	16,943	3,662	28	pediatrics	29,794	25,633

10	family practice	48,193	16,416	29	physical med./rehab.	3,987	1,863
11	forensic pathology	380	134	30	plastic surgery	5,452	529
12	gastroenterology	8,963	763	31	psychiatry	28,159	10,905
13	general practice	14,370	2,465	32	public health	1,237	433
14	general preventive medicine	1,040	510	33	pulmonary diseases	6,088	671
15	general surgery	37,181	3,754	34	radiation oncology	2,880	817
16	internal medicine	95,859	32,576	35	radiology	7,129	1,013
17	medical genetics	147	103	36	thoracic surgery	256	13
18	neurological surgery	4,696	217	37	urological surgery	9,764	264
19	neurology	9,429	2,285				

(a) Lists containing the numbers of men and women physicians in each of the specialties mentioned above have been stored in a file named GENPHYS.83g. Download this file now. The lists, which are now located in the LIST directory of your calculator, are named MEN, WOMEN, and INDEX.

You will use the SortA(option found in your STAT menu to sort the "number of women" variable in ascending order. To use this feature properly you will need to select the WOMEN, MEN, and INDEX lists from the LIST directory and then complete the) and press ENTER :

```
SortA( LWOMEN, LME
N, LINDEX)
```

Sorting these three lists together will allow you to keep track of which specialities the numbers correspond to. Entering the WOMEN list first sorts them by that variable. Now use the SetUpEditor in the STAT menu to place the INDEX and WOMEN lists into the Stat List Editor. The first three indices listed correspond to the speciality rows with the smallest numbers of women. The last three indices listed (use the down arrow to scroll down to the bottom of the lists) correspond to the speciality rows with the largest numbers of women.

(b) Identify the three specialties with the most women and the three specialties with the fewest women; also record those numbers:

<u>most:</u>   <u>fewest:</u>
1.   37.
2.   36.
3.   35.

(c) What aspect of the gender breakdown does the "number of women" variable not take into account?

(d) Use your calculator to determine and store the percentage of women physicians in each specialty in a list named PERCT. You want to divide the number of women by the total number of practitioners in each specialty and then multiply by 100 to form a percentage. You can use the following setup on your home screen:

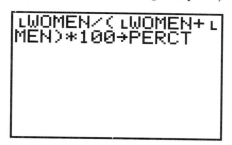

Then sort the "percentage of women" variable (as in (b) above; make sure you enter the PERCT list first and use the list named INDEX when sorting so that you can find the specialty associated with each percentage). Place the sorted PERCT and INDEX lists within your Stat List Editor. Identify the three specialties with the highest percentage of women and the three specialties with the smallest percentage of women; also record those numbers:

<u>highest:</u>   <u>smallest:</u>
1.   37.
2.   36.
3.   35.

(e) Do your lists in (b) and (d) agree exactly? If not, explain (being sure to argue using the data) why they differ.

(f) Use your calculator to produce a dotplot of the distribution of the *number* of women in each specialty. Identify a specialty that seems to have a "typical" number of women physicians. How many women practice this specialty?

(g) Use your calculator to produce a dotplot of the distribution of the *percentages* of women. Identify a specialty that seems to have a "typical" percentage of women physicians. What percentage of physicians in this specialty are women?

(h) Identify a specialty for which many more than a typical *number* of women practice that specialty but for which a much smaller *percentage* of women than is typical practice that specialty. Also record these values for that specialty.

(i) Identify a specialty for which many *fewer* than a typical *number* of women practice that specialty but for which a much *larger percentage* of women than is typical practice that specialty. Also record these values for that specialty.

(j) Based on a casual examination of this dotplot, write a brief paragraph describing key features of the distribution of percentages of women physicians.

(k) Summarize in a sentence or two what this activity reveals about the use of percentages as opposed to counts.

> When a variable involves counting the number of people or objects that belong in categories of different sizes, ***rates*** or ***percentages*** often provide a more appropriate variable to study.

Activity 2-3: Fan Cost Index

The data in the table below were compiled by Team Marketing Research in an effort to measure the cost of attending Major League Baseball games in 1999. The variables recorded were price of adult tickets, price of children's tickets, cost for parking, price of program, price for a medium cap, price for small beer, how many ounces are in a small beer, price for a small soda, how many ounces are in a small soda, and price for a medium hot dog (all in dollars).

	Team	adult	kids	park	pro.	cap	beer	oz	soda	oz	hot dog
1	Anaheim	13.19	12.75	7.00	5.00	12.00	5.00	16	2.00	14	2.75
2	Arizona	16.58	16.58	6.00	3.00	9.00	4.00	14	2.25	16	2.75
3	Atlanta	19.21	19.21	8.00	5.00	12.00	4.75	20	1.50	16	2.50
4	Baltimore	19.82	19.82	5.00	3.00	12.00	4.00	16	1.75	14	2.50
5	Boston	24.05	24.05	12.00	2.00	10.00	4.00	12	2.50	14	2.50
6	Chi. Cubs	17.46	17.46	12.00	3.00	12.00	4.00	16	1.75	14	2.00
7	Chi. White Sox	15.04	15.04	8.00	3.00	12.00	4.00	16	2.00	14	2.25
8	Cincinnati	9.71	9.71	5.00	5.00	10.00	3.75	16	2.00	16	1.75
9	Cleveland	18.43	18.43	10.00	2.50	10.00	3.75	14	2.00	12	2.25
10	Colorado	15.79	15.79	8.00	5.00	12.00	4.00	16	2.75	16	2.75
11	Detroit	12.23	12.23	8.00	4.00	10.00	3.75	14	1.75	14	2.25
12	Florida	12.17	10.84	5.00	4.00	8.00	2.50	12	2.50	20	2.50
13	Houston	13.30	13.10	10.00	5.00	10.00	4.25	22	1.75	16	3.00
14	Kansas City	11.76	11.33	6.00	6.00	12.00	3.00	12	2.00	14	2.00
15	Los Angeles	13.67	13.13	6.00	3.00	15.00	4.00	16	2.00	20	3.00

16	Milwaukee	11.02	10.14	5.00	3.00	10.00	3.75	12	1.75	12	2.00
17	Minnesota	8.46	6.96	5.00	4.00	13.00	4.00	18	2.50	18	2.50
18	Montreal	9.38	8.27	6.61	3.31	9.92	2.81	12	1.65	20	1.82
19	N.Y. Mets	19.89	19.89	7.00	4.00	10.00	5.50	20	3.50	20	3.75
20	N.Y. Yankees	23.33	23.33	6.00	5.00	12.00	4.75	16	2.50	18	3.50
21	Oakland	10.01	8.29	5.00	4.00	10.00	4.00	12	1.50	12	2.25
22	Philadelphia	13.60	13.60	6.00	4.00	14.00	4.75	18	1.25	12	2.75
23	Pittsburgh	10.71	10.06	4.00	3.25	12.00	3.50	16	2.00	16	2.00
24	San Diego	11.92	11.92	6.00	4.00	12.00	4.25	16	2.75	16	2.25
25	San Francisco	12.12	10.39	7.00	4.00	12.00	4.25	14	1.50	12	3.00
26	Seattle	19.01	18.14	11.04	4.00	9.00	3.75	12	2.25	14	2.25
27	St. Louis	16.53	14.88	7.50	4.00	12.00	4.50	20	2.00	14	2.00
28	Tampa Bay	15.08	15.06	6.50	4.00	10.23	3.75	16	2.00	16	2.50
29	Texas	19.93	19.22	8.00	4.00	10.00	4.50	20	2.00	16	2.00
30	Toronto	16.62	16.50	7.93	3.31	7.60	2.48	12	1.37	16	1.41

(a) The "fan cost index" (FCI) is defined to be the cost of two adult tickets, two children's tickets, parking, two programs, two caps, two small beers, four small sodas, and four hot dogs. Download FCI99.83g into your calculator to obtain the TIX, KIDS, PARK, PROG, CAPS, BEER, BSIZE, SODA, SSIZE, and HOTD lists. Use your calculator to calculate and store the FCI for each team:

$2 \times$ TIX $+ 2 \times$ KIDS $+$ PARK $+ 2 \times$ PROG $+ 2 \times$ CAPS $+ 2 \times$ BEER $+ 4 \times$ SODA $+ 4 \times$ HOTD $->$ FCI

Report which team has the highest and the lowest FCI, along with those values.

N.Y. Yankees , Montreal

(b) Use your calculator to look at a dotplot of the distribution of the FCI values. Comment on its features.

2 parallel dotted lines

(c) Report the FCI value for the team that you identified in the "Preliminaries" section as your favorite. Also comment on where it falls relative to the other teams. [*TI Hint:* You may want to place the FCI variable into the Stat List Editor because when a value is selected, the entry number (which is shown at the bottom of the screen) corresponds to the row number.]

(d) If you were to attend a game, who would you go with? Would your group typically be a family of four? Do you think this FCI variable accurately measures how much it would cost *your group* to attend a game?

Family of four, yes

(e) Create a new variable, called the Me Cost Index (MCI), by changing the number of tickets and what is purchased. That is, follow what you did above but change some of the coefficients to better reflect what you think your group would actually spend at a game. Report your new equation and which team has the highest and the lowest MCI, along with those values. Also comment on where your team falls in the distribution of MCI values.

$4 \times tix + 0 \times kids + park + 2 \times prog + 2 \times caps + 4 \times hotd \rightarrow fci$

Yankees, Montreal

(f) Which ballpark(s) has the highest charge for a small soda? Which has the lowest? What are those values?

N.Y. Mets , Philly

$3.50 $1.25

(g) Explain what is misleading about comparing the costs of a small soda or small beer in the different ballparks.

The oz's are different

(h) Use your calculator to create and store a new list for comparing soda prices by converting to a "price per ounce" variable. Which ballpark has the highest soda price per ounce, and which has the lowest? What are those values?

$$SPPO = LSoda / LSSize$$

Activity 2-4: States' SAT Averages

The table below reports the average SAT score for each of the fifty states in 1998 and also the percentage of high school seniors in the state who took the exam.

state	Avg SAT	% take	state	Avg SAT	% take	state	Avg SAT	% take
Alabama	1120	8	Louisiana	1120	8	Ohio	1066	24
Alaska	1041	52	Maine	1005	68	Oklahoma	1132	8
Arizona	1053	32	Maryland	1014	65	Oregon	1056	53
Arkansas	1123	6	Massachusetts	1016	77	Pennsylvania	992	71
California	1013	47	Michigan	1127	11	Rhode Island	996	72
Colorado	1079	31	Minnesota	1183	9	South Carolina	951	61
Connecticut	1019	80	Mississippi	1111	4	South Dakota	1165	5
Delaware	994	70	Missouri	1143	8	Tennessee	1121	13
Florida	1001	52	Montana	1089	24	Texas	995	51
Georgia	968	64	Nebraska	1136	8	Utah	1142	4
Hawaii	996	55	Nevada	1023	33	Vermont	1012	71
Idaho	1089	16	New Hampshire	1043	74	Virginia	1006	66
Illinois	1145	13	New Jersey	1005	79	Washington	1050	53
Indiana	997	59	New Mexico	1105	12	West Virginia	1038	18
Iowa	1194	5	New York	998	76	Wisconsin	1175	7
Kansas	1167	9	North Carolina	982	62	Wyoming	1094	10
Kentucky	1097	13	North Dakota	1189	5			

(a) Which state had the highest average SAT score? Which had the lowest? What do you notice about the percentages of students taking the exam in those states?

(b) We now want to divide the states into two groups: those in which more than 25% of the students took the SAT and those in which 25% or fewer took the SAT. Download SAT98.83g into your calculator. The lists are named SAT (which consists of all the SAT scores), MORE (the SAT scores for states with more than 25% of the students taking the SAT), and FEWER (the SAT scores for the other states). Use your calculator to create dotplots of the distributions of SAT for the two groups, i.e., use the DOTPLOT program and select 2: Compare Plots to compare the lists MORE and FEWER. What do you notice about the distributions of SAT averages for these two groups? Suggest a reasonable explanation for the pattern that is apparent.

(c) Would you conclude that the state with the highest SAT average is doing the best job of preparing its students for the exam and that the state with the lowest SAT average is doing the worst job? In other words, is "SAT average" a good variable for deciding how well a state educates its students? Explain.

(d) How does *your* home state compare to the rest in terms of SAT average and percentage of students taking the test? (Be sure to identify the state also.)

This activity reveals that some properties, such as the effectiveness of a state's educational system, are very difficult to measure.

WRAP-UP

This topic has given you more experience with data and with the ideas of **variability** and **distributions**. You have used your TI calculator to analyze data, and you have also seen that manipulating a variable, for example by converting it to a **rate**, is often necessary. You have begun to consider the conclusions one can draw from statistical studies as related to the variables measured.

In the next topic you will study distributions of data further by considering visual displays other than the dotplot. You will also develop a checklist of features to look for when describing distributions verbally.

HOMEWORK ACTIVITIES

Activity 2-5: Scrabble Names (*cont.*)

Refer to the data on the number of letters and the Scrabble points of you and your classmates collected in the "Preliminaries" section.

(a) Produce (either by hand or using your calculator) dotplots of the distributions of letters and of points. Comment on where you stand relative to your peers with regard to these variables.

(b) Create a new variable: ratio of points to letters. Look at a dotplot of its distribution, and comment on where you stand relative to your peers.

(c) What seems to be a typical, roughly average, value of this ratio?

(d) How do the Scrabble ratios of you and your classmates compare to those of the famous statisticians in Activity 2-1?

Activity 2-6: Fan Cost Index (*cont.*)

Reconsider the FCI data from Activity 2-3 on page 29 and turn your attention to the cost of a small beer at the ballparks (BEER, BSIZE).

(a) Identify the three ballparks with the highest price of a small beer and with the lowest price of a small beer. Also report those prices.

(b) Use your calculator to convert those prices to a "per ounce" basis. Look at a dotplot of the distribution and comment on it.

(c) Identify the three ballparks with the highest price per ounce of a small beer and with the lowest price per ounce of a small beer. Also report those prices. Are these the same ballparks with the highest and lowest prices of a small beer without adjusting for size?

Activity 2-7: Hazardousness of Sports

The following table lists estimates of the number of sports-related injuries treated in U.S. hospital emergency departments in 1997, along with an estimate of the number of participants in the sports taken from *Injury Facts*, National Safety Council, 1999.

sport	injuries	participants	sport	injuries	participants
Basketball	644,921	33,300,000	Skateboarding	48,186	6,300,000
Bicycle riding	544,561	45,100,000	Golf	39,473	26,200,000
Football	334,420	20,100,000	Snowboarding	37,638	2,800,000
Baseball, softball	326,714	30,400,000	Ice skating	25,379	7,900,000
Roller skating	153,023	37,500,000	Bowling	23,317	44,800,000
Soccer	148,913	13,700,000	Tennis	22,294	11,100,000
Weight lifting	86,024	47,900,000	Water skiing	10,657	6,500,000
Swimming	83,772	59,500,000	Racquetball	10,438	4,500,000
Ice hockey	77,491	1,900,000	Billiards, pool	3,685	36,000,000
Fishing	72,598	44,700,000	Archery	3,213	4,800,000
Volleyball	67,340	17,800,000			

(a) If one uses the number of injuries as a measure of the hazardousness of a sport, which sport is more hazardous between bicycle riding and football? between ice hockey and soccer? between swimming and skateboarding?

(b) Use your calculator (SPORTHAZ.83g) to compute each sport's *rate* of injuries (INJUR) *per thousand* participants (PARTI). [*Hint:* A percentage is a rate per hundred, so follow Activity 2-2 on page 25 to figure out how to determine a rate per thousand.]

(c) In terms of the injury rate per thousand participants, which sport is more hazardous between bicycle riding and football? between soccer and ice hockey? between swimming and skateboarding?

(d) How do the answers to (a) and (c) compare to each other? How do they compare to your intuitive perceptions from the "Preliminaries" section?

(e) List the three most and three least hazardous sports according to the injury rate per thousand participants.

(f) Identify some other factors that are related to the hazardousness of a sport. In other words, what information might you use to produce a better measure of a sport's hazardousness that is not already taken into account by the number or rate of injuries?

Activity 2-8: Broadway Shows

The following table lists the 22 shows being produced on Broadway during the week ending September 12, 1999. Also listed are the type of show (play or musical), the gross box office receipts generated by the show that week, the total attendance for the week, and the theater's capacity for attendance that week, as reported by www.buy-broadway.com.

Show	Type	Receipts	Attendance	Capacity	Top Price
Annie Get Your Gun	musical	$572,885	9892	12761	75
Cabaret	musical	$466,670	6647	7360	90
Cats	musical	$350,368	7200	11856	70
Chicago	musical	$536,852	9664	11680	80
Death of a Salesman	play	$351,082	6531	7483	75
Epic Proportions	play	$80,379	2564	4776	65
Footloose	musical	$271,022	6974	10784	75
Fosse	musical	$566,644	8118	9128	80
It Ain't Nothin' But The Blues	musical	$88,014	2366	5500	75
Jekyll & Hyde	musical	$283,674	5613	8512	80
Kat and the Kings	musical	$108,091	3218	8672	75
Les Miserables	musical	$375,318	7677	11336	80
Miss Saigon	musical	$395,522	8558	13945	80
Ragtime	musical	$420,902	8913	14495	75
Rent	musical	$333,248	6219	9448	80
Side Man	play	$106,437	2894	6440	60
Smokey Joe's Cafe	musical	$168,276	3547	9697	75
The Lion King	musical	$880,717	14199	14032	80
The Phantom of the Opera	musical	$601,218	9974	12856	80
The Scarlet Pimpernel	musical	$125,597	2284	5464	80
The Weir	play	$120,605	3450	7576	60
Voices in the Dark	play	$114,607	3848	8688	60

(a) What are the observational units here?

(b) For each of the five variables recorded (type of show, receipts, attendance, capacity, top price), indicate whether it is a quantitative or a categorical variable. If it is a categorical variable, indicate whether or not it is a binary variable.

(c) Use your calculator (BROADWAY.83g) to create a new variable representing attendance (ATTEN) not as a raw total but rather as a percentage of the theater's capacity (CAPAC). Then have your calculator produce a dotplot of the distribution of this new variable. Which show had the *highest* percentage of capacity? What was that value? Which show had the *lowest* percentage of capacity? What was that value?

(d) Write a paragraph describing some features of this distribution. [As always, remember to relate your comments to the context.]

(e) How many productions had a higher attendance figure (in raw numbers) than *Death of a Salesman*? How many had higher receipts than *Death of a Salesman*? How many had a higher percentage of capacity than *Death of a Salesman*?

(f) Suggest in a brief sentence or two an explanation for the discrepancies among the answers to (e).

Activity 2-9: Box Office Blockbusters

For a sample of fifteen popular movies of 1999, the following table (compiled from data at showbizdata.com) lists the number of screens on which each film appeared in its first two weekends of release and also its box office revenue (in millions of dollars) for those weekends:

Movie	first weekend		second weekend	
	number screens	revenue	number screens	revenue
Star Wars: The Phantom Menace	2970	64.8110	3023	51.3999
The Sixth Sense	2161	26.6813	2395	25.7650
Austin Powers II	3312	54.9176	3314	31.4066
The Matrix	2849	27.7883	2903	22.5633
Tarzan	3005	34.2220	3049	24.0558
Big Daddy	3027	41.5364	3121	20.0134
Toy Story 2	3236	57.3888	3238	27.7603
The Mummy	3209	43.3696	3226	24.8563
Runaway Bride	3158	35.0556	3161	20.7727
The Blair Witch Project	27	1.5121	31	1.9782
Notting Hill	2747	21.8112	2752	15.0133
Double Jeopardy	2547	23.1625	2884	17.0188

Wild, Wild West	3342	27.6875	3342	16.8340
Analyze This	2518	18.3835	2537	15.5677
The World is Not Enough	3163	35.5190	3163	23.2377

(a) Which film made the most money in its first weekend of release? How much did it make?

(b) Use your calculator (MOVIES99.83g) to create a dotplot of the first weekend's box office revenue for these films (REV1). Write a few sentences commenting on the distribution. Identify the film for which half of the films made more money and half made less. Also identify any film whose revenue differs markedly from the others. Finally, select two other films (preferably ones that you have seen) and comment on where they fall in the distribution.

(c) Use your calculator to create a new variable: percentage decrease in revenue from week 1 to week 2 (REV2). [*Hint*: Calculate the actual decrease in revenue, then divide by the first weekend's revenue, then multiply by 100 to make it a percentage.] Which film had the highest percentage drop-off? Which had the lowest (which could in fact be an increase in revenue)?

(d) Did the film with the highest first-week revenue have the smallest percentage decrease between the first and second weeks? Explain.

(e) Look at a dotplot of this distribution of percentage decreases and write a few sentences summarizing it.

(f) Suggest an explanation for the film that seems to differ dramatically from the others.

(g) In its third weekend of release, *The Blair Witch Project* played on 1101 screens and generated 29.207 million dollars. Calculate its percentage increase from the second to third week.

Activity 2-10: Box Office Blockbusters (*cont.*)

Reconsider the data on box office revenues presented in Activity 2-9 (MOVIES99).

(a) Use your calculator to compute the average revenue per screen (SCRN1) for the first weekend of release.

(b) Which film had the highest average revenue per screen? Which had the lowest? Which film had half above and half below it in terms of average revenue per screen? Identify all of those values, and be sure to report the units as dollars per screen.

(c) Look at a dotplot of the distribution of average revenue per screen in the first weekend of release, and comment on its key features.

(d) Repeat (a)–(c) for the average revenue per screen in the second weekend of release (SCRN2).

(e) Do the films with the highest revenue per screen in the first weekend tend to be the same films with the highest revenue per screen in the second weekend? Is the same true for those with the lowest per-screen revenues? Explain.

Activity 2-11: Uninsured Americans

The following data are the 1997 populations of each state and the number of the state's residents who do not have health insurance coverage, both reported in thousands from *The 1999 World Almanac and Book of Facts*. Also listed is the region of the country (Northeast, South, Midwest, and West).

State	Region	Population	Uninsured	State	Region	Population	Uninsured
Alabama	S	4,247	659	Montana	W	894	174
Alaska	W	641	116	Nebraska	MW	1,662	180
Arizona	W	4,655	1,141	Nevada	W	1,724	301
Arkansas	S	2,622	639	New Hampshire	NE	1,200	141
California	W	32,987	7,095	New Jersey	NE	7,977	1320
Colorado	W	3,929	592	New Mexico	W	1,827	413
Connecticut	NE	3,299	395	New York	NE	18,143	3,174
Delaware	S	750	98	North Carolina	S	7,352	1,141
Florida	S	14,399	2,817	North Dakota	MW	639	97
Georgia	S	7,647	1,344	Ohio	MW	11,230	1,297
Hawaii	W	1,183	89	Oklahoma	S	3,338	593
Idaho	W	1,257	223	Oregon	W	3,298	440
Illinois	MW	12,098	1,506	Pennsylvania	NE	11,922	1,209
Indiana	MW	5,865	669	Rhode Island	NE	944	96
Iowa	MW	2,830	340	South Carolina	S	3,815	640
Kansas	MW	2,590	304	South Dakota	MW	712	84
Kentucky	S	3,922	587	Tennessee	S	5,542	7,56
Louisiana	S	4,250	827	Texas	S	19,751	4,836
Maine	NE	1,225	182	Utah	W	2,085	280
Maryland	S	5,057	677	Vermont	NE	581	55
Massachusetts	NE	6,004	755	Virginia	S	6,752	854
Michigan	MW	9,794	1,133	Washington	W	5,748	655
Minnesota	MW	4,767	438	West Virginia	S	1,747	300
Mississippi	S	2,737	550	Wisconsin	MW	5,126	409
Missouri	MW	5,322	669	Wyoming	W	491	76

(a) Use your calculator (UNINSURED.83g) to sort these data in terms of the number of uninsured residents. Identify the states with the three highest numbers of uninsured residents and the three states with the lowest numbers.

(b) Create a new variable: percentage of the state's residents who are uninsured. Then sort the data by that variable. Which states have the three highest percentages and which have the three lowest percentages of uninsured residents?

(c) Do the states with the highest uninsured rates also have the most uninsured residents? Do the states with the lowest uninsured rates also have the fewest uninsured residents? Explain.

(d) Write a few sentences explaining the difference between raw numbers (counts) and rates in this context.

Activity 2-12: Signature Measurements (*cont.*)

Refer to the data collected in Topic 1 on students' signatures. Enter the data into a list named SIGS.831. Use your calculator to conduct an analysis of the lengths of signatures and also the lengths adjusted on a per letter basis. Also include in your analysis a comparison of these variables between the two genders. Write a paragraph or two describing your findings.

Activity 2-13: Driver Safety

In 1997 there were 11,012 licensed drivers over age 55 involved in fatal crashes. There were only 7670 licensed drivers between the ages of 16 and 20 involved in fatal crashes. Does this establish that younger drivers are better and safer than older drivers? Explain.

Activity 2-14: Personal Comparison

Think of and describe an example where a rate would provide a much more meaningful comparison than would counts.

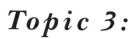

Topic 3:

DISPLAYING AND DESCRIBING DISTRIBUTIONS

OVERVIEW

In the first two topics you discovered the notion of the **distribution** of a set of data. You created visual displays (**bar graphs** and **dotplots**) and wrote verbal descriptions of some distributions. In this topic you will discover some general guidelines to follow when describing the key features of a distribution and also encounter two new types of visual displays for quantitative data: **stemplots** and **histograms**.

OBJECTIVES

- To develop a checklist of important features to look for when describing a distribution of data.
- To anticipate features of data by thinking about the nature of the variable involved.
- To discover how to construct **stemplots** as simple but effective displays of a distribution of data.
- To learn how to interpret the information presented in **histograms**.
- To use your calculator to produce visual displays of distributions.
- To become comfortable and proficient with describing features of a distribution verbally.

PRELIMINARIES

1. What do you think is a typical weight for a male Olympic rower?

180 lbs

41

2. Take a guess as to the length of the longest reign by a British ruler since William the Conqueror. Which ruler do you think reigned the longest?

A foot

3. What do you think is a typical age for an American man to marry? How about for a woman?

25, 22

4. Which type of car would you expect to get the most miles per gallon: large, family, or small?

Small

5. How old was your mother when you were born?

38

6. Think of the salary that you expect to make in the first year of your career after graduation.

$100,000

7. Record the responses of you and your classmates to the question about your mother's age when you were born and your expected first-year salary, along with gender, in a table such as the following:

student	gender	mother's age	expected salary
1			
2			
...			

Activity 3-1: Features of Distributions

Presented below are dotplots of distributions of (hypothetical) exam scores for twelve different classes. The questions following them will lead you to compile a "checklist" of important features to consider when describing distributions of data.

(a) What strikes you as the most distinctive difference among the distributions of exam scores in classes A, B, and C?

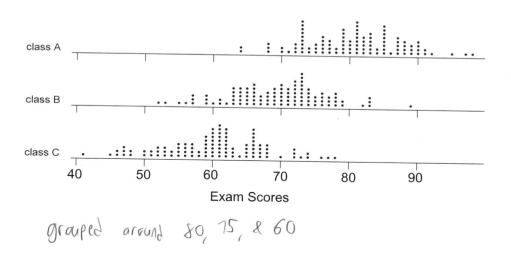

grouped around 80, 75, & 60

(b) What strikes you as the most distinctive difference among the distributions of scores in classes D, E, and F?

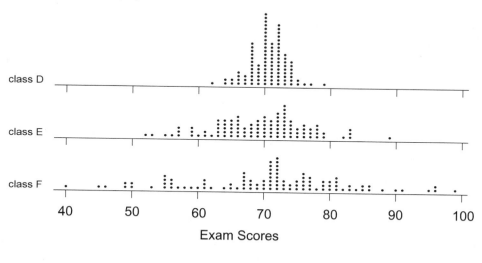

Starts compact & spreads out

(c) What strikes you as the most distinctive difference among the distributions of scores in classes G, H, and I?

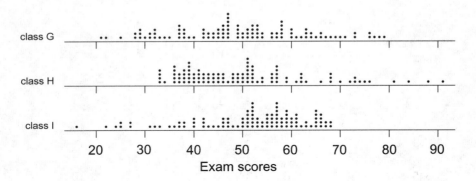

(d) What strikes you as the most distinctive feature of the distribution of scores in class J?

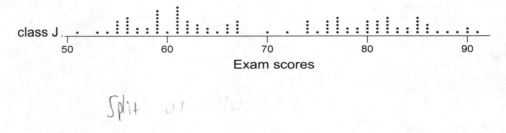

Split *a1*

(e) What strikes you as the most distinctive feature of the distribution of scores in class K?

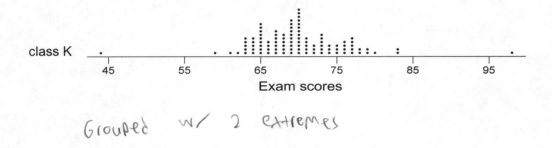

Grouped w/ 2 extremes

(f) What strikes you as the most distinctive feature of the distribution of scores in class L?

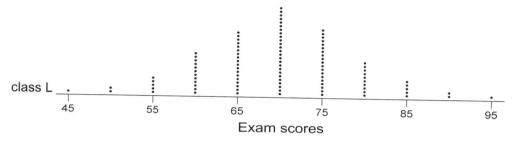

Tappers nicely, goes by 5's

These hypothetical exam scores illustrate six features that are often of interest when one is analyzing a distribution of data:

1. The *center* of a distribution is usually the most important aspect to notice and describe.
2. A distribution's *variability* is a second important feature. *- spread*
3. The *shape* of a distribution can reveal much information. While distributions come in a limitless variety of shapes, certain shapes arise often enough to have their own names. A distribution is **symmetric** if one half is roughly a mirror image of the other, as pictured below:

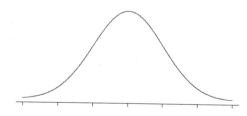

A distribution is **skewed to the right** if it tails off toward larger values:

tail

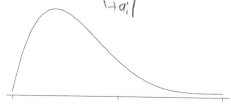

and **skewed to the left** if its tail extends to smaller values:

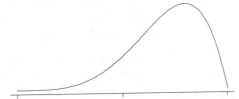

4. A distribution may have *peaks* or *clusters* that indicate that the data fall into natural subgroups.
5. **Outliers**, observations that differ markedly from the pattern established by the vast majority, often arise and warrant close examination.
6. A distribution may display **granularity** if its values occur only at fixed intervals (such as multiples of 5 or 10). Numbered n something other than 1

dots don't fall in middle

Please keep in mind that although we have formed a checklist of important features of distributions, you should not regard these as definitive rules to be applied rigidly to every data set that you consider. Rather, this list should only serve to remind you of some of the features that are *typically* of interest. Every data set is unique and has its own interesting features, some of which may not be covered by the items listed on our checklist.

Activity 3-2: Matching Variables to Dotplots

The dotplots below represent the distributions of eight variables. (The scales have been left off the dotplots intentionally.) For each variable, identify the number of the dotplot that you believe displays its distribution. Also provide a brief explanation of your reasoning in each case.

(a) jersey numbers of 1999 Cal Poly football players
(b) annual snowfall amounts for a sample of cities taken from around the U.S.
(c) margins of victory in a sample of Major League Baseball games
(d) prices of properties on the Monopoly game board
(e) weights of college football players on the 1999 Cal Poly team
(f) ages at which a sample of mothers had their first child
(g) weights of 1999 cars
(h) scores on a statistics exam

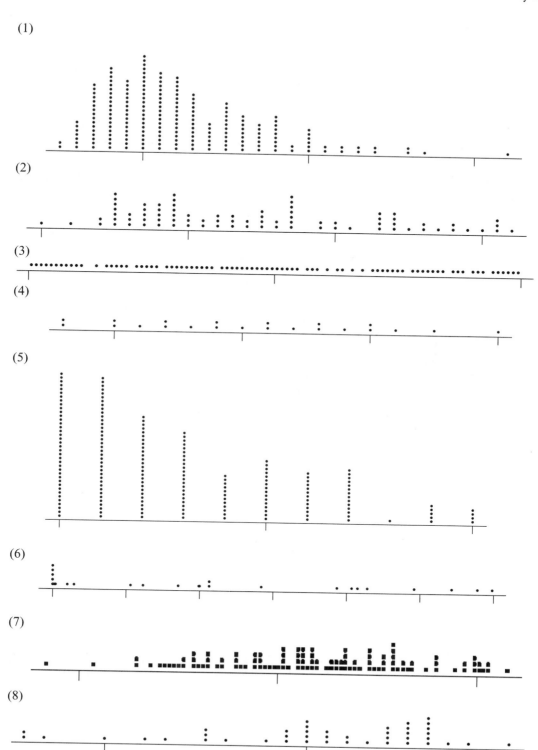

Activity 3-3: Rowers' Weights

The table below records the weight of each rower on the 1996 U.S. Olympic men's rowing team. A dotplot of these weights follows.

name	Weight	event	name	Weight	event
Auth	154	LW double sculls	Klepacki	205	four
Beasley	224	single sculls	Koven	200	eight
Brown	214	eight	Mueller	215	quad
Burden	195	eight	Murray	205	four
Carlucci	160	LW four	Murphy	220	eight
Collins, D	155	LW four	Peterson, M	210	pair
Collins, P	195	eight	Peterson, S	160	LW double sculls
Gailes	205	quad	Pfaendtner	160	LW four
Hall	195	four	Scott	208	four
Holland	195	pair	Schnieder	158	LW four
Honebein	200	eight	Segaloff	121	coxswain
Jamieson	210	quad	Smith	207	eight
Kaehler	210	eight	Young	207	quad

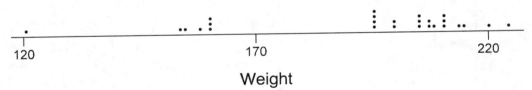

Weight

(a) Which weight has the highest number of rowers at that weight? How can you see this from the dotplot?

195 lbs, tallest stack

(b) Write a paragraph describing key features (e.g., shape, center, spread, outliers) of the distribution of rowers' weights. [As always, be sure to relate your comments to the context.]

Not a lot of variability, one outlier at 121 lbs

(c) Suggest an explanation for the clusters apparent in the distribution.

> The groups of rowers are around the same weight

(d) What is the name of the apparent outlier? Suggest an explanation for why his
weight differs so substantially from the others.

> Segaloff, he doesn't row

Activity 3-4: British Rulers' Reigns

The table below records the lengths of reign (rounded to the nearest year) for the
rulers of England and Great Britain beginning with William the Conqueror in 1066.

ruler	reign	ruler	reign	ruler	reign	ruler	reign
William I	21	Edward III	50	Edward VI	6	George I	13
William II	13	Richard II	22	Mary I	5	George II	33
Henry I	35	Henry IV	13	Elizabeth I	44	George III	59
Stephen	19	Henry V	9	James I	22	George IV	10
Henry II	35	Henry VI	39	Charles I	24	William IV	7
Richard I	10	Edward IV	22	Charles II	25	Victoria	63
John	17	Edward V	0	James II	3	Edward VII	9
Henry III	56	Richard III	2	William III	13	George V	25
Edward I	35	Henry VII	24	Mary II	6	Edward VIII	1
Edward II	20	Henry VIII	38	Anne	12	George VI	15

(a) How long was the longest reign? Who ruled the longest?

> 63 , Victoria

(b) What is the shortest reign? Who ruled the shortest time? What do you think this
value really means?

> 0, Edward V, died before a year

> One can create a visual display of the distribution called a ***stemplot*** by separating each observation into two pieces: a "stem" and a "leaf." When the data consist primarily of two-digit numbers, the natural choice is to make the tens digit the stem and the ones digit the leaf. For example, a reign of 21 years would have 2 as the stem and 1 as the leaf; a reign of 2 years would have 0 as the stem and 2 as the leaf.

(c) Fill in the stemplot below by putting each leaf on the row with the corresponding stem. We have gotten you started by filling in the reign lengths of William I (21 years), William II (13 years), Henry I (35 years), and Stephen (19 years).

length of reign

```
0 | 9 0 2 6 5 3 6 7 9 1
1 | 3 9 0 7 3 3 2 3 0 5
2 | 1 0 2 2 4 2 4 5 5
3 | 5 5 5 9 8 3
4 | 4
5 | 6 0 9
6 | 3
```

(d) The final step to complete the stemplot is to order the leaves (from smallest to largest) on each row. Reproduce the stemplot below with the leaves ordered.

```
0 | 0 1 2 3 5 6 6 7 9 9
1 | 0 0 2 3 3 3 3 5 7 9
2 | 0 1 2 2 2 4 4 5 5
3 | 3 5 5 5 8 9
4 | 4
5 | 0 6 9
6 | 3
```

> The ***stemplot*** is a simple but useful visual display. Its principal virtues are that it is easy to construct by hand (for relatively small sets of data) and that it retains the actual values of the observations. Of particular convenience is the fact that the stemplot also *sorts* the values from smallest to largest.

(e) Write a short paragraph describing the distribution of lengths of reign of British rulers. (Keep in mind the checklist of features that we created above. Also please remember to relate your comments to the context.)

Activity 3-5: Geyser Eruptions

> The ***histogram*** is a visual display similar to a stemplot but one that can be produced even with very large data sets; it also permits more flexibility than does the stemplot. One constructs a histogram by dividing the range of the data into subintervals of equal length, counting the number (also called the ***frequency***) of observations in each subinterval, and constructing rectangles whose heights correspond to the counts in each subinterval. Equivalently, one could make the rectangle heights correspond to the proportions (also called the ***relative frequency***) of observations in the subintervals.

The following histogram displays the distribution of times between eruptions (in minutes) for the Old Faithful geyser in Wyoming in August 1985. Notice that the statistical software reports the midpoint of each subinterval. For example, the first rectangle indicates that there was one inter-eruption time that lasted between 37 and 43 minutes (including the 37 but not the 43).

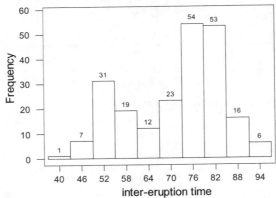

(a) How many of the inter-eruption times lasted between 43 minutes and 49 minutes?

7

(b) How many of the inter-eruption times lasted more than 61 minutes? What proportion of the inter-eruption times were more than 61 minutes?

164   164/222

(c) Can you determine exactly how many inter-eruption times lasted more than 90 minutes? Explain.

No, you don't know where 90 starts & ends

(d) How many distinct clusters (or peaks) can you identify in the distribution? Roughly where do they fall? Can you come up with a reasonable explanation for what might lead to these clusters?

The next basic operation on the TI that you need to understand is how to create visual displays using STAT PLOT. Before attempting to create a histogram (or any other statistical display), you must make sure that your calculator has all graphing functions turned off. To access the STAT PLOT menu, press 2nd Y= . The TI-83 screen should look similar to the following display:

The above picture tells you that Plot 1 is On and is set up to do a histogram of a list named LTV, Plot 2 is off and is set up to do a boxplot (modified) of a list named PROC2, and Plot 3 is off and is set up to do scatterplot of list Y vs. list X (you will learn about boxplots and scatterplots later).

(e) Download the list file GYSR.83l into your calculator. Use your calculator to recreate the histogram of the distribution of times between eruptions (in minutes) for the Old Faithful geyser in Wyoming in August 1985:
- Access the STAT PLOT menu and press ENTER or 1 to begin setting up the first plot.

This screen tells you that Plot1 is on and that you have six choices of graphical displays (scatterplot, xyline, histogram, modified boxplot, boxplot, and normal probability plot), the capability to choose any list, and a choice for frequency.

- Set up your Plot1 as a histogram by pressing the down arrow to move to the Type line and using the right arrow and ENTER to highlight the histogram icon. Then move to the Xlist line and either type the name GYSR using the ALPHA key or select the GYSR list name from the LIST menu.
- Adjust the viewing window for the histogram. For example, after pressing the WINDOW button, input the following numbers:

```
WINDOW
 Xmin=37
 Xmax=97
 Xscl=6
 Ymin=-15
 Ymax=60
 Yscl=10
 Xres=1
```

[*TI Hint:* Make sure you used the (-) key to express the negative sign, not the subtraction key.]

- Press the GRAPH button and write a few sentences describing this distribution.

(f) Now change the WINDOW settings so that the histogram has 5 subintervals rather than 10 by changing the width of the intervals (Xscl) to 12 and pressing the GRAPH button again. Then create another histogram with 20 subintervals. Comment on the different appearances of the distribution revealed by these histograms. Which do you think provides the most useful and informative display? Explain.

WRAP-UP

With this topic you have progressed in your study of distributions of data in many ways. You have created a checklist of various features to consider when describing a distribution verbally: ***center, spread, shape, clusters/peaks, outliers, granularity***. You have encountered three shapes that distributions often follow: ***symmetry, skewness to the left***, and ***skewness to the right***. (Remember that the direction of skewness describes the longer tail of the distribution.) You have also discovered two new techniques for displaying a distribution of quantitative data: ***stemplots*** and ***histograms***.

Even though we have made progress by forming our checklist of features, we have still been somewhat vague to this point about describing distributions. The next two topics will remedy this ambiguity somewhat by introducing you to specific numerical measures of certain features (namely, ***center*** and ***spread***) of a distribution. They will also lead you to studying yet another visual display: the ***boxplot***.

HOMEWORK ACTIVITIES

Activity 3-6: Cars' Fuel Efficiency

(a) The following three dotplots display distributions of EPA highway miles per gallon (MPG) ratings for 1999 cars in three classifications: large, family, and small. Comment on how the *centers* of these three distributions compare.

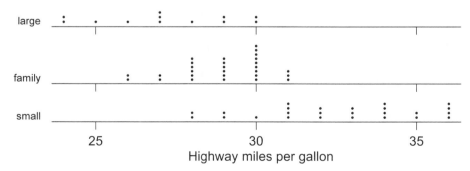

(b) The following three dotplots display distributions of these miles per gallon ratings for cars in the sports, small, and upscale classifications. Comment on how the *spreads* of these distributions compare.

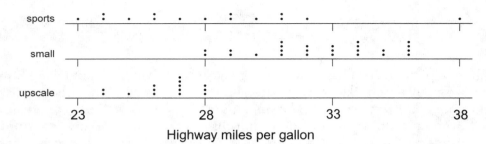

Highway miles per gallon

(c) The following dotplot shows the distribution of weights of cars in certain categories. How many categories do you think are represented here? Explain.

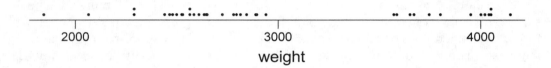

weight

(d) The following data are times that it takes a car to travel a distance of one-quarter mile. What strikes you as a distinctive feature of this distribution?

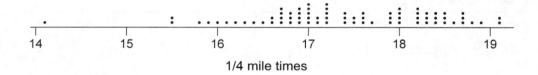

1/4 mile times

Activity 3-7: Features of Distributions (*cont.*)

The following histograms display the distributions of years served by U.S. Senators, percentage of residents living in urban areas for the fifty states, and prices of introductory statistics textbooks, respectively. Comment on how the shapes of these distributions compare. Be sure to identify which is roughly symmetric, which is skewed to the left, and which is skewed to the right.

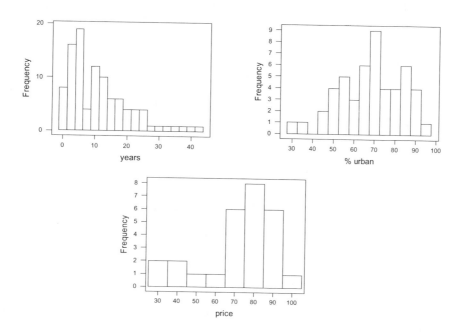

Activity 3-8: Placement Exam Scores

The Department of Mathematics and Computer Science of Dickinson College gives an exam each fall to freshmen who intend to take calculus; scores on the exam are used to determine into which level of calculus a student should be placed. The exam consists of 20 multiple-choice questions. Scores for the 213 students who took the exam in 1992 are tallied in the following table and in PLACE92.831.

score	1	2	3	4	5	6	7	8	9	10
count	1	1	5	7	12	13	16	15	17	32
score	11	12	13	14	15	16	17	18	19	
count	17	21	12	16	8	4	7	5	4	

(a) Use your calculator to produce a dotplot and a histogram of this distribution (PLSCO). Try a few different histograms, with differing numbers of intervals, to find one that best summarizes the distribution.

(b) Write a few sentences describing the shape, center, and spread of this distribution.

(c) What strikes you as an additional distinctive feature of this distribution?

Activity 3-9: Hypothetical Manufacturing Processes

Suppose that a manufacturing process strives to make steel rods with a diameter of 12 centimeters, but the actual diameters vary slightly from rod to rod. Suppose further that rods with diameters within ±0.2 centimeters of the target value are considered to be within specifications (i.e., acceptable). Suppose that 50 rods are collected for inspection from each of four processes and that the dotplots of their diameters are as follows:

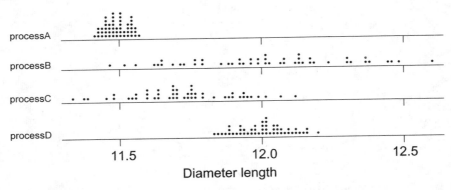

Write a paragraph describing each of these distributions, concentrating on the center and variability of the distributions. Also address each of the following questions in your paragraph:

- Which process is the best as is?
- Which process is the most stable, i.e., has the least variability in rod diameters?
- Which process is the least stable?
- Which process produces rods whose diameters are generally farthest from the target value?

Activity 3-10: College Tuitions

The following histogram displays the distribution of tuition charges (in thousands of dollars) in 1998–99 for four-year colleges in Pennsylvania. Notice that the software reports the midpoint of each subinterval. For example, the second rectangle indicates that 11 colleges had tuitions between $1250 and $3750 (including the $1250 but not the $3750).

(a) How many of the colleges had tuitions between $3750 and $6250?

(b) How many colleges had tuitions greater than $16,250? What proportion of the colleges listed had tuitions greater than $16,250?

(c) Can you determine exactly how many colleges had tuitions above $20,000? Explain.

(d) How many distinct clusters (or peaks) can you identify in the distribution? Roughly where do they fall? Can you come up with a reasonable explanation of which kinds of colleges tend to fall in which clusters?

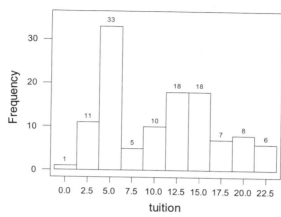

(e) Download TUITS.83l and use your calculator to produce a histogram with 10 subintervals. Then change the histogram to have 5 subintervals and then to have 50 subintervals. Comment on the different appearances of the distribution revealed by these histograms. Which do you think provides the most useful and informative display?

Activity 3-11: Parents' Ages (*cont.*)

The following data are the ages at which a sample of 35 American mothers first gave birth:

20	28	33	23	21	18	24	20	32	16	27	21	17	22
19	40	19	24	24	24	17	31	28	26	18	23	20	18
14	16	21	16	20	20	19							

(a) If you were to create a stemplot of these, how many stems would there be?

(b) In order to create more stems, create a split stemplot in which each stem is listed twice, with leaves of 0–4 appearing on the top stem and leaves of 5–9 on the bottom stem. The stems are drawn here, with the first four leaves (corresponding to 20, 28, 33, and 23) drawn in to get you started:

```
1 |
1 |
2 | 03
2 | 8
3 | 3
3 |
4 |
```

(c) Write a few sentences describing the distribution of these ages.

Activity 3-12: Parents' Ages (*cont.*)

(a) Examine the data collected from you and your classmates on the age of your mother when you were born. Produce a histogram of the distribution, making sure you clearly label both axes.

(b) Write a paragraph describing key features of this distribution.

(c) Identify the observational units here.

(d) Explain how this variable, age of your mother when you were born, differs from the variable in Activity 3-11 on page 59, age at which a mother first gives birth.

Activity 3-13: Marriage Ages

Listed below are the ages of a sample of 24 couples taken from marriage licenses filed in Cumberland County, Pennsylvania, in June and July of 1993.

couple	husband	wife	couple	husband	wife	couple	husband	wife
1	25	22	9	31	30	17	26	27
2	25	32	10	54	44	18	31	36
3	51	50	11	23	23	19	26	24
4	25	25	12	34	39	20	62	60
5	38	33	13	25	24	21	29	26
6	30	27	14	23	22	22	31	23
7	60	45	15	19	16	23	29	28
8	54	47	16	71	73	24	35	36

(a) Select *either* the husbands' ages or the wives' ages, and construct (by hand) a stemplot of their distribution. (Indicate which spouse you are analyzing.)

(b) Write a short paragraph describing the distribution of marriage ages for whichever spouse you chose.

Activity 3-14: Hitchcock Films

The following table lists the running times (in minutes) of the videotape versions of
22 movies directed by Alfred Hitchcock:

film	time	film	time
The Birds	119	Psycho	108
Dial M for Murder	105	Rear Window	113
Family Plot	120	Rebecca	132
Foreign Correspondent	120	Rope	81
Frenzy	116	Shadow of a Doubt	108
I Confess	108	Spellbound	111
The Man Who Knew Too Much	120	Strangers on a Train	101
Marnie	130	To Catch a Thief	103
North by Northwest	136	Topaz	126
Notorious	103	Under Capricorn	117
The Paradine Case	116	Vertigo	128

(a) Construct a stemplot of this distribution.
(b) Comment on key features of this distribution.
(c) One of these movies is particularly unusual in that all of the action takes place in
one room and Hitchcock filmed it without editing. Explain how you might be able
to identify this unusual film based on the distribution of the films' running times.

Activity 3-15: *Jurassic Park* Dinosaur Heights

In the blockbuster movie *Jurassic Park*, dinosaur clones run amok on a tropical island
intended to become mankind's greatest theme park. In Michael Crichton's novel on
which the movie was based, the examination of dotplots of dinosaur heights provides

the first clue that the dinosaurs are not as controlled as the park's creator would like to believe. Here are reproductions of two dotplots presented in the novel:

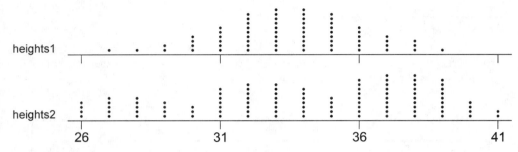

(a) Comment briefly on the most glaring difference in these two distributions of dinosaur heights.

(b) The cynical mathematician Ian Malcolm (a character in the novel) argues that one of these distributions is characteristic of a normal biological population, while the other is what one would have expected from a controlled population that had been introduced in three separate batches (as these dinosaurs had). Identify which distribution corresponds to which type of population.

(c) Take a closer look at the first distribution. There is something suspicious about it that suggests that it does not come from real data but rather from the mind of an author. Can you identify its suspicious quality?

Activity 3-16: Tennis Simulations

As part of a study investigating alternative scoring systems for the sport of tennis, researchers analyzed computer simulations of tennis matches. For 100 simulated games of tennis, the researchers recorded the number of points played in the game. (A game of tennis is won when one player wins four points, with a margin of at least two points over the opponent.) The data are tallied in the following table. As examples of reading this table, 12 games ended after just four points and 2 games required eighteen points to complete.

points in game	4	5	6	8	10	12	14	16	18
tally (count)	12	21	34	18	9	2	1	1	2

(a) Examine a visual display of these data (that you create either by hand or using your calculator; these data are stored in a list named TENN1 in the grouped file TENNSIM.83g), and comment on key features of the distribution.

(b) Describe and explain the unusual granularity that the distribution exhibits.

Activity 3-17: Turnpike Distances

The Pennsylvania Turnpike extends from Ohio in the west to New Jersey in the east. The distances (in miles) between its exits as one travels west to east are listed below.

exit	name	miles	exit	name	miles
1	Ohio Gateway	*	16	Carlisle	25
1A	New Castle	8	17	Gettysburg Pike	9.8
2	Beaver Valley	3.4	18	Harrisburg West Shore	5.9
3	Cranberry	15.6	19	Harrisburg East	5.4
4	Butler Valley	10.7	20	Lebanon–Lancaster	19
5	Allegheny Valley	8.6	21	Reading	19.1
6	Pittsburgh	8.9	22	Morgantown	12.8
7	Irwin	10.8	23	Downingtown	13.7
8	New Stanton	8.1	24	Valley Forge	14.3
9	Donegal	15.2	25	Norristown	6.8
10	Somerset	19.2	26	Fort Washington	5.4
11	Bedford	35.6	27	Willow Grove	4.4
12	Breezewood	15.9	28	Philadelphia	8.4
13	Fort Littleton	18.1	29	Delaware Valley	6.4
14	Willow Hill	9.1	30	Delaware River Bridge	1.3
15	Blue Mountain	12.7			

(a) In preparation for constructing a histogram to display this distribution of distances, count how many values fall into each of the subintervals listed in the table below:

miles	0.1–5.0	5.1–10.0	10.1–15.0	15.1–20.0
tally (count)				
miles	20.1–25.0	25.1–30.0	30.1–35.0	35.1–40.0
tally (count)				

(b) Construct (by hand) a histogram of this distribution, clearly labeling both axes.

(c) Comment in a few sentences on key features of this distribution.

(d) Find a value such that half of the exits are more than this value apart and half are less than this value. Also explain why such a value is not unique.

(e) If a person has to drive between consecutive exits and has only enough gasoline to drive 20 miles, is she very likely to make it? Assume that you do not know *which* exits she is driving between, and explain your answer.

(f) Repeat (e) if she has only enough gasoline to drive 10 miles.

Activity 3-18: ATM Withdrawals

The following table lists both the number and total amount of cash withdrawals by one individual from an automatic teller machine (ATM) during each month of 1994. (For example, January saw nine withdrawals, which totaled $1020.) The data can be downloaded through ATM94.83g.

month	#	total	month	#	total	month	#	total
January	9	$1020	May	8	$980	September	10	$850
February	8	$890	June	13	$1240	October	10	$1010
March	10	$970	July	4	$750	November	7	$860
April	9	$800	August	9	$1130	December	14	$1680

(a) Create a dotplot of the distribution of the *number* of withdrawals in each month, and comment briefly on the distribution (ATMNM).

(b) Create a dotplot of the distribution of the *total* amount withdrawn in each month, and comment briefly on the distribution (ATMTO).

(c) Which month had the most withdrawals and by far the most cash withdrawn? Suggest an explanation for this.

(d) This individual took two extended trips in one of these months. Can you guess which month it was based on these data? Explain.

Activity 3-19: Students' Salary Expectations

Produce a histogram of the responses to the question about the yearly salary at which you expect to begin your career. Then write a paragraph describing key features of the distribution, keeping in mind the checklist of features derived in class.

Activity 3-20: Fan Cost Index (*cont.*)

Consider again the data on the fan cost index from Activity 2-3 on page 29. Below are dotplots of the distributions of prices of a program and of a cap:

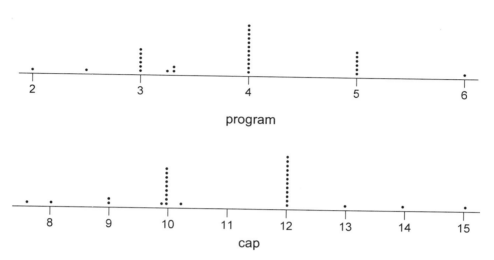

(a) Write a paragraph describing these distributions.

(b) Comment on the granularity in these distributions. Do a few observations violate the pattern of the rest in terms of granularity? Explain.

(c) Look back at the data listing to identify the cities/teams that do not follow the granularity of the others in either cap or program price. Suggest an explanation for why these cities/teams stand out.

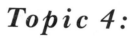

Topic 4:

MEASURES OF CENTER

OVERVIEW

You have been exploring distributions of data, representing them graphically and describing their key features verbally. It is often handy to have a single numerical measure to summarize a certain aspect of a distribution. In this topic you will encounter some of the more common measures of the **center** of a distribution, investigate their properties, apply them to some genuine data, and expose some of their limitations.

OBJECTIVES

- To learn to calculate the **mean**, **median**, and **mode** for summarizing the center of a distribution of data.
- To investigate and discover properties of these summary statistics.
- To explore the property of **resistance** as it applies to these statistics.
- To develop an awareness of situations in which certain measures are and are not appropriate.
- To recognize that these numerical measures do not summarize a distribution completely.
- To acquire the ability to expose faulty conclusions based on misunderstandings of these measures.

PRELIMINARIES

1. Who is the longest-serving justice currently on the Supreme Court? Who is the newest member?

   Alito, Sotomayor

2. Take a guess as to how long a typical member of the U.S. Supreme Court has served.

   30 years

3. Do you suspect that the average tenure of a current Supreme Court justice accurately estimates the average tenure of all Supreme Court justices in history, or do you think it underestimates or overestimates that number?

4. Take a guess concerning the distance from the sun for a typical planet in our solar system.

5. Take a guess concerning the average size (number of students) in a class at your school.

6. Guess the age of your instructor for this course (or, if your instructor prefers, another teacher or administrator at your school), and record the responses of yourself and your classmates.

IN-CLASS ACTIVITIES

Activity 4-1: Supreme Court Service

The table below lists the justices of the Supreme Court of the United States as of October 1999. Also listed is the year of appointment and the tenure (years of service as of 1999) for each.

Justice	year	tenure	Justice	year	tenure	Justice	year	tenure
Rehnquist	1972	27	Scalia	1986	13	Thomas	1991	8
Stevens	1975	24	Kennedy	1988	11	Ginsburg	1993	6
O'Connor	1981	18	Souter	1990	9	Breyer	1994	5

(a) Create a dotplot of the distribution of these years of service.

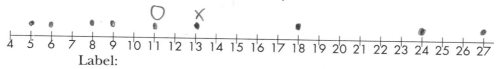

Label:

(b) What number might you choose if you were asked to select a single number to represent the *center* of this distribution? Briefly explain how you arrive at this choice.

11, because there are 4 to the left & 4 to the left

We will consider three commonly used measures of the *center* of a distribution:

- The **mean** is the ordinary arithmetic average, found by adding up the values of the observations and dividing by the number of observations. The mean can be thought of as the balance point of the distribution.
- The **median** is the middle observation (once they are arranged in order).
- The **mode** is the most common value, i.e., the one that occurs most frequently.

If there is a tie b/t modes, they both are
If there is More than 2, then it is no Mode

(c) Calculate the *mean* of these years of service. Mark this value on the dotplot above with an "x".

13.4

(d) How many of the nine justices have served more than the mean number of years? How many have served less than the mean number of years?

3 , 6

(e) Calculate the *median* of these years of service. Mark this value on the dotplot above with an "o".

11

(f) How many of the nine justices have served more than the median number of years? How many have served less than the median number of years?

4, 4

It is easy enough to pick out the median (the middle observation) in a small set of data, but we will try to come up with a general rule for finding the *location* of the median. The first step, of course, is to arrange the observations *in order* from smallest to largest. Let *n* denote the **sample size**, the number of observations in the data set.

(g) With the data you analyzed above (where $n = 9$), the median turned out to be which ordered observation (the second, the third, the fourth, ...)?

(h) Suppose that there had been $n = 5$ observations; the median would have been which (ordered) one? What if there had been $n = 7$ observations? How about if $n = 11$? What about $n = 13$?

$n = 5$: $n = 7$: $n = 9$: $n = 11$: $n = 13$:

n→odd − 1 median
n→even − avg the middle 2

(i) Try to discover the pattern in the question above to determine a general formula (in terms of the sample size n) for finding the *location* of the median of an odd number of ordered observations.

5th data point

$$\frac{n+1}{2}$$

(j) If Justice Rehnquist were to resign suddenly, the Court would consist of eight justices until his replacement was nominated and confirmed. Explain why calculating the median years of service among the remaining eight justices is not as straightforward as it is with nine justices.

If there is an even number of observations, the median is defined to be the average (mean) of the middle two observations.

(k) With this definition in mind, calculate the median years of service of the eight justices excluding Justice Rehnquist.

These activities should have led you to discover that the median of an odd number of observations can be found in position $\frac{(n+1)}{2}$ (once the values are arranged in order), while the median of an even number of observations is the mean of the values occupying positions $\frac{n}{2}$ and $\frac{n}{2}+1$.

(l) Among the 99 justices who served prior to these nine, the years served had a mean of 15.95 and a median of 15. Do the mean and median of the current justices accurately estimate the mean and median for all previous justices? If not, do they underestimate or overestimate? Explain why this makes sense in a sentence or two.

Activity 4-2: Properties of Averages

More important than the ability to calculate these values is understanding their properties and interpreting them correctly.

Refer to Activity 3-6 on page 55, where you studied distributions of highway miles per gallon ratings for 1999 cars classified as small, family, and large. The dotplots are reproduced below:

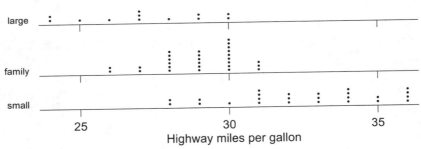

(a) Based on examining those dotplots, which car type do you expect to have the highest mean? the lowest? Take a guess as to what the mean and median MPG will be for each car type. Record your predictions in the table:

	small cars	family cars	large cars
mean			
median			

(b) Download CARSMPG.83g into your calculator. Use your calculator to determine the mean and median values of the highway MPG ratings for these three groups by using the 1-Var Stats option from the STAT CALC menu. For example, for the list SMALL, press the ⬚STAT⬚ button and then use the right arrow to highlight CALC. From this menu, choose 1-Var Stats and press ⬚ENTER⬚. At the prompt in the home screen, choose the SMALL list from the LIST directory. You should see the following on your home screen, and then press ⬚ENTER⬚:

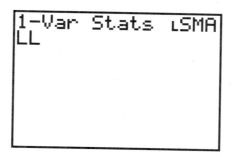

Notice the arrow in the lower left corner of your screen. This arrow lets you know that you can scroll down to view more information. Use the scroll buttons on your calculator to view the rest of the information. Write the mean $\bar{x}$ and median (MED) below in the "small cars" column. Repeat these steps for the family cars (FAMLY) and the large cars (LARGE).

	small cars	family cars	large cars
mean			
median			

(c) Were your predictions about the categories with the highest and lowest means and medians correct?

(d) How accurate were your guesses for the means and medians?

Now consider a dotplot of U.S. Senators' years of service:

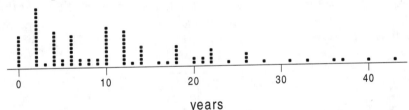

years

(e) Based on the shape of the distribution, do you expect the mean and median of this distribution to be close together, do you expect the mean to be noticeably higher than the median, or do you expect the median to be noticeably higher than the mean?

Median higher than mean

(f) Answer question (e) with respect to the distribution of states' percentages of urban residents, displayed below:

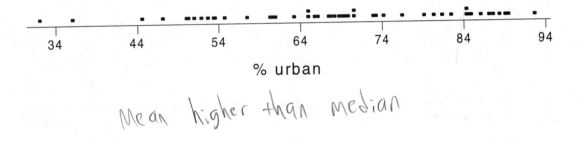

% urban

Mean higher than median

(g) Answer question (e) with respect to the distribution of textbook prices, displayed below:

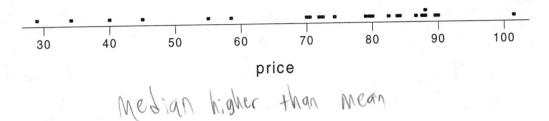

price

Median higher than mean

(h) Download AVERAGES.83g into your calculator and compute the mean and median values for these three variables (SRVC, URBAN, and PRICE). Record the results below, and indicate whether the calculations confirm your intuitions expressed in the three previous questions.

	Senators' service	% urban residents	textbook prices
mean	11.09	68.2	73.3
median	10	68.8	79.5

$\overline{X}$ = Mean

Medx = Median

(i) Summarize what the previous question reveals about how the shape of the distribution (symmetric or skewed in one direction or the other) relates to the relative location of the mean and median.

When something is skewed to the right, then the mean is to the right (toward the tail)

> These data should have revealed that the mean is close to the median with symmetric distributions, while the mean is greater than the median with skewed right distributions and the mean is less than the median with skewed left distributions. In other words, when one tail of the distribution is longer, the mean follows the tail.

(j) Does it make sense to talk about the mean gender or the median party of the Senators? How about the mode gender or the mode party? Explain.

No, you need #'s to calculate things

One can calculate the mean only with quantitative variables. The median can be found with quantitative variables and with categorical variables for which a clear ordering exists among the categories. The mode applies to all categorical variables but is only useful with some quantitative variables.

Activity 4-3: Rowers' Weights (*cont.*)

Reconsider the data from Activity 3-3 on page 48 on weights of rowers on the 1996 U.S. Olympic men's rowing team.

(a) The data on weights of rowers on the 1996 U.S. Olympic men's rowing team are stored in a list file named ROWWT.83l. Download this file into your calculator. Use your calculator to determine the mean and median of these weights. Record the results in the table below.

(b) If you were told only the mean and median weights, but you were not given the individual weights or shown a visual display of the weights, would you have a complete understanding of the distribution of rowers' weights? Explain.

No, you need the weights

(c) In what direction do you expect the mean and median to change if the coxswain is removed from the analysis? Explain briefly.

Go up b/c you are getting rid of the outlier

(d) Now remove the coxswain's weight and use your calculator to recalculate the mean and median. To do this, place your list into the Stat List Editor and then use your DEL button to remove the coxswain. Again record the results in the table. Was your prediction about the direction of the change correct?

(e) Which measure (mean or median) do you expect to change more if all of the lightweight rowers are removed from the analysis? Explain briefly.

The mean

(f) Now remove the lightweight rowers (in addition to the coxswain) from the analysis, and use your calculator to recalculate the mean and median. Record these in the table below. Which measure was more affected? Was your prediction correct?

(g) Now change the weight of the heaviest rower to 324 pounds instead of 224. Use your calculator to determine the mean and median after making this change (still with the lightweights and coxswain removed). Record the results in the table.

(h) Finally, suppose that the heaviest rower's weight had inadvertently been recorded as 2224 rather than 224. Recalculate the mean and median with this change. How many rowers weigh less than the mean? Do you think that these values are extreme enough to draw attention to the typographical error?

	whole team	without coxswain	also without lightweights	with max at 324	with max at 2224
mean	191.8	194.7	206.3	211.6	312
median	202.5	205	207	207	207

(handwritten column labels above table: A, D, F, G, H)

> A measure whose value is relatively unaffected by the presence of outliers in a distribution is said to be **resistant**.

(i) Based on these calculations, would you say that the mean or the median is resistant? Explain why this makes sense, basing your argument on the definition of each.

(j) Is there any limit in principle to how large or small the mean can become just by changing *one* of the values in the distribution?

Activity 4-4: Readability of Cancer Pamphlets

Researchers in Philadelphia investigated whether pamphlets containing information for cancer patients are written at a level that the cancer patients can comprehend. They applied tests to measure the reading levels of 63 cancer patients and also the readability levels of 30 cancer pamphlets (based on such factors as the lengths of sentences and number of polysyllabic words). These numbers correspond to grade levels, but patient reading levels of under grade 3 and above grade 12 are not determined exactly.

The following tables indicate the number of patients at each reading level and the number of pamphlets at each readability level. The dotplots reveal the distributions on the same scale (with "below 3" appearing at level 2 and "above 12" at level 13 for convenience).

patients' reading level	< 3	3	4	5	6	7	8	9	10	11	12	>12
count	6	4	4	3	3	2	6	5	4	7	2	17

pamphlets' readability	6	7	8	9	10	11	12	13	14	15	16
count	3	3	8	4	1	1	4	2	1	2	1

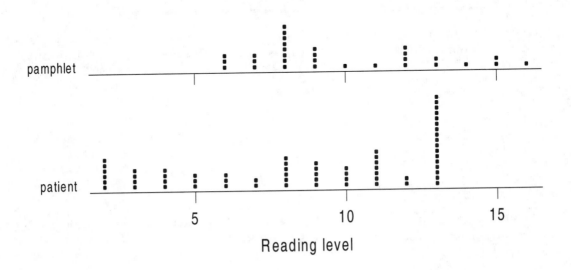

(a) Explain why the form of the data does not allow one to calculate the *mean* reading skill level of a patient.

(b) Determine the *median* reading level of a patient. (Be sure to consider the counts.)

(c) Determine the median readability level of a pamphlet.

(d) How do these medians compare? Are they fairly close?

(e) Does the closeness of these medians indicate that the pamphlets are well matched to the patients' reading levels? Compare the dotplots above to guide your thinking.

(f) What proportion of the patients do not have the reading skill level necessary to read even the simplest pamphlet in the study? (Examine the dotplots above to address this question.)

This activity illustrates that while measures of center are often important, they do not summarize all aspects of a distribution.

WRAP-UP

You have explored in this topic how to calculate a number of measures of the center of a distribution. You have discovered many properties of the **mean**, **median**, and **mode** (such as the important concept of **resistance**) and discovered that these statistics can produce very different values with certain data sets. Most importantly, you have learned that these statistics measure only *one* aspect of a distribution and that you must combine these numerical measures with what you already know about displaying distributions visually and describing them verbally.

In the next topic you will discover similar measures of another aspect of a distribution of data: its variability.

HOMEWORK ACTIVITIES

Activity 4-5: Planetary Measurements

The following table lists the average distance from the Sun (in millions of miles), diameter (in miles), and period of revolution around the Sun (in Earth days) for the nine planets of our solar system.

planet	distance	diameter	period
Mercury	36	3,030	88
Venus	67	7,520	225
Earth	93	7,926	365
Mars	142	4,217	687
Jupiter	484	88,838	4,332
Saturn	887	74,896	10,760
Uranus	1,765	31,762	30,684
Neptune	2,791	30,774	60,188
Pluto	3,654	1,428	90,467

(a) Calculate (by hand) the median value of each of these variables.

(b) If a classmate uses the $(n + 1)/2$ formula and obtains a median diameter of 88,838 miles, what do you think would be the most likely cause of his or her mistake? (Notice that this is Jupiter's diameter.)

(c) Without doing any calculations, would you expect the mean values of these variables to be close to, less than, or greater than the medians? Explain.

Activity 4-6: Age Guesses

(a) Produce a dotplot of the guesses of your instructor's age collected in the "Preliminaries" section (save the list as AGEGU).

(b) Write a paragraph describing the distribution of these guesses.

(c) If your instructor is willing to reveal her or his actual age, determine how many guesses are correct/above/below. Comment on the accuracy of the guesses.

(d) Based on this distribution, do you expect the mean to be greater than the median, less than the median, or very close to the median?

(e) Calculate (either by hand or using your calculator) the mean and median of these age guesses. Was your prediction about the relative values of the mean and median confirmed?

(f) If the data contain any obvious outliers, remove them and recalculate the mean and median. Then comment on how much these values changed.

(g) If your instructor is willing to reveal her or his actual age, comment on how well the mean and median of the guesses approximate the actual value.

Activity 4-7: Students' Data (*cont.*)

Refer back to any of the data collected in class, such as states visited or signature lengths or Scrabble points or mothers' ages.

(a) Judging from a dotplot of the distribution, would you expect the mean to be greater than the median, to be close to the median, or to be less than the median? Explain.

(b) Calculate the mean and median of the distribution, and comment on whether your answer to (a) is confirmed or refuted.

Activity 4-8: Salary Expectations (*cont.*)

Reconsider the data collected in Topic 3 on students' expectations for the first year's salary of their career.

(a) Calculate the mean and median of these salary expectations.

(b) Suppose that another student had been in class who expected to begin her career with a salary of $1,000,000. Perform the appropriate calculation to determine how much this student would have affected the mean and median.

(c) Which is affected more by the inclusion of the student in (b), the mean or the median? Explain why this makes sense.

Activity 4-9: Tennis Simulations (*cont.*)

Refer back to the data presented in Activity 3-16 on page 62 concerning lengths (as measured in points) of simulated tennis games.

(a) Determine the mean, median, and mode of the game lengths. (You may use your calculator.)

The experiment also simulated tennis games using the "no-ad" scoring system, which awards the game to the first player to reach four points. The data are tallied below:

points in game	4	5	6	7
tally (count)	13	22	33	32

(b) Enter these numbers into your calculator, putting the first row in a list named POINT and the second row in a list named TALLY. To determine the mean, median, and mode of the game lengths with no-ad scoring, use the 1-Var Stats feature of your calculator found in the STAT CALC menu and select both the POINT and TALLY lists and then press the ENTER key: `1-Var Stats LPOINT, LTALLY`. Write a sentence or two comparing these values with their counterparts from conventional scoring.

Activity 4-10: ATM Withdrawals (*cont.*)

Reconsider the data presented in Activity 3-18 on page 64 concerning ATM withdrawals. There was a total of 111 withdrawals during the year.

(a) Do the data as presented in that activity enable you to identify the mode value of those 111 withdrawal amounts? If so, identify the mode. If not, explain.

(b) Do the data as presented in that activity enable you to determine the median value of those 111 withdrawal amounts? If so, determine the median. If not, explain.

(c) Do the data as presented in that activity enable you to calculate the mean value of those 111 withdrawal amounts? If so, calculate the mean. If not, explain.

(d) The following table summarizes the individual amounts of the 111 withdrawals. For example, 17 withdrawals were for the amount of $20 and 3 were for the amount of $50; no withdrawals were made of $40. Use this new information to calculate whichever of the mode, median, and mean you could not calculate before.

amount	$20	$50	$60	$100	$120	$140	$150	$160	$200	$240	$250
count	17	3	7	37	3	8	16	10	8	1	1

(e) Create (by hand) a dotplot of the distribution of these 111 withdrawals. Comment on key features of the distribution, including its unusual granularity.

Activity 4-11: Gender of Physicians (*cont.*)

Consider the data from Activity 2-2 on page 25 about the numbers of male and female physicians of different specialties in the U.S. (GENPHYS.83g).

(a) Use your calculator to produce a dotplot of the distribution of percentage of women in the various specialties. [*TI Hint*: You stored this variable as PERCT in Activity 2-2.] Use your calculator to compute the mean and median of this distribution.

(b) Suppose that ten doctors are gathered for a meeting in room A, while twenty have convened in room B and fifty in room C. If six of the doctors in room A are women, compared to eight in room B and eleven in room C, calculate the percentage of women in each room.

(c) Find the mean of the three percentages found in (b). Does this equal the overall percentage of women among the doctors in the three rooms combined? Explain.

(d) Does the mean that you calculated in (a) equal the percentage of women among all specialties of U.S. physicians? If not, use your calculator to compute this percentage from the data given.

Activity 4-12: Creating Examples

For each of the following properties, try to construct a data set of ten hypothetical exam scores that satisfies the property. Also produce a dotplot of your hypothetical data in each case. Assume that the exam scores are integers between 0 and 100, inclusive. You may use your calculator.

(a) 90% of the scores are greater than the mean

(b) the mean is greater than twice the mode

(c) the mean is less than two-thirds the median

(d) the mean equals the median but the mode is greater than twice the mean

(e) the mean does not equal the median and none of the scores are between the mean and the median

Activity 4-13: Incorrect Conclusions

For each of the following arguments, explain why the conclusion drawn is not valid. Also include a simple hypothetical example that illustrates that the conclusion drawn need not follow from the information.

(a) A real estate agent notes that the mean housing price for an area is $125,780 and concludes that half of the houses in the area cost more than that.

(b) A businesswoman calculates that the median cost of the five business trips that she took in a month is $600 and concludes that the total cost must have been $3000.

(c) A company executive concludes that an accountant must have made a mistake because she prepared a report stating that 90% of the company's employees earn less than the mean salary.

(d) A restaurant owner decides that more than half of her customers prefer chocolate ice cream because chocolate is the mode when customers are offered their choice of chocolate, vanilla, and strawberry.

Activity 4-14: Properties of Averages (*cont.*)

Suppose that an instructor is teaching two sections of a course and that she calculates the mean exam score to be 60 for section 1 and 90 for section 2.

(a) Do you have enough information to determine the mean exam score for the two sections combined? Explain.

(b) What can you say with certainty about the value of the overall mean for the two sections combined?

(c) Without seeing all of the individual students' exam scores, what information would you need to be able to calculate the overall mean?

(d) Suppose that section 1 contains 20 students and section 2 contains 30 students. Calculate the overall mean exam score. Is the overall mean closer to 60 or to 90?

(e) Give an example of sample sizes for the two sections for which the overall mean turns out to be less than 65.

(f) If you do not know the numbers of students in the sections but do know that there is the same number of students in the two sections, can you determine the overall mean? If not, explain. If so, do.

(g) Explain how it could happen that a student could transfer from section 1 to section 2 and cause the mean score for each section to decrease. What would have to be true about that student's score?

Activity 4-15: Geyser Eruptions (*cont.*)

Refer again to the data displayed in Activity 3-5 on page 51 on the times between eruptions of Old Faithful geyser. The mean of these times turns out to be 71 minutes, and the median is 75 minutes.

(a) Are many of the inter-eruption times very close to 71 or 75 minutes? In other words, does either do a good job of indicating where most of the data values fall? Explain.

(b) Considering the shape of this distribution, can one number do a thorough job of describing its center? Explain.

Activity 4-16: Properties of Averages (*cont.*)

In this activity you will investigate the effects on measure of center of adding a constant to a set of data and of multiplying by a constant.

Suppose that the Supreme Court justices listed in Activity 4-1 on page 69 are still serving in the year 2004, so each of their years of service at that time would be five years more than listed in Activity 4-1.

(a) Record these new data on years of service.

(b) Determine the mean and median of these new years of service at that point.

(c) How do the measures of center for the new data compare with those for the original data?

Now consider the placement exam scores in Activity 3-8 on page 57. Suppose that every score is multiplied by 5 in order to put it on a percentage correct scale.

(d) Determine the mean and median for these new scores.

(e) How do these statistics compare with those for the original data?

(f) If temperature data such as those in Activity 3-5 on page 51 were converted from Fahrenheit to Celsius, they would first have 32 subtracted and then be multiplied by 5/9. Without actually doing the conversion, what do you think the effect would be on measures of center? [*Hint*: First consider the effect of subtracting a constant, then consider the effect of multiplying by a constant.]

Topic 5:

MEASURES OF SPREAD

OVERVIEW

In the previous topic you explored important numerical measures of the center of a distribution. In this topic you will investigate similar numerical measures of a distribution's **variability**. These measures will also lead you to discover another visual display (the **boxplot**) and a very important technique (**standardization**) that will appear throughout the course.

OBJECTIVES

- To learn to calculate certain statistics (**range, interquartile range, standard deviation**) for summarizing the variability of a distribution of data.
- To discover the **five-number summary** of a distribution.
- To explore the **boxplot** as another convenient and informative visual display of a distribution.
- To investigate and determine properties of these summary statistics.
- To understand the **empirical rule** as a means for interpreting the value of standard deviation for certain types of distributions.
- To appreciate the applicability of calculating **z-scores** for comparing distributions of different variables.
- To recognize some of the limitations of these measures of variability.

5

PRELIMINARIES

1. Take a guess concerning the average temperature in January in Raleigh. Do the same for the average temperature in San Francisco in January.

   Raleigh:

   San Francisco:

2. Take a guess concerning the average temperature in July in Raleigh. Do the same for the average temperature in San Francisco in July.

   Raleigh:

   San Francisco:

3. Think about the average temperature for Raleigh for each of the twelve months of the year and the average temperature for San Francisco for the twelve months. For which city do you expect to see more variability in those twelve temperatures?

4. Think about the average high temperatures in January for selected cities from across the United States. Would you expect this distribution to be more or less variable than the average high temperatures in July for those same cities?

5. Think of what you consider the ideal outdoor temperature: If you could make today's high temperature anything you wish, what temperature would you choose (in degrees Fahrenheit)?

6. Gather and record these data on ideal temperature from yourself and your classmates.

IN-CLASS ACTIVITIES

Activity 5-1: City Temperatures

The following table reports the average monthly temperatures for San Francisco, California, and for Raleigh, North Carolina. Dotplots of these twelve temperatures for each city appear below.

	Jan	Feb	Mar	Apr	May	Jun	Jul	Aug	Sep	Oct	Nov	Dec
Raleigh	39	42	50	59	67	74	78	77	71	60	51	43
San Francisco	49	52	53	56	58	62	63	64	65	61	55	49

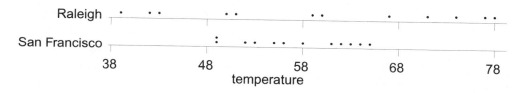

(a) Calculate the median value of these temperatures for each city. (Remember to arrange the data in order first.) Are these medians fairly close?

Raleigh median: $\dfrac{74 + 78}{2} = 76$

San Francisco median: $\dfrac{62 + 63}{2} = 62.5$

Close? No! really

(b) Since the centers of these distributions are very close, can one conclude that there is no difference between these two cities with regard to monthly temperatures? Explain.

No

(c) Which city has more variability in its monthly temperatures?

Raleigh

(d) What is the highest monthly temperature for Raleigh? What is the lowest? Calculate the difference between them.

highest: 78   lowest: 39   *(range)* difference: 39

(e) What is the highest monthly temperature for San Francisco? What is the lowest? Calculate the difference between them.

highest: 65   lowest: 49   *(range)* difference: 16

A very simple, but not particularly useful, measure of variability is the **range**, calculated as the difference between the maximum and minimum values in the data.

Another measure of variability is the **interquartile range** (IQR), the difference between the **upper quartile** and the **lower quartile** of the distribution. The *lower quartile* (or the 25th percentile) is the value such that about 25% of the observations fall below it and 75% above it, while the *upper quartile* (or 75th percentile) is the value such that about 75% of the observations fall below it and 25% above it. Thus, the IQR is the range of the middle 50% of the data. To find the lower quartile, one finds the median of those observations falling below the location of the actual median. Similarly, the upper quartile is the median of those observations falling above the location of the actual median. (When there is an odd number of observations, one does *not* include the actual median in either group.)

(f) To find the lower quartile for Raleigh, first list the observations that fall below the location of the median. Then determine the median of this list.

(g) Similarly, to find the upper quartile of Raleigh's distribution, list the observations that fall above the location of the median. Then determine the median of this list.

(h) Verify that one-quarter of the values fall below the lower quartile, one-quarter above the upper quartile, and half in between them.

(i) Calculate the interquartile range (IQR) of Raleigh's distribution by determining the difference between the quartiles.

(j) Is the IQR of Raleigh's monthly temperatures greater or less than the IQR for San Francisco's monthly temperatures, which turns out to be 10? Is this what your answer to (c) would have predicted?

> The median, quartiles, and extremes (minimum and maximum) of a distribution constitute its ***five-number summary***, so called because it provides a quick and convenient description of where the four quarters of the data fall.

5

(k) The table below reports the five-number summary for San Francisco's monthly temperatures. Fill in this summary for Raleigh's monthly temperatures.

	minimum	lower quartile	median	upper quartile	maximum
Raleigh					
San Francisco	49	52.5	57	62.5	65

> The five-number summary forms the basis for a visual display called a **boxplot**. To construct a boxplot, one draws a "box" between the quartiles, thus indicating where the middle 50% of the data fall. Horizontal lines called "whiskers" are then extended from the middle of the sides of the box to the minimum and to the maximum. The median is then marked with a vertical line inside the box.

(1) Construct a boxplot of the distribution of Raleigh's monthly temperatures below the boxplot for San Francisco's:

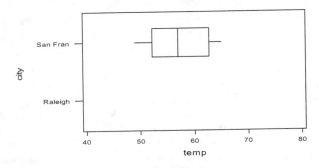

Activity 5-2: City Temperatures (*cont.*)

Other measures of variability examine observations' deviations from the mean.

(a) The mean of Raleigh's monthly temperatures is 59.25 degrees Fahrenheit. Complete the two missing entries (for March and for July) in the "deviation from mean" row below.

(b) Add those deviations and then calculate their sum. What value should this sum equal (if there are no rounding discrepancies)?

(c) Since a measure of spread is concerned with distances from the mean rather than direction from the mean, work with the absolute values of the deviations by completing the missing entries in that column of the table. Then add those absolute deviations.

month	temperature	deviation from mean	absolute deviation	squared deviation
January	39	–20.25	20.25	410.06
February	42	–17.25	17.25	297.56
March	50	(a)	(c)	(e)
April	59	–0.25	0.25	0.06
May	67	7.75	7.75	60.06
June	74	14.75	14.75	217.56
July	78	(a)	(c)	(e)
August	77	17.75	17.75	315.06
September	71	11.75	11.75	138.06
October	60	0.75	0.75	0.56
November	51	–8.25	8.25	68.06
December	43	–16.25	16.25	264.06
sum	711	(b)	(c)	(e)

(d) Calculate the average (mean) of these twelve absolute deviations from the overall mean temperature.

The measure of variability that you have just calculated is the ***mean absolute deviation*** (MAD). It is an intuitively sensible but not widely used measure of variability.

(e) An alternative to working with absolute values is to square the deviations from the mean. Complete the missing entries in the "squared deviations" column of the table, and then sum those squared deviations.

(f) Divide the sum of squared deviations by 11 (one less than the sample size).

(g) To convert back to the original units rather than square units, take the square root of your answer to (f).

> You have just calculated the **standard deviation** for these data, a widely used measure of variability. To compute the standard deviation, one calculates the difference between the mean and each observation and then squares that value for each observation. These squares are then added and the sum is divided by $n-1$ (one less than the sample size). The standard deviation is the square root of the result.

(h) Use your calculator to verify your computation of the standard deviation of the Raleigh monthly temperatures. Then use your calculator to determine the standard deviation of the San Francisco monthly temperatures. Which is larger? Is this what you expected?

5.75

It might be helpful to introduce some notation at this point. If we let x_i denote the ith observation and continue to let n denote the number of observations, then we can express the formula for calculating the sample mean (which we will denote by $\bar{x}$) by

$$\bar{x} = \frac{\sum\limits_{i=1}^{n} x_i}{n} = \frac{x_1 + x_2 + \cdots + x_n}{n},$$

where the Σ symbol means to add up what follows it. Similarly, we can express the formula for calculating the sample standard deviation (which we will denote by s_x or s) by

$$s_x = \sqrt{\frac{\sum\limits_{i=1}^{n} (x_i - \bar{x})^2}{n-1}}$$

Activity 5-3: Interpreting Spread and Boxplots

As with measures of center, the abilities to understand the properties of these measures of spread and to interpret them correctly are more important than being able to perform the calculations.

The following dotplots show the distributions of highway miles per gallon ratings for cars in the sports, small, and upscale classifications:

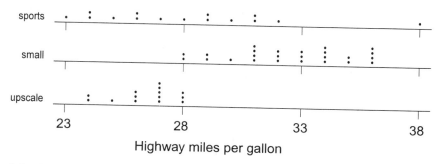

(a) The tables below report the interquartile range and standard deviation of these MPG ratings for each group. Your task is to use the dotplots (do not perform any calculations) to identify which pair of statistics (A, B, C) goes with which car classification (sports, small, upscale).

A:

IQR	std. dev.
1.5	1.34

car type:

upscale

B:

IQR	std. dev.
6.0	3.91

car type:

sports

C:

IQR	std. dev.
3.5	2.52

car type:

small

(b) Recall that dotplots for the data on Senators' years of service, states' percentages of urban residents, and textbook prices appear in Activity 4-2 on page 72. The table below reports the five-number summaries for these data:

	minimum	lower quartile	median	upper quartile	maximum
Senators' service	0	4	10	16.75	43
states' % urban	32.10	56.72	68.80	81.57	92.60
textbook prices	28.95	69.95	79.50	86.55	101.50

Boxplots of these variables appear below. (These vertical boxplots are just like horizontal ones but have lower values at the bottom and higher ones at the top.) Use the information in the dotplots and in the five-number summaries to identify which variable goes with which boxplot. Also write a sentence or two explaining your choices.

(i)   (ii)   (iii)

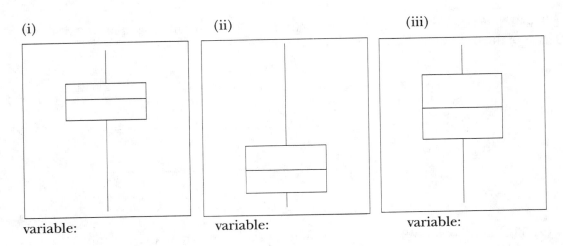

variable: variable: variable:

Activity 5-4: Supreme Court Justices (*cont.*)

Reconsider the data on Supreme Court justices' years of service as reported in Activity 4-1 on page 69.

(a) Enter the data into your calculator as a list named SPMCT and use the 1-Var Stats command in the STAT CALC menu to compute the standard deviation, quartiles, minimum, and maximum of the years of tenure. Use this information to calculate the interquartile range and range of the years of tenure and record these results in the first row of the table below. [*TI Hint*: On the calculator Q_1 represents the lower quartile and Q_3 represents the upper quartile.]

(b) Now imagine that Justice Rehnquist has served for 47 years rather than 27. Use the Stat List Editor to change his value to 47. Recalculate the standard deviation, IQR, and range in this case; record the values in the table.

(c) Finally, suppose that Justice Rehnquist's 27 years of service had been mistakenly recorded as 227 years. Again change this value and recalculate the standard deviation, IQR, and range, recording the values in the table.

	std. dev.	IQR	range
Justices			
Justices with "big" outlier			
Justices with "huge" outlier			

(d) Based on these calculations, which of these measures of spread would you say is *resistant?* Explain. (Recall the definition of resistance from Activity 4-3 on page 76.)

Activity 5-5: Placement Exam Scores (*cont.*)

Refer back to Activity 3-8 on page 57, which reported scores on a mathematics placement exam. The mean score on this exam is $\bar{x} = 10.221$. The standard deviation of the exam scores is $s_x = 3.859$. A histogram of this distribution follows:

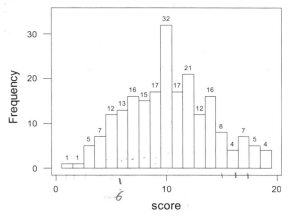

(a) Does this distribution appear to be roughly symmetric and mound-shaped?

(b) Consider the questions of how many scores fall within one standard deviation of the mean (denoted by $\bar{x} \pm s$). Determine the upper endpoint of this interval by adding the value of the standard deviation to that of the mean. Then determine the interval's lower endpoint by subtracting the value of the standard deviation from that of the mean.

$\bar{x} + s:$

$\bar{x} - s:$

(c) Look back at the table of tallied scores to determine how many of the 213 scores fall within *one* standard deviation of the mean. What proportion of the 213 scores is this?

(d) Determine how many of the 213 scores fall within *two* standard deviations of the mean, which turns out to be between 2.503 and 17.939. What proportion is this?

(e) Determine how many of the 213 scores fall within *three* standard deviations of the mean, which turns out to be between −1.356 and 21.798. What proportion is this?

You have discovered what is sometimes called the **empirical rule**. It turns out that with mound-shaped distributions, about 68% of the observations fall within one standard deviation of the mean, about 95% fall within two standard deviations of the mean, and virtually all of the observations fall within three standard deviations of the mean.

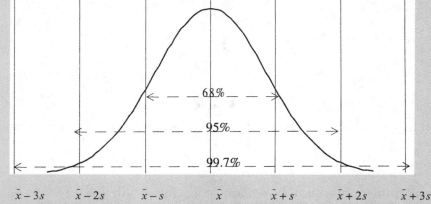

$$\bar{x} - 3s \qquad \bar{x} - 2s \qquad \bar{x} - s \qquad \bar{x} \qquad \bar{x} + s \qquad \bar{x} + 2s \qquad \bar{x} + 3s$$

Notice that this applies to mound-shaped distributions but not necessarily to distributions of other shapes. This rule provides a way of interpreting the value of a distribution's standard deviation. Note how this compares with the IQR, which reveals the range of the middle 50% of the data, and with the quartiles themselves, which specify the values between which that middle 50% falls.

Activity 5-6: SATs and ACTs

Suppose that a college admissions office needs to compare scores of students who take the Scholastic Aptitude Test (SAT) with those who take the American College Test (ACT). Suppose that among the college's applicants who take the SAT, scores have a mean of 896 and a standard deviation of 174. Further suppose that among the college's applicants who take the ACT, scores have a mean of 20.6 and a standard deviation of 5.2.

(a) If applicant Bobby scored 1080 on the SAT, how many points above the SAT mean did he score?

(b) If applicant Kathy scored 28 on the ACT, how many points above the ACT mean did she score?

(c) Is it sensible to conclude that since your answer to (a) is greater than your answer to (b), Bobby outperformed Kathy on the admissions test? Explain.

(d) Determine how many standard deviations above the mean Bobby scored by dividing your answer to (a) by the standard deviation of the SAT scores.

(e) Determine how many standard deviations above the mean Kathy scored by dividing your answer to (b) by the standard deviation of the ACT scores.

This activity illustrates the use of the standard deviation to make comparisons of individual values from different distributions. One calculates a **z-score**, or ***standardized score***, by subtracting the mean from the value of interest and then dividing by the standard deviation. These z-scores indicate how many standard deviations above (or below) the mean a particular value falls. One should use z-scores only when working with mound-shaped distributions, however.

(f) Which applicant has the higher z-score for his or her admissions test score?

(g) Explain in your own words which applicant performed better on his or her admissions test.

(h) Calculate the z-score for applicant Peter, who scored 740 on the SAT, and for applicant Kelly, who scored 19 on the ACT.

(i) Which of Peter and Kelly has the higher z-score?

(j) Under what conditions does a z-score turn out to be negative?

Activity 5-7: Value of Statistics (*cont.*)

Recall Activity 1-4 on page 9, where you examined ratings of the value of statistics on a 1–9 scale for five hypothetical classes in addition to your own (real) class. Now consider the ratings of five more hypothetical classes, where the data are given in the table and displayed in the histograms on the following page:

rating	1	2	3	4	5	6	7	8	9
class F count	0	3	1	5	7	2	4	2	0
class G count	1	2	3	4	5	4	3	2	1
class H count	1	0	0	0	22	0	0	0	1
class I count	12	0	0	0	1	0	0	0	12
class J count	2	2	2	2	2	2	2	2	2

(a) Judging from the tables and histograms, take a guess as to which has more variability between classes F and G.

(b) Judging from the tables and histograms, which would you say has the most variability among classes H, I, and J? Which would you say has the least variability?

most variability: least variability:

(c) Download VALUESFJ.83g and use your calculator to determine the range, interquartile range, and standard deviation of the ratings for each class. Record the results in the table:

	class F	class G	class H	class I	class J
range					
interquartile range					
standard deviation					

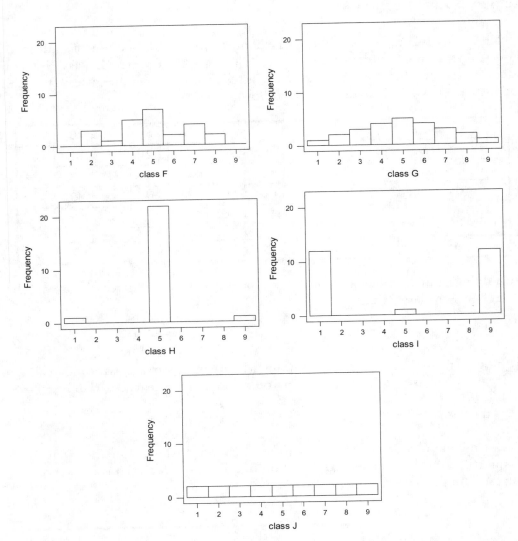

(d) Judging from these statistics that measure spread, does class F or G have more variability? Was your expectation in (a) correct?

(e) Judging from these statistics that measure spread, which among classes H, I, and J has the most variability? Which has the least? Was your expectation in (b) correct?

(f) Between classes F and G, which has more "bumpiness" or unevenness? Does that class have more or less variability than the other?

(g) Among classes H, I, and J, which distribution has the most distinct values? Does that class have the most variability of the three?

(h) Based on the previous two questions, does either "bumpiness" or "variety" relate directly to the concept of variability? Explain.

A common misconception about variability is to believe that a "bumpier" histogram indicates a more variable distribution, but this is not the case. Similarly, the number of distinct values represented in a histogram does not necessarily indicate greater variability.

WRAP-UP

In this topic you have learned to calculate and studied properties of the **range**, **interquartile range**, and **standard deviation** as measures of the variability of a distribution. You have also discovered a new visual display, the **boxplot**, and studied the **five-number summary** on which it is based. In addition, you have explored the **empirical rule** and **z-scores** as applications of the standard deviation.

To this point you have dealt primarily with one distribution at a time. Often it is more interesting to compare distributions between two or more groups. In the next topic you will discover some new techniques and also apply what you have learned so far to that task.

HOMEWORK ACTIVITIES

Activity 5-8: City Temperatures (*cont.*)

Average monthly temperatures for selected cities appear in the table below.
(a) Before looking closely at the table's entries, order the cities from the one that you expect to have the most variability in these monthly temperatures to the one that you expect to have the least variability.

month	Atlanta	Chicago	Honolulu	Minneapolis	Phoenix	Pittsburgh	San Antonio	San Diego
January	41	21	73	12	54	26	49	57
February	45	25	73	18	58	29	54	59
March	54	37	74	31	62	39	62	60
April	62	49	76	46	70	50	69	62
May	69	59	78	59	79	60	76	64
June	76	69	79	68	88	68	82	67
July	79	73	81	74	94	72	85	71
August	78	72	81	71	92	71	85	73
September	73	64	81	61	86	64	79	71
October	62	53	80	49	75	52	70	68
November	53	40	77	33	62	42	60	62
December	45	27	74	18	54	32	52	57

(b) Use your calculator (CITYTEMPS.83g) to determine the standard deviations of these monthly temperatures for each city. Report these in order from largest to smallest.

(c) How closely does the ordering of cities in (b) match your expectation from (a)? Explain.

(d) Calculate the range of these temperatures for each city. Does the ordering of the cities by the magnitude of their ranges correspond with the ordering by the magnitude of their standard deviations? Calculate the interquartile range of these temperatures for each city. Does the ordering of the cities by the magnitude of their ranges correspond with the ordering by the magnitude of their standard deviations?

(e) Select one city that you expect to have a large mean absolute deviation and one that you expect to have a small MAD. Calculate the MAD for each of these two cities. Do they compare as you expected?

Activity 5-9: Hypothetical Manufacturing Processes (*cont.*)

Look back at the dotplots from Activity 3-9 on page 58 of data from hypothetical manufacturing processes. The following table lists the means and standard deviations of these processes. Match each dotplot (processes A, B, C, and D) with its numerical statistics (process 1, 2, 3, or 4). Explain your answers.

	process 1	process 2	process 3	process 4
mean	12.008	12.004	11.493	11.723
std. dev.	0.274	0.089	0.041	0.18

Activity 5-10: Climatic Conditions

The following table lists average high temperatures in January and in July for selected cities from across the U.S.

city	Jan hi	July hi	city	Jan hi	July hi	city	Jan hi	July hi
Atlanta	50.4	88.0	Kansas City	34.7	88.7	Phoenix	65.9	105.9
Baltimore	40.2	87.2	Los Angeles	65.7	75.3	Pittsburgh	33.7	82.6
Boston	35.7	81.8	Miami	75.2	89.0	St. Louis	37.7	89.3
Chicago	29.0	83.7	Minneapolis	20.7	84.0	Salt Lake City	36.4	92.2
Cleveland	31.9	82.4	Nashville	45.9	89.5	San Diego	65.9	76.2
Dallas	54.1	96.5	New Orleans	60.8	90.6	San Francisco	55.6	71.6
Denver	43.2	88.2	New York	37.6	85.2	Seattle	45.0	75.2
Detroit	30.3	83.3	Philadelphia	37.9	82.6	Washington	42.3	88.5
Houston	61.0	92.7						

(a) Calculate (by hand or using your calculator and CLIMATE.83g) the interquartile range for the January high temperatures and for the July high temperatures.

(b) Use your calculator to determine the standard deviations for both variables.

(c) Which variable has greater variability in its distribution? Comment on the accuracy of your guess from the "Preliminaries" section.

(d) Which variable generally has higher temperatures, January or July high temperatures?

(e) Do you think that if one variable tends to cover larger values than another, then the former variable must have more variability in its values as well? Explain.

Activity 5-11: Planetary Measurements (*cont.*)

Refer back to the data presented in Activity 4-5 on page 80 concerning planetary measurements.
(a) Calculate (by hand) the five-number summary of distance from the sun.
(b) Draw (by hand) a boxplot of the distribution of distances from the sun.
(c) Would you classify this distribution as roughly symmetric, skewed left, or skewed right?

Activity 5-12: Word Lengths (*cont.*)

Reconsider the data collected in Topic 1 concerning the lengths of your words. Calculate (by hand) the five-number summary of this distribution and draw (also by hand) a boxplot. Comment on what the boxplot reveals about the distribution of your word lengths.

Activity 5-13: Tennis Simulations (*cont.*)

Consider the data from Activity 3-16 on page 62 and Activity 4-9 on page 82 concerning tennis simulations. Use your calculator to determine the standard deviation of the game lengths with conventional scoring (TENN1) and with no-ad scoring (TENN2). Write a sentence or two describing your findings.

Activity 5-14: Rowers' Weights (*cont.*)

Recall from Activity 3-3 on page 48 the data on the weights of rowers on the 1996 U.S. Olympic men's team.
(a) Calculate (by hand) the five-number summary for these weights.
(b) Draw (by hand) a boxplot of the distribution of rowers' weights.
(c) Is the boxplot as informative as the dotplot in this situation? If so, explain your answer. If not, describe what the dotplot reveals that the boxplot does not.

Activity 5-15: Students' Data (*cont.*)

Refer back to any of the data collected in class, such as states visited or signature lengths or Scrabble points.
(a) Create a dotplot of this distribution and comment on its features, particularly on the amount of variability.
(b) Calculate the range, IQR, MAD, and standard deviation of these data. Explain what each of these represents in the context of the data.

(c) Report the five-number summary of these data, draw a boxplot, and comment on its features.

Activity 5-16: Mothers' Ages (*cont.*)

Refer back to the data on the ages at which a sample of 35 mothers had their first child, as reported in Activity 3-11 on page 59. The list MMAGE can be found in AGE-CHILD.83g.

(a) Look at a dotplot of the distribution of these ages. Based on the shape of this dotplot, would you expect the empirical rule to hold in this case? Explain.

(b) Use your calculator to determine the mean and standard deviation of these mothers' ages. Report their values.

(c) Determine what proportion of the 35 ages falls within *one* standard deviation of the mean. How closely does this proportion match what the empirical rule would predict?

(d) Determine what proportion of the 35 ages falls within *two* standard deviations of the mean. How closely does this proportion match what the empirical rule would predict?

(e) Remove the outlier from the analysis, answer these questions again, and comment on how your answers change.

Activity 5-17: Ideal Temperatures

Analyze the data collected in the "Preliminaries" section concerning the ideal temperature in the minds of you and your classmates. Write a paragraph describing the distribution. Be sure to include graphical displays and numerical summaries. Also comment on where your ideal temperature falls in relation to those of your classmates.

Activity 5-18: GPAs

Suppose the distribution of GPAs at Jefferson High School has a mean of 2.7 and a standard deviation of 0.37 points. The GPAs at Washington High School have a mean of 2.8 and a standard deviation of 0.33 points.

(a) Ted, a student at Washington High School, has a GPA of 3.25, and Frank, a student at Jefferson High School, has a GPA of 3.17. Calculate the *z*-score for Ted and Frank and comment on which of them has the higher GPA relative to his peers.

(b) What GPA would Ted need to have the same *z*-score as Frank?

(c) Torsten, another student at Jefferson High, has a GPA of 3.07. Assuming that these GPAs follow a mound-shaped distribution, approximately what proportion of Jefferson High School students have a larger GPA than Torsten?

Activity 5-19: Heights of Volleyball Players

Suppose the average height of women collegiate volleyball players is 5'9", with a standard deviation of 2.1". Assume that heights among these players follow a mound-shaped distribution.

(a) According to the empirical rule, about 95% of women collegiate volleyball players have heights between what two values?

(b) What does the empirical rule say about the proportion of players who are between 62.7 inches and 75.3 inches?

(c) Reasoning from the empirical rule, what is the tallest we would expect a woman collegiate volleyball player to be?

Activity 5-20: Guessing Standard Deviations

Notice that each of the following hypothetical distributions is roughly symmetric and mound-shaped.

(a) Use the empirical rule that you discovered above to make an educated guess about the mean and standard deviation of each distribution.

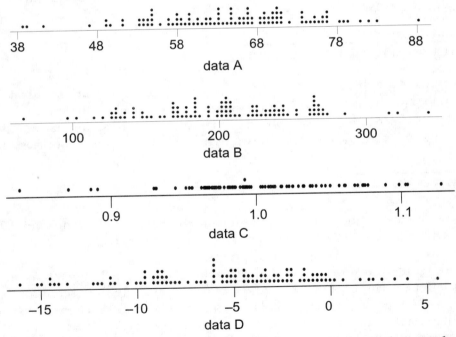

(b) Download HYPOSTDV.83g and use your calculator to compute the actual means and standard deviations. Comment on the accuracy of your guesses.

Activity 5-21: Limitations of Boxplots

Consider the hypothetical exam scores presented below for three classes of students. Dotplots of the distributions are also presented.

class A:	50	50	50	63	70	70	70	71	71	72	72	79	91	91	92
class B:	50	54	59	63	65	68	69	71	73	74	76	79	83	88	92
class C:	50	61	62	63	63	64	66	71	77	77	77	79	80	80	92

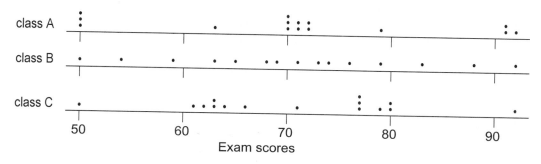

(a) Do these dotplots reveal differences among the three distributions of exam scores? Explain briefly.

(b) Find (by hand) five-number summaries of the three distributions. (Notice that the scores are already arranged in order.)

(c) Create boxplots (on the same scale) of these three distributions.

(d) If you had not seen the actual data and had been shown only the boxplots, would you have been able to detect the differences in the three distributions?

Activity 5-22: Creating Examples (*cont.*)

For each of the following properties, try to construct a data set of ten hypothetical exam scores that satisfies the property. Also, produce a dotplot of your hypothetical data in each case. Assume that the exam scores are integers between 0 and 100, inclusive. You may use your calculator.

(a) fewer than half of the exams fall within one standard deviation of the mean

(b) the standard deviation is positive but the interquartile range equals 0

(c) all of the exams fall within one standard deviation of the mean

(d) the standard deviation is as large as possible

(e) the interquartile range is as large as possible but the standard deviation is not as large as possible

(f) the mean absolute deviation is exactly half as large as the range

Activity 5-23: More Measures of Center

Consider the following three numerical measures of the center of a distribution:
- *trimmed mean*: the mean of the observations after removing the largest 5% and smallest 5%
- *midrange*: the average of the minimum and maximum
- *midhinge*: the average of the lower and upper quartiles

(a) Which of these three measures of center is resistant to outliers and which is not resistant? Explain your answers based on how the measures are defined.

(b) Calculate the midrange and midhinge of the Supreme Court justices' years of service from Activity 4-1 on page 69 and Activity 5-4 on page 96 (SPMCT.831). Comment on how these compare to the mean and median found above.

(c) Calculate the midrange and midhinge of the data collected in Topic 3 on salaries expected by students. Comment on how these compare to the mean and median found in Activity 4-8 on page 81.

Activity 5-24: Hypothetical ATM Withdrawals

Suppose that a bank wants to monitor the withdrawals that its customers make from automatic teller machines at three locations. Suppose further that they sample 50 withdrawals from each location and tally the data in the following table. All missing entries are zeros. The "raw" (untallied) data are stored in HYPOATM.83g, containing lists ATM1, ATM2, and ATM3.

cash amount	20	30	40	50	60	70	80	90	100	110	120
machine 1 tally			25						25		
machine 2 tally	2	8	1	9	2	6	2	9	1	8	2
machine 3 tally	9					32					9

(a) Use your calculator to produce visual displays of the distributions of cash amounts at each machine. Is each distribution perfectly symmetric?

(b) Use your calculator to determine the mean and standard deviation of the cash withdrawal amounts at each machine. Are they identical for each machine?

(c) Are the *distributions* themselves identical? Would you conclude that the mean and standard deviation provide a complete summary of a distribution of data?

Activity 5-25: Five on Five

Suppose that one's data consists of the numbers 1, 2, 3, 4, and 5.

(a) Calculate the five-number summary.

(b) Does the five-number summary consist of these five numbers?

Activity 5-26: Properties of Measures of Spread (*cont.*)

In this activity you will investigate the effects on measure of spread of adding a constant to a set of data and of multiplying by a constant.

Suppose that the Supreme Court justices listed in Activity 5-4 on page 96 are still serving in the year 2004, so each of their years of service at that time would be five years more than listed in Activity 4-1 on page 69.

(a) Record these new data on years of service.

(b) Determine the range, IQR, MAD, and standard deviation of these new years of service at that point.

(c) How do the measures of spread for the new data compare with those for the original data?

Now consider the placement exam scores in Activity 5-5 on page 97. Suppose that every score is multiplied by 5 in order to put it on a percentage correct scale.

(d) Determine the range, IQR, and standard deviation for these new scores.

(e) How do these statistics compare with those for the original data?

(f) If the temperatures in Activity 5-1 on page 89 were converted from Fahrenheit to Celsius, they would first have 32 subtracted and then be multiplied by 5/9. Without actually doing the conversion, what do you think the effect would be on measures of spread? [*Hint:* First consider the effect of subtracting a constant, then consider the effect of multiplying by a constant.]

5

Unit II

Exploring Data: Comparisons and Relationships

Topic 6:

COMPARING DISTRIBUTIONS I: QUANTITATIVE VARIABLES

OVERVIEW

You have been analyzing distributions of data by constructing various graphical displays (dotplot, histogram, stemplot, boxplot), by calculating numerical measures of various aspects of that distribution (mean, median, and mode for center; range, interquartile range, and standard deviation for spread), and by commenting verbally on features of the distribution revealed in those displays and statistics. Thus far you have primarily concentrated on one distribution at a time. With this topic you will apply these techniques in the more interesting case of analyzing, comparing, and contrasting distributions from two or more groups simultaneously.

OBJECTIVES

- To see how to construct **side-by-side stemplots** to compare two distributions of data.
- To understand the meaning of the phrase **statistical tendency**.
- To learn a formal rule of thumb for determining whether an observation is an **outlier**.
- To discover **modified boxplots** as displays conveying more information than ordinary boxplots.
- To acquire extensive experience with using graphical, numerical, and verbal means of comparing and contrasting distributions from two or more groups.
- To use your calculator as an important tool for performing such analyses.

PRELIMINARIES

1. Which state would you guess has the highest birth rate in the U.S.?

California

2. Would you expect to find much of a difference in birth rates between eastern and western states? If so, which region would you expect to have higher birth rates?

East Coast

3. Consider the assertion that "men are taller than women." Does this mean that every man is taller than every woman? If not, write one sentence indicating what you think the assertion does mean.

No,

4. Take a guess as to the most money won by a female professional golfer in 1999. Do the same for a male golfer.

$200,000 $1 Mill

5. Make a guess as to the year in which you expect a woman to become President of the U.S. for the first time.

4280

6. In 1998 the American Film Institute (AFI) selected the top 100 American films of all time. Place a check beside each of these films that you have seen, and then count up how many you have seen.

Number seen:

rank	title	O?	yr	rank	title	O?	yr
1	Citizen Kane	n	41	2	Casablanca	y	42
3	The Godfather	y	72	4	Gone with the Wind	y	39
5	Lawrence of Arabia	y	62	6	The Wizard of Oz	n	39
7	The Graduate	n	67	8	On the Waterfront	y	54
9	Schindler's List	y	93	10	Singin' in the Rain	n	52
11	It's A Wonderful Life	n	46	12	Sunset Boulevard	n	10
13	The Bridge on the River Kwai	y	57	14	Some Like It Hot	n	59
15	Star Wars	n	77	16	All About Eve	y	10
17	The African Queen	n	51	18	Psycho	n	60
19	Chinatown	n	74	20	One Flew over the Cuckoo's Nest	y	75
21	The Grapes of Wrath	n	40	22	2001: A Space Odyssey	n	68
23	The Maltese Falcon	n	41	24	Raging Bull	n	10
25	E.T the Extra-Terrestrial	n	82	26	Dr. Strangelove	n	64
27	Bonnie and Clyde	n	67	28	Apocalypse Now	n	79
29	Mr. Smith Goes to Washington	n	39	30	The Treasure of the Sierra Madre	n	48
31	Annie Hall	y	77	32	The Godfather Part II	y	74
33	High Noon	n	52	34	To Kill a Mockingbird	n	62
35	It Happened One Night	y	34	36	Midnight Cowboy	y	69
37	The Best Years of Our Lives	n	46	38	Double Indemnity	n	44
39	Doctor Zhivago	n	65	40	North by Northwest	n	59
41	West Side Story	y	61	42	Rear Window	n	54
43	King Kong	n	33	44	The Birth of a Nation	n	15
45	A Streetcar Named Desire	n	51	46	A Clockwork Orange	n	71
47	Taxi Driver	n	76	48	Jaws	n	75
49	Snow White and the Seven Dwarfs	n	37	50	Butch Cassidy and the Sundance Kid	n	69
51	The Philadelphia Story	n	40	52	From Here to Eternity	y	53
53	Amadeus	y	84	54	All Quiet on the Western Front	y	30
55	The Sound of Music	y	65	56	M*A*S*H	n	70
57	The Third Man	n	49	58	Fantasia	n	40
59	Rebel Without a Cause	n	55	60	Raiders of the Lost Ark	n	81
61	Vertigo	n	58	62	Tootsie	n	82
63	Stagecoach	n	39	64	Close Encounters of the Third Kind	n	77
65	The Silence of the Lambs	y	91	66	Network	n	76
67	The Manchurian Candidate	n	62	68	An American in Paris	y	51
69	Shane	n	53	70	The French Connection	y	71
71	Forrest Gump	y	94	72	Ben-Hur	y	59
73	Wuthering Heights	n	39	74	The Gold Rush	n	25
75	Dances with Wolves	y	90	76	City Lights	n	31
77	American Graffiti	n	73	78	Rocky	y	76
79	The Deer Hunter	y	78	80	The Wild Bunch	n	69
81	Modern Times	n	36	82	Giant	n	56
83	Platoon	y	86	84	Fargo	n	96
85	Duck Soup	n	33	86	Mutiny on the Bounty	y	35
87	Frankenstein	n	31	88	Easy Rider	n	69
89	Patton	y	70	90	The Jazz Singer	n	27
91	My Fair Lady	y	64	92	A Place in the Sun	n	51
93	The Apartment	y	60	94	Goodfellas	n	90
95	Pulp Fiction	n	94	96	The Searchers	n	56
97	Bringing up Baby	n	38	98	Unforgiven	y	92
99	Guess Who's Coming to Dinner	n	67	100	Yankee Doodle Dandy	n	42

6

7. Record the following variables for you and your classmates: predicted year of a female President, number of "top 100" films seen, gender, and political inclination (liberal, moderate, conservative) as in Topic 1.

IN-CLASS ACTIVITIES

Activity 6-1: Birth and Death Rates

The following table lists the birth rates and death rates (per 1000 residents) for each of the 50 states in 1997.

(a) Indicate in the table whether the state lies mostly to the east (E) or to the west (W) of the Mississippi River.

state	birth	death	e/w	state	birth	death	e/w
Alabama	14.1	9.9	E	Montana	11.9	8.8	W
Alaska	15.9	4	W	Nebraska	14	9.2	W
Arizona	18.9	8.2	W	Nevada	16.1	7.7	W
Arkansas	14.7	10.1	M	New Hampshire	12.3	8.1	E
California	16.3	7.1	W	New Jersey	14.1	9	E
Colorado	13.4	6.6	W	New Mexico	15.4	7.4	w
Connecticut	13.1	8.9	E	New York	15.6	8.9	E
Delaware	13.9	8.8	E	North Carolina	14.2	8.9	E
Florida	13.1	10.6	E	North Dakota	13	9.4	w
Georgia	15.8	7.9	E	Ohio	13.6	9.4	E
Hawaii	14.4	6.5	w	Oklahoma	14.5	10.2	W
Idaho	14.9	7.4	W	Oregon	13.4	8.8	W
Illinois	15.2	8.6	E	Pennsylvania	11.8	10.6	E
Indiana	11.8	8	E	Rhode Island	12.5	9.9	E
Iowa	12.9	9.2	W	South Carolina	13.6	8.8	E
Kansas	14.6	9.3	W	South Dakota	13.7	9.5	W
Kentucky	13.5	9.8	E	Tennessee	13.9	9.7	
Louisiana	14.8	8.7	W	Texas	16.3	6.9	
Maine	10.9	8.9	E	Utah	21.2	5.6	
Maryland	13	7.9	E	Vermont	11.4	9	
Massachusetts	13.5	8.5	E	Virginia	13.2	7.9	
Michigan	13.7	8.5	E	Washington	14.2	7.7	
Minnesota	13.8	7.9	E	West Virginia	11.3	11.5	
Mississippi	15.6	10	E	Wisconsin	12.8	8.7	
Missouri	13.8	10.1	W	Wyoming	13.3	7.8	

> In a **side-by-side stemplot**, a common set of stems is used in the middle of the display with leaves for each category branching out in either direction, one to the left and one to the right. The convention is to order the leaves from the middle out toward either side.

(b) Construct a side-by-side stemplot of these birth rates according to the state's region; use the stems listed below. (Alabama's, Alaska's, and Arizona's values have already been placed on the plot to get you started.)

```
        west              east
                   10
                   11
                   12
           4       13 | 1      9
        7          14 | 1
              9    15          8
              3    16
                   17
              9    18
                   19
                   20
                   21
```

(c) Remember to arrange the leaves in order from the inside out:

```
        west              east
                   10
                   11
                   12
                   13
                   14
                   15
                   16
                   17
                   18
                   19
                   20
                   21
```

6

(d) Calculate the median value of the states' birth rates for each region.

East:

West:

(e) Identify your home state and comment on where it fits into the distribution.

(f) Does one region (east or west) *tend* to have higher birth rates than the other? Explain.

(g) Is it the case that every state from one region has a higher birth rate than every state from the other? If not, identify a pair such that the eastern state has a higher birth rate than the western state.

(h) If you were to randomly pick one state from each region, which would you expect to have the higher birth rate? Explain.

You have discovered an important (if somewhat obvious) concept in this activity— that of **statistical tendency**. You found that western states *tend* to have higher birth rates than do eastern states. It is certainly not the case, however, that *every* western state has a higher birth rate than *every* eastern state.

Similarly, men *tend* to be taller than women, but there are certainly some women who are taller than most men. Statistical tendencies pertain to average or typical cases but not necessarily to individual cases. Just as Geena Davis and Danny DeVito do not disprove the assertion that men are taller than women, the cases of Montana and Georgia do not contradict the finding that western states *tend* to have higher birth rates than eastern states.

Activity 6-2: Professional Golfers' Winnings

The following table presents the winnings (in thousands of dollars) of the 30 highest money winners on each of the three professional golf tours (PGA for males, LPGA for females, and Seniors for males over 50 years of age) in 1999.

rank	PGA	winnings	LPGA	winnings	Senior	winnings
1	Woods	6617	Webb	1592	Fleisher	2516
2	Duval	3642	Inkster	1337	Irwin	2025
3	Love	2475	Pak	957	Doyle	1912
4	Singh	2283	Sorenstam	864	Nelson	1514
5	Perry	2146	Kane	758	Morgan	1493
6	Sutton	2128	Mallon	680	Quigley	1328
7	Stewart	2078	Steinhauer	663	Jenkins	1167
8	Leonard	2021	Kim	584	Summerhays	1118
9	Maggert	2016	Jones	584	Fernandez	1108
10	Toms	1960	Pepper	578	Canizares	1087
11	Franco	1865	Hjorth	573	Inman	1051
12	Furyk	1828	Hetherington	538	Marsh	1039
13	Pate	1756	Lunn	512	Jacobs	997
14	Mickelson	1723	Davies	502	McCord	993
15	Els	1711	Fukushima	498	Mahaffey	989
16	Funk	1639	Kuehne	485	Dougherty	951
17	Sluman	1621	Moodie	448	Eichelberger	883
18	Price	1572	Robbins	440	Graham	870
19	Geiberger	1541	Barrett	411	Thorpe	858
20	Tryba	1534	Neumann	405	Hall	816
21	Huston	1519	Matthew	370	Baiocchi	754
22	Herron	1511	Alfredsson	369	Snead	744
23	Weir	1491	Klein	367	Archer	738
24	Lehman	1436	Daniel	356	Duval	727
25	Appleby	1360	McCurdy	354	Dent	715
26	Estes	1358	Scranton	322	O'Connor	711
27	Paulson	1314	Dobson	304	Floyd	683
28	Waldorf	1303	Turner	301	Colbert	639
29	Hart	1268	Iverson	298	Thompson	635
30	Roberts	1259	Stephenson	296	Green	631

6

(a) Notice that one cannot construct side-by-side stemplots to compare these distributions, since there are three groups and not just two to be compared. One can, however, construct comparative boxplots of the distributions. Start by calculating (by hand) five-number summaries for each of the three groups, recording them below. [*Hint:* Notice that the observations have already been arranged in order and that they are numbered, which should greatly simplify your calculations.]

tour	minimum	lower quartile	median	upper quartile	maximum
PGA					
LPGA					
Senior					

(b) Construct (by hand) boxplots of these three distributions on the same scale; an axis has been drawn for you below.

PGA

LPGA

Senior

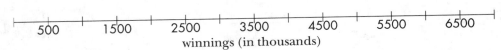

500   1500   2500   3500   4500   5500   6500

winnings (in thousands)

(c) The boxplot for the PGA golfers reveals one of the weaknesses of boxplots as visual displays. Clearly, Tiger Woods was an outlier, earning almost $3,000,000 more than his nearest competitor. Would the boxplot look any different, however, if his nearest competitor had won only $5 less than him?

> One way to address this problem with boxplots is to construct what are sometimes called **modified boxplots**. These treat outliers differently by marking them with a special symbol (*) and then extending the box-plot's "whiskers" only to the most extreme non-outlying value. To do this requires an explicit rule for identifying outliers. The rule that we will use regards outliers as observations lying more than 1.5 times the interquartile range away from the nearer quartile.

With the PGA winnings the lower quartile is Q_1 = 1491 and the upper quartile is Q_3 = 2021, so the interquartile range is IQR = 2021 − 1491 = 530. Thus, 1.5 x IQR = 1.5 (530) = 795, so any observation more than 795 away from its nearer quartile will be considered an outlier. To look for such observations, we add 795 to Q_3, obtaining 2021 + 795 = 2816. Since Woods' 6617 and Duval's 3642 exceed this, they are outliers. On the other end, we subtract 795 from 1491, obtaining 696. Since nobody on the list earned less than that, there are no outliers on the low end.

(d) Use this rule to check for and identify outliers on the LPGA and Senior lists.

new MiN   new MaX

LPGA	Q_1	Q_3	IQR=Q_3−Q_1	1.5IQR	Q_1−1.5IQR	Q_3+1.5IQR

Outliers?

Senior	Q_1	Q_3	IQR=Q_3−Q_1	1.5IQR	Q_1−1.5IQR	Q_3+1.5IQR

Outliers?

Modified boxplots are constructed by marking outliers with an "*" and extending "whiskers" only to the most extreme (i.e., largest on the high end, smallest on the low end) *non*outlier.

(e) Construct modified boxplots of the three distributions below:

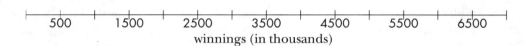

winnings (in thousands)

(f) Lists containing the professional golfer's winnings have been stored in a grouped file named GOLFWINN.83g. Download this grouped file into your calculator (the lists are named PGA, LPGA, and SEN) and create side-by-side modified boxplots on your calculator by using Plot1, Plot2, and Plot3 simultaneously. For example, you might want to set Plot1 up as follows:

Note that you have two choices for boxplots. The one chosen above is the modified boxplot. Once you have set up Plot1, Plot2, and Plot 3 press $\boxed{\text{ZOOM}}$ $\boxed{9}$. By pressing the $\boxed{\text{TRACE}}$ button and then using the left and right scroll buttons to move the trace along the horizontal axis you can view the five-number summary of the boxplot. Compare the modified boxplots from the calculator to your hand-drawn modified boxplots above.

Do the modified boxplots provide more visual information than the "unmodified" ones? Explain.

(g) Write a paragraph comparing and contrasting key features of the distributions of the earnings of the top money winners on these three golf tours.

Activity 6-3: Variables of State (*cont.*)

The following boxplots display distributions of variables with states in the U.S. as the cases. Each graph contains four boxplots, corresponding to the state's region of the country. These regions are midwest, northeast, south, and west, but they are not necessarily displayed in that order. Your task is to try to identify which region is labeled number 1, which is number 2, and so on. The boxplots are of these variables:
- the percentage of a state's residents who have a college degree
- the percentage of a state's residents who have never been married
- the percentage of a state's births that are to mothers under age 20
- the percentage of a state's residents who are of Mexican descent

(See Activity 6-18 on page 135 for a listing of the regions into which each state is classified.)

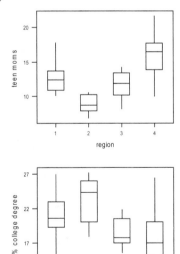

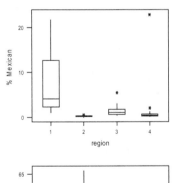

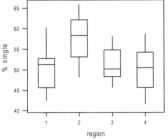

Based on examining these boxplots, make your educated guess as to which region is which, and write a paragraph describing how you came to arrive at this guess.

1:                    2:   3:   4:

Explanation:

Activity 6-4: Top American Films

Consider the American Film Institute's list of 100 top films from the "Preliminaries" section.

(a) Lists containing the American Film Institute's list of 100 top films have been stored in a grouped file named FILMS.83g. Download this file along with the program SPRT.83p. The names of the lists are AGE and OSCAR. To separate the films into those that won the Academy Award for Best Picture and those that did not, run the program SPRT and enter the AGE list at the DATA? prompt and the OSCAR list at the CATEGORY? prompt. The two lists will be named GRP1 and GRP2. Use the 1-Var Stats feature of your calculator to determine the five-number summary for each group. Record these statistics, as well as the sample sizes, the sample means, and the sample standard deviations, in the table below:

	number	mean	std. dev.	min.	Q_1	median	Q_3	max.
won BP		34.1	18.2	6	22.5	33	46.5	70
did not		43.6	18.0	4	30.5	44.5	59.5	85

(b) Use your calculator to produce comparative boxplots (on the same scale) of the ages of these films, again separated by whether or not the film won an Academy Award for Best Picture. To do this you will need to set up Plot1 and Plot2 as in Activity 6-2, but you will need to turn off Plot3. Write a paragraph comparing and contrasting the two distributions. [As always, remember to relate your comments to the context of the data. Also, do not forget about the checklist of features to look for when describing a distribution.]

(c) Enter into your calculator a new categorical variable indicating whether or not you have seen the film (use 0's and 1's for N's and Y's, respectively). Use your calculator and the SPRT program to determine numerical summaries for the distribution of the ages of these films, separated by whether or not you have seen it (the lists will be named GRP0 and GRP1). Record these statistics in the table:

	number	mean	std. dev.	minimum	Q_1	median	Q_3	maximum
seen it								
have not								

(d) Use your calculator to produce comparative boxplots (on the same scale) of the ages of these films, again separated by whether or not you have seen the film. Write a paragraph comparing and contrasting the two distributions.

WRAP-UP

You have been introduced in this topic to methods of **comparing** distributions between two or more groups. This task has led you to the very important concept of statistical **tendencies**. You have also expanded your knowledge of visual displays of distributions by encountering **side-by-side stemplots** and **modified boxplots**. Another object of your study has been a formal test for determining whether an unusual observation is an outlier.

The methods of this topic apply to quantitative variables. In the next topic you will encounter two-way tables as you study methods for comparing distributions of *categorical* variables.

HOMEWORK ACTIVITIES

Activity 6-5: Birth and Death Rates (*cont.*)

Refer to the data in Activity 6-1 on page 118 on birth rates and death rates among the states.

(a) Construct a side-by-side stemplot of the death rates.

(b) Comment on the distributions of death rates between eastern and western states.

(c) Ignoring east/west distinctions, use your calculator to compute five-number summaries of the birth rates and of the death rates (BIRTHDEATH.83g). Report these five-number summaries.

(d) Use the five-number summaries to conduct (by hand) outlier tests for the two variables.

(e) Write a few sentences identifying and commenting on any outliers that you find.

Activity 6-6: Rowers' Weights (*cont.*)

Recall from Activity 3-3 on page 48 the data on the weights of rowers on the 1996 U.S. Olympic team.

(a) Identify the rower who appeared to be a clear outlier, and also report his weight.

(b) Conduct (by hand) an outlier test on the weights. (You may have already calculated the five-number summary in Activity 5-14.)

(c) Look carefully at the calculations involved to explain why the obvious outlier is not declared an outlier by this test.

Activity 6-7: Top American Films (*cont.*)

Continuing your analysis of Activity 6-4 on page 126, study the "rank" variable rather than the "age" variable (FILMS.83g). Do the distributions of movie ranks differ between those that have won an Oscar for Best Picture and those that have not? Do the distributions of ranks differ between the films you have seen and the ones you have not? Use your calculator to produce graphical displays and numerical summaries to address these questions. Write a paragraph or two reporting your findings.

Activity 6-8: Signature Measurements (*cont.*)

Refer back to the data collected in Topic 1 on signature measurements. Compare men's and women's distributions of signature lengths by answering the following questions, supplying displays and/or calculations to support your answers:

(a) Does one gender *tend* to have larger values of this variable than the other gender? If so, by about how much do the genders differ on average with respect to this variable? Does *every* member of one gender have a larger value of this variable than *every* member of the other gender?

(b) Does one gender have more variability in the distribution of these measurements than the other gender?

(c) Are the shapes of both genders' distributions fairly similar?

(d) Does either gender have outliers in their distribution of this variable?

Activity 6-9: Female President

Consider the data collected on the year in which people expect a woman to become President of the United States.

(a) Convert the data into years from now that the person expects a woman to become President.

(b) Compare the distributions of men and women. Examine appropriate visual displays and calculate relevant summary statistics. Write a paragraph or two reporting your findings.

(c) Compare the distributions of the three political categories. Examine appropriate visual displays and calculate relevant summary statistics. Write a paragraph or two reporting your findings.

Activity 6-10: Tennis Simulations (*cont.*)

Reconsider the data from Activity 3-16 on page 62 and Activity 4-9 on page 82 concerning tennis simulations. The experiment also analyzed a "handicap" scoring method that uses no-ad scoring and also awards weaker players bonus points at the start of a game. The lengths of 100 simulated games with this handicap scoring system are tallied below and the raw data are stored in TENN3.

points in game	1	2	3	4	5	6	7
tally (count)	3	4	12	18	28	25	10

(a) Use your calculator and determine the means and standard deviations of the game lengths for each of the three scoring systems.

(b) Write a paragraph addressing whether one method tends to produce longer or shorter games than the others and whether one method produces more or less variability in game lengths than the others.

Activity 6-11: Parents' Ages (*cont.*)

Reconsider Activity 3-11 on page 59, where you created a split stemplot for the ages at which a sample of mothers had their first child. The ages at which a sample of fathers had their first child appears here:

23	32	35	19	20	13	23	23	26	25	25	25	24	29	24	26	39	31
22	24	22	21	25	29	25	28	19	30	23	30	30	26	26	32	16	

(a) Create a split side-by-side stemplot to compare the distributions of these ages between fathers and mothers. [Refer to Activity 3-11 on page 59 for a description of what "split" means.]

(b) Use the stemplots to calculate the five-number summary for each group.

(c) Compare the distributions of these ages between fathers and mothers. Address the question of whether one or the other tends to have their first child at a younger age.

Activity 6-12: States' Population Percentages

Which states have higher percentages of female residents than others? Does one part of the country tend to have higher percentages of females than another? The table below reports the percentage of female residents in each of the fifty states, separated by states east and west of the Mississippi River. (Also reported in the table is an automobile theft rate described in the next activity.)

Eastern state	% female	theft rate	Western state	% female	theft rate
Alabama	52.1	348	Alaska	47.3	565
Connecticut	51.5	731	Arizona	50.6	863
Delaware	51.5	444	Arkansas	51.8	289
Florida	51.6	826	California	49.9	1016
Georgia	51.5	674	Colorado	50.5	428
Illinois	51.4	643	Hawaii	49.1	381
Indiana	51.5	439	Idaho	50.2	165
Kentucky	51.6	199	Iowa	51.6	170
Maine	51.3	177	Kansas	51.0	335
Maryland	51.5	709	Louisiana	51.9	602
Massachusetts	52.0	924	Minnesota	51.0	366

Michigan	51.5	714	Missouri	51.8	539
Mississippi	52.2	208	Montana	50.5	243
New Hampshire	51.0	244	Nebraska	51.3	178
New Jersey	51.7	940	Nevada	49.1	593
New York	52.1	1043	New Mexico	50.8	337
North Carolina	51.5	284	North Dakota	50.2	133
Ohio	51.8	491	Oklahoma	51.3	602
Pennsylvania	52.1	506	Oregon	50.8	459
Rhode Island	52.0	954	South Dakota	50.8	110
South Carolina	51.6	386	Texas	50.7	909
Tennessee	51.8	572	Utah	50.3	238
Vermont	51.0	208	Washington	50.4	447
Virginia	51.0	327	Wyoming	50.0	149
West Virginia	52.0	154			
Wisconsin	51.1	416			

(a) Create a split side-by-side stemplot of the female percentages according to whether the state is east or west of the river. (The term "split" means to break each stem into two rows, with 0–4 leaves placed on one row and 5–9 leaves on the next. For example, Illinois's 51.4 and Indiana's 51.5 should be represented on different rows. Also see the description in Activity 3-11 on page 59.)

(b) Write a paragraph commenting on similarities and differences between the eastern and western states with regard to this variable. Be sure to address whether one region tends to have a higher percentage of female residents than the other, but also comment on spread, shape, and outliers.

(c) Which state is an obvious outlier? What percentage of its residents are female? Suggest a plausible explanation for this unusual percentage.

Activity 6-13: Automobile Theft Rates

Investigate whether states in the eastern or western part of the U.S. tend to have higher rates of motor vehicle thefts. The table above (Activity 6-12) separates states according to whether they lie east or west of the Mississippi River and lists their 1990 rate of automobile thefts per 100,000 residents.

(a) Create a side-by-side stemplot to compare these distributions. Ignore the last digit of each state's rate; use the hundreds digit as the stem and the tens digit as the leaf.

(b) Calculate (by hand) the five-number summary for each distribution of automobile theft rates. [Make sure that you consider the last digit again even though it does not appear in your stemplot.]

(c) Conduct (by hand) the outlier test for each distribution. If you find any outliers, identify the outlying states.

(d) Construct (by hand) modified boxplots to compare the two distributions.

(e) Write a paragraph describing your findings about whether motor vehicle theft rates tend to differ between eastern and western states.

Activity 6-14: Video Rental Times

The following boxplots reveal the distributions of running times of the 512 movie videos found in the "new releases" section of Blockbuster Video in June of 1999. They are grouped according to Blockbuster's "type" category, where the types (in alphabetical order) are action, comedy, drama, family, special.

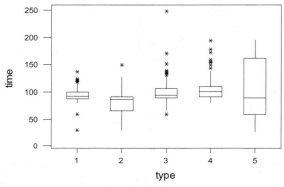

(a) Which type (by number) has the highest median running time? Estimate that median as accurately as you can from the graph.

(b) Which type (by number) has the shortest median running time? Estimate that median as accurately as you can from the graph.

(c) Which type (by number) has the movie with the longest running time? Estimate that time as accurately as you can from the graph.

(d) Which type (by number) has the movie with the shortest running time? Estimate that time as accurately as you can from the graph.

(e) Which type (by number) has the largest interquartile range of running times? Estimate that IQR as accurately as you can from the graph.

(f) Which type (by number) has the smallest interquartile range of running times? Estimate that IQR as accurately as you can from the graph.

(g) Which type (by number) has the largest lower quartile of running times? Estimate that lower quartile as accurately as you can from the graph.

(h) Which type (by number) has the smallest upper quartile of running times? Estimate that upper quartile as accurately as you can from the graph.

(i) Give a guess as to which "type" categories correspond to which numbers. Write a paragraph explaining the reasons behind your guesses.

Activity 6-15: College Tuitions (*cont.*)

Consider the data on the 1997–98 tuition charges and the founding years of Pennsylvania institutions of higher learning. The colleges are classified by type as private four-year, public four-year, private two-year, and public two-year. The following tables provide five-number summaries of the tuition charges and founding dates for the four types of schools. Use this information to produce comparative boxplots of the tuition charges of these colleges. Comment on similarities and differences in the distributions among the four types, being sure to address issues of center, spread, and shape. Then do the same for the founding years. Prepare a well-written, 2–3 paragraph report of your findings.

tuition	n	minimum	lower quartile	median	upper quartile	maximum
private 2-year	5	7112	7211	10305	11994	13080
private 4-year	67	6100	12640	14380	18000	22664
public 2-year	20	1400	2072	2211	5654	6540
public 4-year	25	4023	4211	4404	5832	6164

founding date	n	minimum	lower quartile	median	upper quartile	maximum
private 2-year	5	1865	1879	1921	1956	1966
private 4-year	67	1740	1847	1870	1924	1965
public 2-year	20	1923	1942	1965	1967	1971
public 4-year	25	1787	1855	1871	1926	1966

Activity 6-16: Lifetimes of Notables

The 1991 World Almanac and Book of Facts contains a section in which it lists "noted personalities." These are arranged according to a number of categories, such as "noted writers of the past" and "noted scientists of the past." One can calculate (approximately, anyway) the lifetimes of these people by subtracting their year of birth from their year of death. Distributions of the lifetimes of the people listed in nine different categories have been displayed in boxplots below:

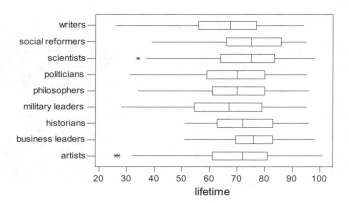

Use the information contained in these boxplots to answer the following questions. (In cases where the boxplots are not clear enough to make a definitive determination, you will have to make educated guesses.)

(a) Which group has the *individual* with the longest lifetime of anyone listed; about how many years did he or she live?

(b) Which group has the largest *median* lifetime; what (approximately) is that median value?

(c) Which group has the largest *range* of lifetimes; what (approximately) is the value of that range?

(d) Which group has the largest *interquartile range* of lifetimes; what (approximately) is the value of that IQR?

(e) Which group has the smallest *interquartile range* of lifetimes; what (approximately) is the value of that IQR?

(f) Which group has the smallest *median* lifetime; what (approximately) is the value of that median?

(g) Describe the general shape of the distributions of lifetimes.

(h) Suggest an explanation for writers tending to live shorter lives than scientists.

Activity 6-17: Hitchcock Films (*cont.*)

Reconsider the data from Activity 3-14 on page 61 concerning running times of movies directed by Alfred Hitchcock. Perform (by hand) the outlier test to determine whether any of the films constitute outliers in terms of their running times. Comment on your findings in light of your analysis in Activity 3-14.

Activity 6-18: Voting for Perot

The following table lists by state (and D.C.) the percentage of the popular vote received by Ross Perot in the 1992 Presidential election (PEROT92.83g).

state	region	% Perot	state	region	% Perot
Alabama	S	10.8	Alaska	W	27.6
Arizona	W	24.1	Arkansas	S	10.6
California	W	20.8	Colorado	W	23.5
Connecticut	NE	21.7	Delaware	S	20.6
Dist. of Colum.	S	4.2	Florida	S	19.9
Georgia	S	13.3	Hawaii	W	14.4
Idaho	W	27.8	Illinois	MW	16.9
Indiana	MW	19.9	Iowa	MW	18.8
Kansas	MW	27.4	Kentucky	S	13.8
Louisiana	S	12.0	Maine	NE	30.4
Maryland	S	14.4	Massachusetts	NE	22.9
Michigan	MW	19.2	Minnesota	MW	24.1
Mississippi	S	8.8	Missouri	MW	21.8
Montana	W	26.4	Nebraska	MW	23.7
Nevada	W	26.7	New Hampshire	NE	22.8
New Jersey	NE	15.9	New Mexico	W	16.4
New York	NE	15.8	North Carolina	S	13.7
North Dakota	MW	23.2	Ohio	MW	21.0
Oklahoma	S	23.1	Oregon	W	25.1
Pennsylvania	NE	18.3	Rhode Island	NE	22.8
South Carolina	S	11.6	South Dakota	MW	21.8
Tennessee	S	10.1	Texas	S	22.2
Utah	W	28.8	Vermont	NE	22.6
Virginia	S	13.7	Washington	W	24.3
West Virginia	S	15.9	Wisconsin	MW	21.7
Wyoming	W	25.6			

Analyze possible differences in the vote percentage distributions among the four regions of the country. Write a paragraph summarizing your findings, paying particular attention to the question of whether Perot tended to receive more support in certain regions and less in others.

Activity 6-19: Cars' Fuel Efficiency

As you saw in Activity 4-2 on page 72, the *Consumer Reports 1999 New Car Buying Guide* reported on the EPA highway miles per gallon ratings for a variety of cars classified in six categories. The following boxplots display the distributions:

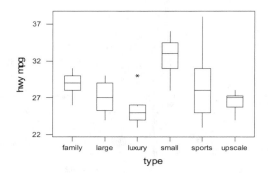

(a) Which category of car tends to get the best fuel efficiency? second best? worst?
(b) Does every car in the best fuel efficiency group have a higher miles per gallon rating than every car in the second best group?
(c) Does every car in the best fuel efficiency group have a higher miles per gallon rating than every car in the worst fuel efficiency group?
(d) Which group has the most variability in miles per gallon ratings?

Activity 6-20: Word Lengths (*cont.*)

The following table lists the lengths of the 26 words that we used in the first sentence of the "Overview" section of Activity 1-1 on page 5:

10	2	3	7	2	9	4	4	2	1	7	5	2
5	4	5	2	2	9	4	2	5	2	3	4	4

(a) Compare the distribution of *our* word lengths with the data that you collected in Topic 1 on the lengths of *your* words. Write a brief paragraph describing your findings; include any visual displays and numerical calculations that you find informative.
(b) Select a sentence or two by a favorite writer of yours and analyze her or his word length distribution in comparison with yours and ours. Again write a brief paragraph describing your findings.

Activity 6-21: Students' Data (*cont.*)

Refer back to any quantitative variable on which you collected class data (signature lengths, states visited, etc.) for which you also have data on a categorical variable (gender, political leaning, penny opinion, etc.). Analyze the data using graphical and numerical techniques to investigate potential differences in the distributions of the quantitative variable among the groups represented by the categorical variable. Write a paragraph summarizing your findings, referring to displays and numbers as appropriate.

Topic 7:

COMPARING DISTRIBUTIONS II: CATEGORICAL VARIABLES

OVERVIEW

In the previous topic you encountered the notion of a statistical tendency and studied techniques for comparing distributions of quantitative variables. In this topic you will study some basic techniques for comparing distributions of **categorical variables**. (Remember that a categorical variable is one that records simply that category into which a person or thing falls on the characteristic in question.) These techniques involve the analysis of two-way tables of counts. You will use no more complicated mathematical operations than addition and calculation of proportions, but you will acquire some very powerful analytical tools.

OBJECTIVES

- To learn to produce a **two-way table** as a summary of the information contained in a pair of categorical variables.
- To develop skills for interpreting information presented in two-way tables of counts.
- To become familiar with the concepts of **marginal** and **conditional distributions** of categorical variables.
- To discover **segmented bar graphs** as visual representations of the information contained in two-way tables.
- To acquire the ability to understand, recognize, and explain the phenomenon of **Simpson's paradox** as it relates to interpreting and drawing conclusions from two-way tables.
- To explore and understand the concepts of **independence** and **relative risk**.
- To gain experience in applying techniques of analyzing two-way tables to genuine data.

PRELIMINARIES

1. Do you think that a student's political inclination has any bearing on whether he or she wants to see the penny retained or abolished?

2. Do you think that Americans tend to be more interested in politics when they are older or when they are younger, or do you suspect that there is no relationship between age group and political interest?

3. Is there a difference between the proportion of American men who are U.S. Senators and the proportion of U.S. Senators who are American men? If so, which proportion is greater?

4. Do you think it would be more common to see a toy advertisement in which a boy plays with a traditionally female toy or one in which a girl plays with a traditionally male toy?

5. Do you suspect that it is any more or less common for a physician to be a woman if she is in a certain age group? If so, in what age group would you expect to see the highest proportion of woman physicians?

6. If you could choose one of the following accomplishments for your life, which would you choose:
   - to win an Olympic gold medal
   - to win a Nobel Prize
   - to win an Academy Award
   - to become President of the United States

7. Think about whether you agree or disagree with the following statement: "Most men are better suited emotionally for politics than most women are."

8. Would you expect men and women to differ in terms of the proportions who agree with the statement in question 7? If so, how?

9. Would you expect different political inclinations to differ in terms of the proportions who agree with the statement in question 7? If so, how?

10. Record anonymously the responses of your classmates and yourself to the "lifetime achievement" and the "men better suited" questions. Also record each student's gender and political inclination, classified as in Topic 1 as liberal, moderate, or conservative. Create a table such as the following:

student	gender	politics	lifetime achievement	"men better suited"?
1				
2				
...				

IN-CLASS ACTIVITIES

Activity 7-1: Suitability for Politics

Consider two of the variables on which you have collected class data: political inclination and agreement/disagreement that "most men are better suited emotionally for politics than are most women."

(a) Identify each variable as either quantitative or categorical.

political inclination:

suitability opinion:

> Often we are interested in considering one variable (the **response variable**) as being affected or predicted by the other variable (the **explanatory variable**).

(b) Which of these variables would you consider as the explanatory variable and which as the response variable? In other words, which makes more sense as a predictor or indicator of the other?

explanatory:

response:

(c) Count how many students fall into each of the six possible pairs of responses to these questions. Record these counts in the appropriate cells of the table below. For example, the number that you place in the upper left cell of the table should be the number of students who classified themselves as liberal and also agree that men are better suited for politics.

	liberal	moderate	conservative
agree with statement			
disagree with statement			

> This table is called a **two-way table** since it classifies each person according to two variables. In particular, it is a 2 x 3 table; the first number represents the number of categories of the row variable (opinion about statement), and the second number represents the number of categories of the column variable (political inclination). The explanatory variable should be in columns and the response variable in rows.

Activity 7-2: Age and Political Interest

In a national survey of adult Americans in 1998, people were asked to indicate their age and to classify their interest in politics as very much, somewhat, or not much. While age is typically a quantitative variable, it was categorized into three groups for this analysis: 18–35, 36–55, and 56–94 (the oldest subject in the survey). The results are summarized in the following table of counts; notice that row and column totals are also provided:

	18–35	36–55	56–94	total
not much	146	146	89	381
somewhat	192	260	154	606
very much	47	125	106	278
total	385	531	349	1265

(a) What proportion of the survey respondents were between ages 18 and 35?

$$385 / 1265 = 0.304$$

(b) What proportion of the survey respondents were between 36 and 55 years of age?

$$531 / 1265 = 0.420$$

(c) What proportion of the survey respondents were over age 55?

$$349 / 1265 = 0.280$$

total

> You have calculated the **marginal distribution** of the age variable. When analyzing two-way tables, one typically starts by considering the marginal distribution of each of the variables by itself before moving on to explore possible relationships between the two variables. You saw in Topic 1 that marginal distributions can be represented graphically in **bar graphs**.
>
> To study possible relationships between two categorical variables, one examines **conditional distributions**, i.e., distributions of one variable for *given* categories of the other variable.

marginal is   cat. total / total total

(d) Restrict your attention (for the moment) to just the respondents under 35 years of age. What proportion *of these young respondents* classify themselves as having not much interest in politics?

$$146/385 = 0.379$$

(e) What proportion *of the young respondents* classify themselves as somewhat interested in politics?

$$192/385 = 0.498$$

(f) What proportion *of the young respondents* classify themselves as having very much interest in politics?

$$47/385 = 0.122$$

(g) Record the conditional distribution that you have just calculated in the "18–35" column of the table below:

	18–35	36–55	56–94
not much	0.379	.275	.255
somewhat	0.498	.490	.441
very much	0.122	.235	.304
total	1.000	1.000	1.000

Conditional distributions can be represented visually with **segmented bar graphs**. The rectangles in a segmented bar graph all have a height of 100%, but they contain *segments* whose lengths correspond to the conditional proportions.

(h) Complete the segmented bar graph below by constructing the conditional distribution of political interest among those aged 18–35.

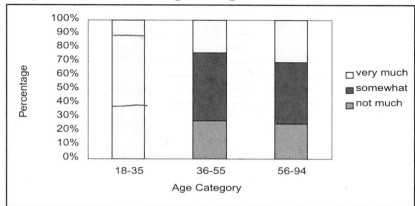

(i) Based on the calculations that you have performed and the display that you have created above, write a few sentences commenting on whether there seems to be any relationship between age and political interest. In other words, does the distribution of political interest seem to differ among the three age groups? If so, describe key features of the differences.

you become More interested as you
get older.

In dealing with conditional proportions, it is very important to keep straight which category is the one being conditioned on. For example, the proportion *of American males* who are U.S. Senators is very small, yet the proportion *of U.S. Senators* who are American males is very large.

Refer to the original table of counts to answer the following:

(j) What proportion of respondents aged 36–55 classified themselves as not much interested in politics?

146 / 531 = 0.275

(k) What proportion of those with not much interest in politics are of age 36–55?

$$\frac{146}{381} = 0.383$$

(l) What proportion of the people surveyed identified themselves as being both between ages 36–55 and having not much political interest?

$$146/1265 = 0.115$$

Activity 7-3: Pregnancy, AZT, and HIV

In an experiment reported in the March 7, 1994 issue of *Newsweek*, 164 pregnant, HIV-positive women were randomly assigned to receive the drug AZT during pregnancy, and 160 such women were randomly assigned to a control group that received a placebo ("sugar" pill). The following segmented bar graph displays the conditional distributions of the child's HIV status (positive or negative) for mothers who received AZT and for those who received a placebo.

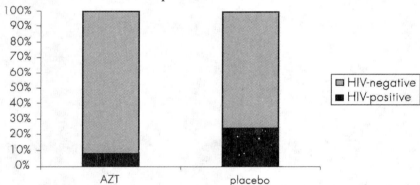

(a) Use the graph to estimate the proportion of AZT-receiving women who had HIV-positive babies and the proportion of placebo-receiving women who had HIV-positive babies.

The actual results of the experiment were that 13 of the mothers in the AZT group had babies who tested HIV-positive, compared to 40 HIV-positive babies in the placebo group.

(b) Use this information to calculate the proportions asked for in (a). Compare your calculations to your estimates based on the graph.

(c) The proportion of HIV-positive babies among placebo mothers is how many times greater than the proportion of HIV-positive babies among AZT mothers? In other words, what is the *ratio* of the proportions of HIV-positive babies between the AZT group and the placebo group?

> You have calculated the ***relative risk*** of having an HIV-positive baby between the AZT and placebo groups. If the response variable categories are incidence and nonincidence of a disease, then the relative risk is the ratio of the proportions having the disease between the two groups of the explanatory variable.

(d) Comment on whether the difference between the two groups appears to be important. What conclusion would you draw from the experiment?

Activity 7-4: Hypothetical Hospital Recovery Rates

The following two-way table classifies hypothetical hospital patients according to the hospital that treated them and whether they survived or died:

	survived	died	total
hospital A	800	200	1000
hospital B	900	100	1000

(a) Calculate the proportion of hospital A's patients who survived and the proportion of hospital B's patients who survived. Which hospital saved the higher percentage of its patients?

A) 800/1000 = 80%.   B) 900/1000 = 90%.

Hospital B

Suppose that when we further categorize each patient according to whether they were in fair condition or poor condition prior to treatment we obtain the following two-way tables:

fair condition:

	survived	died	total
hospital A	590	10	600
hospital B	870	30	900

poor condition:

	survived	died	total
hospital A	210	190	400
hospital B	30	70	100

(b) Convince yourself that when the "fair" and "poor" condition patients are combined, the totals are indeed those given in the table above.

(c) Among those who were in *fair* condition, compare the recovery rates for the two hospitals. Which hospital saved the greater percentage of its patients who had been in fair condition?

A) 98%.   B) 96%.   Hospital A

    590   870
    ───   ───
    600   900

(d) Among those who were in *poor* condition, compare the recovery rates for the two hospitals. Which hospital saved the greater percentage of its patients who had been in poor condition?

A) $\dfrac{210}{400} = 53\%$ B $\dfrac{30}{100} = 30\%$

Hospital A

> The phenomenon that you have just discovered is called **Simpson's paradox**, which refers to the fact that aggregate proportions can reverse the direction of the relationship seen in the individual pieces. In this case, hospital B has the higher recovery rate overall, yet hospital A has the higher recovery rate for each type of patient.

(e) Write a few sentences explaining (arguing from the data given) how it happens that hospital B has the higher recovery rate overall, yet hospital A has the higher recovery rate for each type of patient. [*Hints*: Do fair or poor patients tend to survive more often? Does one type of hospital tend to treat most of one type of patient? Is there any connection here?]

Fair & poor patients survive more often in hospital A, Hospital B tends to help more people who are in fair condition

(f) Which hospital would you rather go to if you were ill? Explain.

I would rather go to hospital A b/c the survival rates are higher

Activity 7-5: Women Senators

Refer to the data on 1999 U.S. Senators provided in Activity 12-3 on page 251.

(a) Create a two-way table of gender by party. Fill in the results in the following table.

	Republicans	Democrats	row total
men	52	39	91
women	3	6	9
column total	55	45	100

(b) Calculate the conditional distribution of gender for each party. Which party has a higher proportion of women among its Senators?

$$\frac{52}{55} = 95\%. \qquad \frac{39}{45} = 87\%. \qquad \frac{91}{100} = 91\%.$$

$$\frac{3}{55} = 5\%. \qquad \frac{6}{45} = 13\%. \qquad \frac{9}{100} = 9\%.$$

(c) Construct a segmented bar graph showing the conditional distributions of gender within each party.

(d) Write a sentence or two summarizing the relationship between gender and party among members of the 1999 U.S. Senate.

More males in general. The biggest proportion of women is in the Democrat Party

> Two categorical variables are said to be **independent** if the conditional distributions of one variable are identical for every category of the other variable.

(e) Are the variables gender and party independent among members of the 1999 U.S. Senate? Explain.

No, the totals are different

Suppose that at some point in the future the numbers of Senators break down as follows:

	Republicans	Democrats	row total
men	48	32	80
women	12	8	20
column total	60	40	100

$\frac{60}{100} \rightarrow .6$ $.4 \leftarrow \frac{40}{100}$

(f) Fill in the empty cells in such a way that gender and party are independent. [*Hints:* What proportion of the Senators are women? If that same proportion of the Republicans were women, how many women Republican senators would there be? In order for the marginal totals to be correct, is there any more freedom in filling in the remaining empty cells?]

WRAP-UP

With this topic we have concluded our investigation of distributions of data. This topic has differed from earlier ones in that it has dealt exclusively with categorical variables. The most important technique that this topic has covered has involved interpreting information presented in two-way tables. You have encountered the ideas of marginal distributions and conditional distributions, and you have learned to draw bar graphs and segmented bar graphs to display these distributions. You have explored the notion of relative risk, and you have discovered and explained the phenomenon known as Simpson's paradox, which raises interesting issues with regard to analyzing two-way tables.

Comparing distributions of categorical variables can also be thought of as exploring relationships between those variables. The next unit will be devoted to exploring relationships between quantitative variables. You will find that our approach will again involve starting with graphical displays (**scatterplots**) and proceeding to numerical summaries (**correlation**). You will then study a technique for predicting one variable from the value of another (**regression**).

HOMEWORK ACTIVITIES

Activity 7-6: Suitability for Politics (*cont.*)

Consider the data collected in class on whether people agree with the assertion that men are better suited emotionally for politics than women. Refer to the table you constructed in Activity 7-1 on page 139.

(a) For each of the three political categories, calculate the proportion of students who agree with the assertion that men are better suited for politics.
(b) Create a segmented bar graph to display these results.
(c) Comment on any association that you find between political inclination and reaction to this statement.
(d) Noting the method of selecting this sample, how broadly do you think your findings can be generalized? Explain.
(e) Repeat (a)–(c) using gender as the explanatory variable rather than political inclination. [The table will have to be 2×2, of course.]

Activity 7-7: Suitability for Politics (*cont.*)

The 1998 General Social Survey, a large-scale survey of randomly selected adult Americans, asked the same questions about gender, political inclination, and reaction to the statement that men are better suited emotionally for politics than are women. The results are summarized in the following tables:

	liberal	moderate	conservative
agree	74	139	169
disagree	410	471	422

	men	women
agree	169	236
disagree	565	777

Analyze these data to address the question of whether there is a relationship between political inclination and reaction to the statement or between gender and reaction to the statement. Write a paragraph or two describing your findings. Include appropriate calculations and displays in your write-up.

Activity 7-8: Gender-Stereotypical Toy Advertising

To study whether toy advertisements tend to picture children with toys considered typical of their gender, researchers examined pictures of toys in a number of children's catalogs. For each picture, they recorded whether the child pictured was a boy or girl. (We will ignore ads in which boys and girls appeared together.) They also recorded whether the toy pictured was a traditional "male" toy (like a truck or a toy soldier) or a traditional "female" toy (like a doll or a kitchen set) or a "neutral" toy (like a puzzle or a toy phone). Their results are summarized in the following two-way table:

	boy shown	girl shown
traditional "boy" toy	59	15
traditional "girl" toy	2	24
neutral gender toy	36	47

(a) Calculate the marginal totals for the table.
(b) What proportion of the ads showing *boys* depicted traditionally male toys? traditionally female toys? neutral toys?
(c) Calculate the conditional distribution of toy types for ads showing *girls*.
(d) Construct a segmented bar graph to display these conditional distributions.
(e) Based on the segmented bar graph, comment on whether the researchers' data seem to suggest that toy advertisers do indeed tend to present pictures of children with toys stereotypical of their gender.

Activity 7-9: Gender-Stereotypical Toy Advertising (*cont.*)

Reconsider the data concerning toy advertising presented in Activity 7-8. Let us refer to ads that show boys with traditionally "female" toys and ads that show girls with traditionally "male" toys as "crossover" ads.
(a) What proportion of the ads under consideration are "crossover" ads?
(b) What proportion *of the crossover ads* depict girls with traditionally male toys?
(c) What proportion *of the crossover ads* depict boys with traditionally female toys?
(d) When toy advertisers do defy gender stereotypes, in which direction does their defiance tend?

Activity 7-10: Gender of Physicians (*cont.*)

The following data address the question of whether percentages of women physicians are changing with time. The table classifies physicians according to their gender and age group.

	under 35	35–44	45–54	55–64
male	93,287	153,921	110,790	80,288
female	40,431	44,336	18,026	7,224

(a) For each age group, calculate the proportion of its physicians who are women.
(b) Construct a segmented bar graph to represent these conditional distributions.
(c) Comment on whether your analysis reveals any connection between gender and age group. Suggest an explanation for your finding.

Activity 7-11: Children's Living Arrangements

The following table classifies the living arrangements of American children (under 18 years of age) in 1993 according to their race/Hispanic origin and which parent(s) they live with. Analyze these data to address the issue of whether a relationship exists between race/Hispanic origin and parental living arrangements. Write a paragraph reporting your findings, supported by appropriate calculations and visual displays.

	white	black	Hispanic
both	40,842,340	3,833,640	4,974,720
just mom	9,017,140	5,750,460	2,176,440
just dad	2,121,680	319,470	310,920
neither	1,060,840	745,430	310,920

Activity 7-12: Baldness and Heart Disease

To investigate a possible relationship between heart disease and baldness, researchers asked a sample of 663 male heart patients to classify their degree of baldness on a 5-point scale. They also asked a control group (not suffering from heart disease) of 772 males to do the same baldness assessment. The results are summarized in the following table:

	none	little	some	much	extreme
heart disease	251	165	195	50	2
control	331	221	185	34	1

(a) What proportion of these men identified themselves as having little or no baldness?

(b) Of those who had heart disease, what proportion claimed to have some, much, or extreme baldness?

(c) Of those who declared themselves as having little or no baldness, what proportion were in the control group?

(d) Construct a segmented bar graph to compare the distributions of baldness ratings between subjects with heart disease and those from the control group.

(e) Summarize your findings about whether a relationship seems to exist between heart disease and baldness.

(f) Consider the "none" or "little" baldness categories as one group and the other three categories as another group. Calculate the relative risk of heart disease between these two groups.

Activity 7-13: Driver Safety (*cont.*)

Recall from Activity 2-13 on page 40 that in 1997 there were 46,568,949 licensed drivers over age 55, with 11,012 involved in fatal crashes. There were 12,587,060 licensed drivers between the ages of 16 and 20, with 7670 drivers of them involved in fatal crashes. Calculate the relative risk of being involved in a fatal crash between younger and older drivers.

Activity 7-14: Gender and Lung Cancer

Researchers in New York investigated a possible gender difference in incidence of lung cancer among smokers. They screened 1000 people who were smokers of age 60 or higher. They found 459 women and 541 men, with 19 of the women and 10 of the men suffering from lung cancer.

(a) Identify the explanatory and response variables in this study.

(b) Construct a 2 × 2 table for these data. Also include the marginal totals in the table.

(c) Calculate the conditional distributions of lung cancer for each gender.

(d) Construct a segmented bar graph to display these distributions.

(e) Calculate the relative risk of having lung cancer between women and men.

(f) Write a sentence or two interpreting the relative risk value in this context.

Activity 7-15: Friendly Observers (*cont.*)

Refer to Activity 13-10 on page 293, which discusses a psychology experiment in which subjects were randomly assigned to one of two groups. Those in group A were told that an observer would share in a prize with them if they could navigate a video game's obstacle course in a certain length of time. Those in group B were not told anything about the observer sharing in the prize. The researchers had conjectured that those in group B would perform better on the task, since they were not under as much peer pressure. It turned out that 3 of the 12 subjects in group A won the game, while 8 of 11 subjects in group B achieved success.

(a) Identify the explanatory and response variables in this study.
(b) Construct a two-way table to present these data.
(c) Calculate the conditional distributions of performance for each of the two groups.
(d) Construct a segmented bar graph to display these distributions.
(e) Write a sentence or two comparing the two distributions, addressing whether the data seem to support the researchers' conjecture.

Activity 7-16: Top American Films (*cont.*)

Refer again to the data on AFI's top 100 American films, and consider three categorical variables:

whether or not the film won an Academy Award for Best Picture
whether or not the film was produced after 1960
whether or not you have seen the film

(a) Produce two-way tables to represent all three pairs of variables.
(b) Analyze all three tables. Write a paragraph describing your findings about relationships among these three variables. Include calculations and displays to support your findings.

Activity 7-17: Penny Thoughts (*cont.*)

Consider the data collected in Topic 1 on the question of whether students favor retaining or abolishing the penny as a unit of U.S. currency. Create a two-way table classifying students according to their opinion about the U.S. penny and their political inclination. Analyze this table to address the question of whether a relationship exists between these two variables. Write a paragraph summarizing your findings.

Activity 7-18: Lifetime Achievements (*cont.*)

Consider the data collected in class on which achievement (Olympic medal, Nobel Prize, etc.) students would most like to fulfill.

(a) Examine the marginal distribution of this variable, construct a bar graph of it, and comment on the distribution.
(b) Create a two-way table of this "achievement" variable by gender, and examine the conditional distributions of this variable for each gender. Construct a segmented bar graph, and comment on your findings.

Activity 7-19: Hypothetical Hospital Recovery Rates (*cont.*)

Reconsider the hypothetical data in Activity 7-4 on page 146. Calculate the relative risk of dying between hospital A and hospital B for patients:
(a) overall;
(b) in fair condition;
(c) in poor condition.

Activity 7-20: Graduate Admissions Discrimination

The University of California at Berkeley was charged with having discriminated against women in their graduate admissions process for the fall quarter of 1973. The table below identifies the number of acceptances and denials for both men and women applicants in each of the six largest graduate programs at the institution at that time.

	men accepted	men denied	women accepted	women denied
program A	511	314	89	19
program B	352	208	17	8
program C	120	205	202	391
program D	137	270	132	243
program E	53	138	95	298
program F	22	351	24	317
total				

(a) Start by ignoring the program distinction, collapsing the data into a two-way table of gender by admission status. To do this, find the total number of men accepted and denied and the total number of women accepted and denied. Construct a table such as the one below:

	admitted	denied	total
men			
women			
total			

(b) Consider for the moment just the *men* applicants. Of the men who applied to one of these programs, what proportion were admitted? Now consider the *women* applicants; what proportion of them were admitted? Do these proportions seem to support the claim that men were given preferential treatment in admissions decisions?

(c) To try to isolate the program or programs responsible for the alleged mistreatment of women applicants, calculate the proportion of men and the proportion of women *within each program* who were admitted. Record your results in a table such as the one below.

	proportion of men admitted	proportion of women admitted
program A		
program B		
program C		
program D		
program E		
program F		

(d) Does it seem as if any program is responsible for the large discrepancy between men and women in the overall proportions admitted?

(e) Reason from the data given to explain how it happened that men had a much higher rate of admission overall even though women had higher rates in most programs and no program favored men very strongly.

Activity 7-21: Softball Batting Averages

Construct your own hypothetical data to illustrate Simpson's paradox in the following context. Show that it is possible for one softball player (Amy) to have a higher proportion of hits than another (Barb) in the first half of the season and in the second half of the season and yet to have a lower proportion of hits for the season as a whole. We will get you started: suppose that Amy has 100 at-bats in the first half of the season and 400 in the second half, and suppose that Barb has 400 at-bats in the first half and 100 in the second half. You are to make up how many hits each player had in each half of the season, so that the above statement holds. See how large you can make the differences in the players' proportions of hits. (The proportion of hits is the number of hits divided by the number of at-bats.)

	first half of season	second half of season	season as a whole
Amy's hits	30	100	130
Amy's at-bats	100	400	500
Amy's proportion of hits	30%	25%	26%
Barb's hits	117	24	141
Barb's at-bats	400	100	500
Barb's proportion of hits	29%	24%	28%

Activity 7-22: Employee Dismissals

Suppose that you are asked to investigate the practices of a company that has recently been forced to dismiss many of its employees. The company figures indicate that of the 1000 men and 1000 women who worked there a month ago, 300 of the men and 200 of the women were dismissed. The company employs two types of employees, professional and clerical, so you ask to see the breakdown for each type. Even though the company dismissed a higher percentage of men than women, you know that it is possible (Simpson's paradox) for the percentage of women dismissed within each employee type to exceed that of the men within each type. The company representative does not believe this, however, so you need to construct a hypothetical example to convince him of the possibility. Do so, by constructing and filling in tables such as the following.

overall (professional and clerical combined):

	dismissed	retained
men	300	700
women	200	800

professional only:

	dismissed	retained
men		
women		

clerical only:

	dismissed	retained
men		
women		

Activity 7-23: Hypothetical Employee Retention Predictions

Suppose that an organization is concerned about the number of its new employees who leave the company before they finish one year of work. In an effort to predict whether a new employee will leave or stay, they develop a standardized test and apply it to 100 new employees. After one year, they note what the test had predicted (stay or leave) and whether the employee actually stayed or left. They then compile the data into the following table:

	actually stays	actually leaves	row total
predicted to stay	63	12	75
predicted to leave	21	4	25
column total	84	16	100

(a) Of those employees predicted to stay, what proportion actually left?

(b) Of those employees predicted to leave, what proportion actually left?

(c) Is an employee predicted to stay any less likely to leave than an employee predicted to leave?

(d) Considering your answers to (a)–(c), does the test provide any helpful information about whether an employee will leave or stay?

(e) Is the employee's action *independent* of the test's prediction? Explain.

(f) Create a segmented bar graph to display these conditional distributions.

(g) Sketch what the segmented bar graph would look like if the test were *perfect* in its predictions.

(h) Sketch what the segmented bar graph would look like if the test were *very useful* but not quite perfect in its predictions.

Activity 7-24: Women Senators (*cont.*)

Refer to the table of gender vs. party breakdown among 1999 U.S. Senators you created in Activity 7-5 on page 148.

(a) What proportion of the Senators are women?

(b) What proportion of the Senators are Democrats?

(c) Is it fair to say that most Democratic Senators are women? Support your answer with an appropriate calculation.

(d) Is it fair to say that most women Senators are Democrats? Support your answer with an appropriate calculation.

Activity 7-25: Politics and Ice Cream

Suppose that 500 college students are asked to identify their preferences in political affiliation (Democrat, Republican, or Independent) and in ice cream (chocolate, vanilla, or strawberry). Fill in the following table in such a way that the variables political affiliation and ice cream preference turn out to be completely *independent*. In other words, the conditional distribution of ice cream preference should be the same for each political affiliation, and the conditional distribution of political affiliation should be the same for each ice cream flavor.

	chocolate	vanilla	strawberry	row total
Democrat	108	96 .48	3 6 .48	240
Republican	8 1	72 .36	27 .36	18 0
Independent	3 6	32	12	80
column total	225	200	75	500

240/500 = .48
180/500 = .36
80/500
.5 00

Activity 7-26: Variables of Personal Interest (*cont.*)

Think of a pair of categorical variables that you would be interested in exploring the relationship between. Describe the variables in as much detail as possible and indicate how you would present the data in a two-way table.

7

Topic 8:

GRAPHICAL DISPLAYS OF ASSOCIATION

OVERVIEW

To this point you have been investigating and analyzing distributions of a single variable. In the previous two topics you compared distributions among groups. This topic introduces you to another issue that plays an important role in statistics: exploring relationships between variables. You will investigate the concept of **association** and explore the use of graphical displays (namely, **scatterplots**) as you begin to study relationships between quantitative variables.

OBJECTIVES

- To understand the concept of **association** between two variables and the notions of the **direction** and **strength** of the association.
- To learn to construct and to interpret **scatterplots** as graphical displays of the relationship between two variables.
- To discover the utility of including a 45° ($y = x$) line on a scatterplot when working with **paired data**.
- To become familiar with **labeled scatterplots** as devices for including information from a categorical variable into a scatterplot.
- To use your calculator to explore association between variables of genuine data in a variety of applications.

8

PRELIMINARIES

1. Do you expect that there is a tendency for heavier cars to get worse fuel efficiency (as measured in miles per gallon) than lighter cars?

2. Do you think that if one car is heavier than another, it must always be the case that its gets worse fuel efficiency?

3. Take a guess as to a typical atmospheric temperature for a space shuttle launch in Florida.

4. What do you think is a typical salary for the governor of a state? The highest governor's salary? The lowest governor's salary?

   typical: highest: lowest:

5. Record the foot length and height of you and your classmates. Also record the person's gender.

IN-CLASS ACTIVITIES

Activity 8-1: Cars' Fuel Efficiency (*cont.*)

The following table reports the EPA's city miles per gallon rating and the weight (in pounds) for the sports cars described in *Consumer Reports 1999 New Car Buying Guide*. (The EPA rating for the Audi TT was not reported.)

Model	city MPG	weight	model	city MPG	weight
Audi TT	*	2655	Mazda MX-5 Miata	25	2365
BMW 318Ti	23	2790	Mercedes/Benz SLK	22	3020
BMW Z3	19	2960	Mercury Cougar	20	3140
Chevrolet Camaro	17	3545	Mitsubishi Eclipse	23	3235
Chevrolet Corvette	17	3295	Pontiac Firebird	18	3545
Ford Mustang	17	3270	Porsche Boxster	19	2905
Honda Prelude	22	3040	Saturn SC	27	2420
Hyundai Tiburon	22	2705	Toyota Celica	22	2720

(a) What are the observational units here?

(b) How many variables are reported in the table for each observational unit? What type (categorical or quantitative) is each variable?

> The simplest graph for displaying two quantitative variables simultaneously is a **scatterplot**, which uses a vertical axis for one of the variables and a horizontal axis for the other. A dot is placed for each observational pair at the intersection of its two values. If you are interested in using the value of one variable to predict the value of another variable, the convention is to place the variable to be predicted (the **response variable**) on the vertical axis and the variable to do the predicting (the **explanatory variable**) on the horizontal axis.

8

(c) Use the axes below to construct a scatterplot of miles per gallon vs. weight.

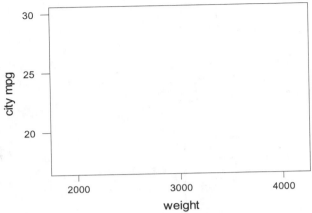

(d) Does the scatterplot reveal any relationship between a car's weight and its fuel efficiency? In other words, does knowing a car's weight reveal any information about its fuel efficiency? Write a few sentences about the relationship between the two variables.

Two variables are said to be ***positively associated*** if larger values of one variable tend to occur with larger values of the other variable; they are said to be ***negatively associated*** if larger values of one variable tend to occur with smaller values of the other. The strength of the association depends on how closely the observations follow that relationship. In other words, the strength of the association reflects how accurately one could predict the value of one variable based on the value of the other variable.

(e) Is fuel efficiency positively or negatively associated with weight?

(f) Can you find an example where one car weighs more than another and still manages to have better fuel efficiency than that other car? If so, identify such a pair and circle them on the scatterplot above.

Clearly, the concept of **association** is an example of a **statistical tendency**. It is *not always* the case that a heavier car is less fuel efficient, but heavier cars certainly do *tend* to be.

tends

Activity 8-2: Guess the Association

The data on 1999 cars reported by *Consumer Reports* included not just sports cars but also classifications of small, family, large, luxury, and upscale. The following nine scatterplots display pairs of variables for these cars. The variables are:

city MPG rating	highway MPG rating
weight	% front weight
time to accelerate from 0 to 60 miles per hour	time to cover 1/4 mile
page number on which the car appeared	fuel capacity

(a) Evaluate the direction and strength of the association between the variables in each graph. Do this by arranging the associations revealed in the scatterplots from those that reveal the most strongly positive association to those that reveal virtually no association to those that reveal the most strongly negative association. Arrange them by letter in the table below. (Since you are to use each letter only once, you should probably look through all nine plots first.)

	strongly negative		mildly negative		virtually none		mildly positive		strongly positive
letter:	D	G	A	H	C	E	I	F	B

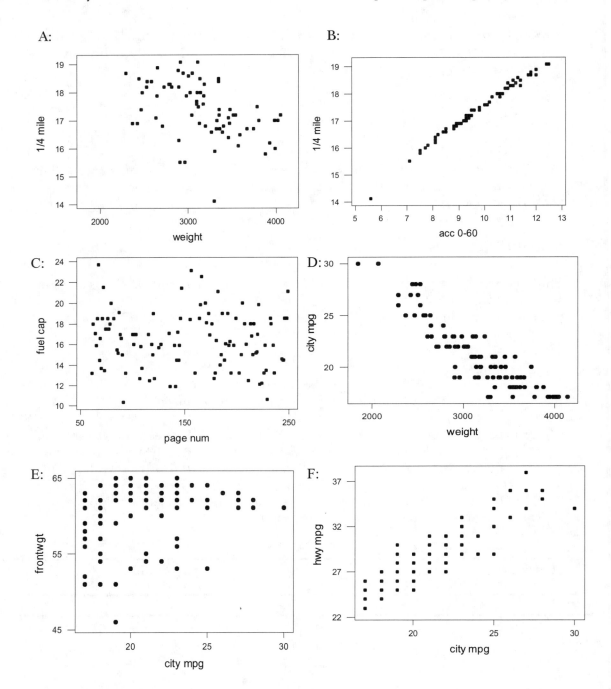

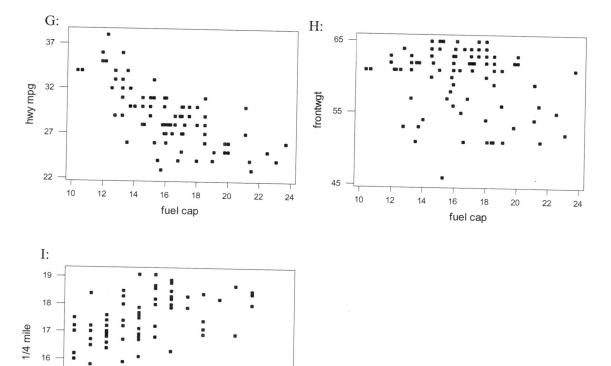

(b) Indicate what you would expect for the direction (positive, negative, or none at all) and strength (none, weak, moderate, or strong) of the association between the pairs of variables listed below.

pair of variables	direction	strength
length and width of signatures	positive	
SAT score & college G.P.A.	positive	
latitude & avg. January temperature of U.S. cities	negative	
lifetime & weekly cigarette consumption		
serving size & calories of fast food sandwiches		
airfare & distance to destination		
birth rates & death rates of U.S. states		
foot length & height		

∞

Activity 8-3: Marriage Ages (*cont.*)

Refer to the data from Activity 3-13 on page 60 concerning the ages of 24 couples applying for marriage licenses. The following scatterplot displays the relationship between husband's age and wife's age. The line drawn on the scatterplot is the "$y = x$" line, where the husband's age would equal the wife's age.

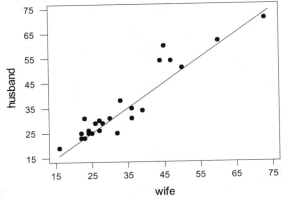

(a) Does there seem to be an association between husband's age and wife's age? If so, is it positive or negative? Would you characterize it as strong, moderate, or weak? Explain.

(b) Look back at the original listing of the data to determine how many of the 24 couples' ages fall exactly on the line. In other words, how many couples listed the same age for both the man and the woman on their marriage license?

(c) Again looking back at the data, for how many couples is the husband older than the wife? Do these couples fall above or below the line drawn in the scatterplot?

(d) For how many couples is the husband younger than the wife? Do these couples fall above or below the line drawn in the scatterplot?

(e) Summarize what one can learn about the ages of marrying couples by noting that the majority of couples produce points that fall above the "$y = x$" line.

> This activity illustrates that in working with paired data, including a "$y = x$" line on a scatterplot can provide valuable information about whether one member tends to have a larger value of the variable than the other member of the pair.

Activity 8-4: Space Shuttle O-Ring Failures

The following data were obtained from 23 shuttle launches prior to *Challenger*'s fatal launch. Each of four joints in the shuttle's solid rocket motor is sealed by two O-ring joints. After each launch, the reusable rocket motors were recovered from the ocean. This table lists the number of O-ring seals showing evidence of thermal distress and the launch temperature for each of the 23 flights.

flight date	O-ring failures	temperature	flight date	O-ring failures	temperature
4/12/81	0	66	11/8/84	0	67
11/12/81	1	70	1/24/85	3	53
3/22/82	0	69	4/12/85	0	67
11/11/82	0	68	4/29/85	0	75
4/4/83	0	67	6/17/85	0	70
6/18/83	0	72	7/29/85	0	81
8/30/83	0	73	8/27/85	0	76
11/28/83	0	70	10/3/85	0	79
2/3/84	1	57	10/30/85	2	75
4/6/84	1	63	11/26/85	0	76
8/30/84	1	70	1/12/86	1	58
10/5/84	0	78			

(a) Lists containing the information in the table above have been stored in a grouped file named CHALLENG.83g. Download this grouped file into your calculator. The lists have been named ORING and TEMP. To create a scatterplot of O-ring vs. temperature with your calculator, set up Plot1 as:

The "mark" prompt offers you three choices for the style of point to be displayed. To view the graph, press ZOOM 9 .

Write a few sentences commenting on whether the scatterplot reveals any association between O-ring failures and temperature.

(b) The forecasted low temperature for the morning of the fateful launch was 31°F. What does the scatterplot reveal about the likelihood of O-ring failure at such a temperature?

(c) Place the two lists into the Stat List Editor and delete temperature and O-ring values for those flights that had no O-ring failures. Considering just the remaining seven flights, use your calculator to construct a new scatterplot of O-ring failures vs. temperature. Does this scatterplot reveal association between O-ring failures and temperature? If so, does it make the case as strongly as the previous scatterplot did?

(d) NASA officials argued that flights on which no failures occurred provided no information about the issue. They therefore examined only the second scatterplot and concluded that there was little evidence of an association between temperature and O-ring failures. Comment on the wisdom of this approach and specifically on the claim that flights on which no failures occurred provided no information.

Activity 8-5: Fast Food Sandwiches

A *categorical* variable can be incorporated into a scatterplot by constructing a **labeled scatterplot**, which assigns different labels to the dots in a scatterplot.

The following table lists some nutritional information reported in 1999 about sandwiches offered by the fast food chain Arby's. Serving sizes are in ounces, fat and protein in grams, and sodium in milligrams.

sandwich	serv oz	calories	total fat	protein	sodium	meat
Arby's Melt with Cheddar	5.2	368	18	18	937	roast beef
Arby-Q	6.4	431	18	22	1321	roast beef
Bac'n Cheddar Deluxe	8.1	539	34	22	1140	roast beef
Beef'n Cheddar	6.7	487	28	25	1216	roast beef
Giant Roast Beef	8.1	555	28	35	1561	roast beef
Junior Roast Beef	4.4	324	14	17	779	roast beef
Regular Roast Beef	5.4	388	19	23	1009	roast beef
Super Roast Beef	8.7	523	27	25	1189	roast beef
Breaded Chicken Fillet	7.2	536	28	28	1016	chicken
Chicken Cordon Bleu	8.5	623	33	38	1594	chicken
Grilled Chicken BBQ	7.1	388	13	23	1002	chicken
Grilled Chicken Deluxe	8.1	430	20	23	848	chicken
Roast Chicken Club	8.5	546	31	31	1103	chicken
Roast Chicken Deluxe (ssb)	7.6	433	22	24	763	chicken
Roast Chicken Santa Fe	6.4	436	22	29	818	chicken

French Dip	6.8	475	22	30	1411	roast beef
Hot Ham 'n Swiss	9.3	500	23	30	1664	ham
Italian Sub	10.1	675	36	30	2089	ham
Philly Beef 'n Swiss	10.4	755	47	39	2025	roast beef
Roast Beef Sub	10.8	700	42	38	2034	roast beef
Triple Cheese Melt	8.4	720	45	37	1797	*
Turkey Sub	9.8	550	27	31	2084	turkey
Roast Beef Deluxe	6.4	296	10	18	826	roast beef
Roast Chicken Deluxe	6.8	276	6	20	777	chicken
Roast Turkey Deluxe	6.8	260	7	20	1262	turkey
Fish Fillet	7.7	529	27	23	864	fish
Ham 'n Cheese	5.9	359	14	24	1283	ham
Ham 'n Cheese Melt	4.9	329	13	20	1013	ham

(a) Download the file ARBYS.83g. The data on calories and serving ounces are stored in lists named CALOR and SERV. Create a scatterplot of calories vs. serving ounces. Does the scatterplot reveal an association between a sandwich's serving size and its calories? Explain.

(b) Now use your calculator to produce a labeled scatterplot of calories vs. serving size, using different labels for the meat, poultry and other (Fish Fillet and Triple Cheese Melt). Create a new list MEAT that distinguishes these groupings. Since your calculator accepts only numeric values in lists, you need to use a 0 to represent meat, 1 to represent poultry, and 2 to represent other. Download and run the program LBLSCAT.83p using SERV as the X-LIST, CALOR as the Y-LIST, and MEAT as the CATEGORICAL VARIABLE LIST. Also, you will need to type in names for the different categories, i.e., you could use M for meat, P for poultry, and O for other.

Comment on tendencies you might observe with regard to the type of meat (including cheese). Does one type of meat tend to have more or fewer calories than other types of similar serving size? Elaborate.

WRAP-UP

This topic has introduced you to an area that occupies a prominent role in statistical practice: exploring relationships between variables. You have discovered the concept of **association** and learned how to construct and to interpret **scatterplots** and **labeled scatterplots** as visual displays of association.

Just as we moved from graphical displays to numerical summaries when describing distributions of data, we will consider in the next topic a numerical measure of the degree of association between two quantitative variables: the **correlation coefficient.**

HOMEWORK ACTIVITIES

Activity 8-6: Height and Foot Length

Enter the data collected on the heights and foot lengths into your calculator as lists named HTS and FEET. Suppose that your goal is to predict a person's height from his or her foot length (perhaps a footprint was found at a crime scene and you are asked to estimate the height of the perpetrator).

(a) Use your calculator to produce the appropriate scatterplot.

(b) Comment on the direction and strength of the association, keeping in mind the goal of predicting height from foot length.

(c) Use your calculator to produce a labeled scatterplot of height vs. foot length, using gender as the labels. Comment on any gender differences that you notice.

Activity 8-7: Broadway Shows (*cont.*)

The following scatterplots present all pairs of variables among the four related to Broadway shows: gross receipts, attendance, capacity, and top ticket price.

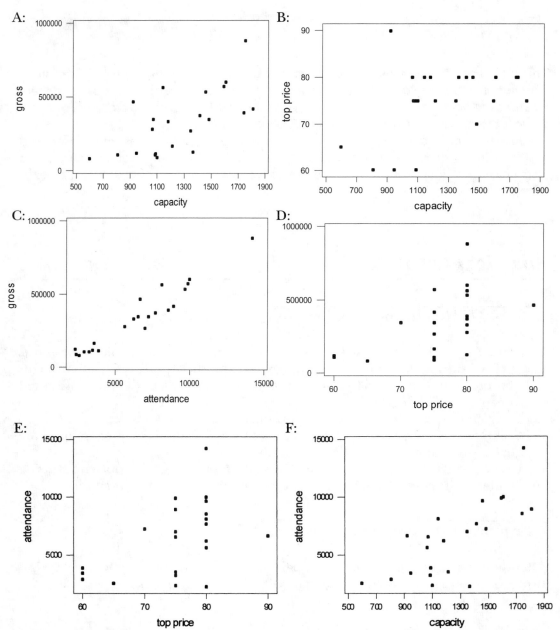

(a) In what direction is the association between the variables in all of these scatter-plots?   positive

(b) Arrange these in order from the strongest to the weakest association.

C, F, D, A, E, B

Activity 8-8: College Football Players

Refer to Activity 12-16 on page 265 for data on Cal Poly's 1999 football team. The following scatterplots display the relationship between a player's weight and his jersey number. In the first plot the labels show different class years, while in the second plot they indicate whether the player plays primarily offense, defense, or special teams.

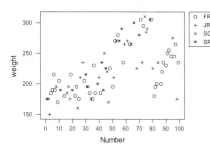 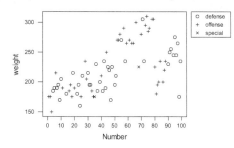

(a) For each statement, decide whether it is true or false based on the scatterplots above.
- There is a positive association between weight and number.
- There is a strong positive association between weight and number.
- The association is stronger for numbers below 70 than for all numbers.
- Most defensive players weigh more than most offensive players.
- Numbers above 80 are split roughly evenly between offensive and defensive players.
- Defensive players generally weigh more than offensive players with similar numbers.
- Most numbers above 80 belong to freshmen.
- Most freshmen have numbers above 80.
- Most of the players are freshmen.

(b) For each statement that is false, draw a rough sketch of a labeled scatterplot that would make the statement true.

Activity 8-9: Birth and Death Rates (*cont.*)

Refer to the data on states' birth and death rates presented in Activity 6-1 on page 118 (BIRTHDEATH.83g).
(a) Use your calculator to produce a scatterplot with birth rate on the vertical axis and death rate on the horizontal. ✗
(b) How would you characterize the direction and strength of the association?
(c) Does this agree with your prediction in Activity 8-2 on page 165? Explain.
(d) How many states have a higher death rate than birth rate? Identify them by name and also list their birth rates and death rates.
(e) Where do their points fall relative to a "$y = x$" line on the scatterplot?

(f) Identify at least three other states whose rates stand out on the scatterplot, and explain what is unusual about them.

Activity 8-10: City Temperatures (*cont.*)

Recall from Activity 5-1 on page 89 your analysis of the average monthly temperatures for the cities of Raleigh and San Francisco. Following are two scatterplots of the data with a"$y = x$" line drawn on plot B:

A: B:

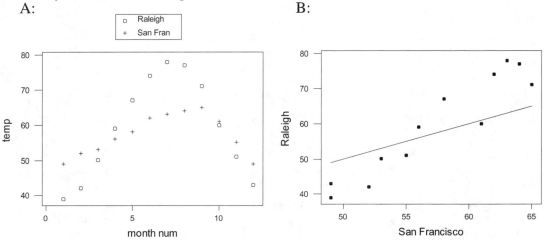

(a) In how many of the 12 months is Raleigh's monthly temperature higher than San Francisco's? Explain how you could use either graph to discern this.

(b) In which of the twelve months is Raleigh's monthly temperature higher than San Francisco's? With which graph can you discern this? Explain.

(c) How would you characterize the strength of the association between the two cities' monthly temperatures? With which graph is it easier to judge this? Explain.

Activity 8-11: Fast Food Sandwiches (*cont.*)

Consider again the data on Arby's sandwiches from Activity 8-4 on page 169. Use your calculator to create labeled scatterplots of fat vs. serving size, protein vs. serving size, and sodium vs. serving size. In each case use "type of meat" as the labels. Comment on differences between meat and poultry sandwiches as revealed in these scatterplots.

Activity 8-12: Broadway Shows (*cont.*)

Reconsider the data from Activity 2-8 on page 36 concerning Broadway shows (BROADWAY.83g). Notice that there are two different types of shows listed: plays and musicals.

(a) Produce a labeled scatterplot of receipts vs. *percentage* capacity using different labels to distinguish plays from musicals.

(b) Ignoring for the moment the distinction between plays and musicals, comment on the association between receipts and percentage capacity.

(c) Now comment on differences in the relationship between receipts and percentage capacity based on whether the show is a play or a musical. For shows with similar percentage capacities, does one type of show tend to take in more money than the other type? Explain.

Activity 8-13: Scrabble Names (*cont.*)

Reconsider the data collected in Topic 2 concerning numbers of letters and Scrabble points in students' names.

(a) Use your calculator to create a scatterplot of Scrabble points vs. number of letters.

(b) Comment on the association between Scrabble points and number of letters as revealed in the scatterplot.

(c) Can you find examples of pairs of students where the student with the longer name has fewer Scrabble points? If so, identify such a pair and circle them on the scatterplot.

(d) Do any points fall below the "$y = x$" line on this plot (assuming that Scrabble points are on the vertical axis and number of letters on the horizontal)? Explain why this makes sense.

(e) What can you conclude about any points that fall on the "$y = x$" line in this scatterplot? (Be sure to relate your answer to the context.)

(f) Use your calculator to create a new variable: *ratio* of Scrabble points to letters.

(g) Use your calculator to create a scatterplot of this ratio vs. Scrabble points. Does it appear that names with more points tend to have higher ratios or lower ratios than names with fewer points? Elaborate.

(h) Use your calculator to produce a scatterplot of ratio vs. letters. Comment on any association between these variables that is revealed.

Activity 8-14: Signature Measurements (*cont.*)

Reconsider the data collected in Topic 1 concerning students' signatures.

(a) Use your calculator to create a labeled scatterplot of signature length vs. height, using different labels for each gender.

(b) Disregarding gender for the moment, write a sentence or two to describe the association between signature length and height.

(c) Comment on any gender differences that the scatterplot reveals.

(d) Does the manner in which we collected the data allow you to examine the relationship between the number of letters in a person's name and the length of her or his signature? Explain.

Activity 8-15: College Alumni Donations

For the graduating classes of 1961–1998 at Harvey Mudd College, the following table lists the percentage of alumni who made a financial contribution to the college during 1998–99. Also listed is the average amount of a contribution among those who made one (HMCDONORS.83g).

year	% give	avg gift	year	% give	avg gift	year	% give	avg gift	year	% give	avg gift
1961	67	773	1971	65	228	1981	45	339	1991	50	111
1962	81	12187	1972	60	349	1982	40	339	1992	44	142
1963	64	2951	1973	50	399	1983	51	239	1993	30	83
1964	51	2453	1974	63	2567	1984	36	255	1994	32	80
1965	58	2476	1975	56	406	1985	48	172	1995	51	58
1966	64	723	1976	47	444	1986	41	157	1996	44	66
1967	60	614	1977	54	344	1987	50	178	1997	32	124
1968	61	298	1978	53	295	1988	40	234	1998	35	236
1969	54	4945	1979	49	3106	1989	43	156			
1970	61	402	1980	48	201	1990	46	133			

(a) Ignore the class year information for the moment, and use your calculator to analyze the distribution of donor percentages (PERGV). Write a few sentences commenting on key features of this distribution. (Consider this your occasional reminder *always* to relate your comments to the context at hand.)

(b) Continue to ignore the class year information for the moment, and use your calculator to analyze the distribution of average gifts (AVGGV). Write a few sentences commenting on key features of this distribution.

(c) Now use your calculator to produce a scatterplot of donor percentage vs. class year. Comment on any patterns revealed in the scatterplot.

(d) Use your calculator to produce a scatterplot of average gift vs. class. Comment on any patterns revealed in the scatterplot.

(e) Does one class stick out as deviating from the pattern established by the majority in either scatterplot? If so, identify it and comment on its oddity.

Activity 8-16: Peanut Butter

The September 1990 issue of *Consumer Reports* rated thirty–seven varieties of peanut butter. Each variety was given an overall sensory quality rating, based on taste tests by a trained sensory panel. Also listed was the cost (per three tablespoons, based on the average price paid by *CR* shoppers) and sodium content (per three tablespoons, in milligrams) of each product. Finally, each variety was classified as creamy (cr) or chunky (ch), natural (n) or regular (r), and salted (s) or unsalted (u). The results are given below and in PBUTTER.83g (with list names COST, SODIU, QUAL, CRCH, RN, and SU).

brand	cost	sodium	quality	cr/ch	r/n	s/u
Jif	22	220	76	cr	r	s
Smucker's Natural	27	15	71	cr	n	u
Deaf Smith Arrowhead Mills	32	0	69	cr	n	u
Adams 100% Natural	26	0	60	cr	n	u
Adams	26	168	60	cr	n	s
Skippy	19	225	60	cr	r	s
Laura Scudder's All Natural	26	165	57	cr	n	s
Kroger	14	240	54	cr	r	s
Country Pure Brand	21	225	52	cr	n	s
NuMade	20	187	43	cr	r	s
Peter Pan	21	225	40	cr	r	s
Peter Pan	22	3	35	cr	r	u
A&P	12	225	34	cr	r	s
Hollywood Natural	32	15	34	cr	n	u
Food Club	17	225	33	cr	r	s
Pathmark	9	255	31	cr	r	s
Lady Lee	16	225	23	cr	r	s
Albertsons	17	225	23	cr	r	s
Shur Fine	16	225	11	cr	r	s
Smucker's Natural	27	15	89	ch	n	u
Jif	23	162	83	ch	r	s
Skippy	21	211	83	ch	r	s
Adams 100% Natural	26	0	69	ch	n	u
Deaf Smith Arrowhead Mills	32	0	69	ch	n	u
Country Pure Brand	21	195	67	ch	n	s
Laura Scudder's All Natural	24	165	63	ch	n	s
Smucker's Natural	26	188	57	ch	n	s
Food Club	17	195	54	ch	r	s
Kroger	14	255	49	ch	r	s
A&P	11	225	46	ch	r	s
Peter Pan	22	180	45	ch	r	s
NuMade	21	208	40	ch	r	s

Health Valley 100% Natural	34	3	40	ch	n	u
Lady Lee	16	225	34	ch	r	s
Albertsons	17	225	31	ch	r	s
Pathmark	9	210	29	ch	r	s
Shur Fine	16	195	26	ch	r	s

(a) Select any *pair* of quantitative variables that you would like to examine. Use your calculator to produce a scatterplot of these variables, and write a paragraph of a few sentences commenting on the relationship between them.

(b) Now select *one* of the three categorical variables, and use your calculator to produce a labeled scatterplot of your two variables from (a) using this new variable for the labels. Comment on any effect of this categorical variable as well as on any other features of interest in the plot.

Activity 8-17: States' SAT Averages (*cont.*)

Recall the data from Activity 2-4 on page 32, which listed the average SAT score for each of the fifty states and the percentage of high school seniors in the state who took the SAT test (SAT98.83g).

(a) Use your calculator to look at a scatterplot of these data. Write a paragraph describing the relationship between average SAT score and percentage of students taking the test. Include a reasonable explanation for the type of association that is apparent.

(b) Which state has the highest SAT average? Would you conclude that this state does the best job of educating its students? Which state has the lowest SAT average? Would you conclude that this state does the worst job of educating its students? Explain.

(c) How does *your* home state compare to the rest in terms of SAT average and percentage of students taking the test? (Identify the state also.)

Activity 8-18: Governors' Salaries

The following table lists each state's per capita income (as of 1997) and governor's salary (as of 1998) and also indicates the state's region of the country (Northeast, Midwest, South, or West).

state	region	income	gov sal	state	region	income	gov sal
Alabama	S	20,842	87,643	Montana	W	20,046	78,246
Alaska	W	25,305	81,648	Nebraska	MW	23,803	65,000
Arizona	W	22,364	75,000	Nevada	W	26,791	90,000

Arkansas	S	19,585	60,000	New Hampshire	NE	28,047	90,547
California	W	26,570	131,000	New Jersey	NE	32,654	85,000
Colorado	W	27,051	70,000	New Mexico	W	19,587	90,000
Connecticut	NE	36,263	78,000	New York	NE	30,752	130,000
Delaware	S	29,022	107,000	North Carolina	S	23,345	107,132
Florida	S	25,255	110,962	North Dakota	MW	20,271	75,372
Georgia	S	24,061	115,939	Ohio	MW	24,661	115,762
Hawaii	W	26,034	94,780	Oklahoma	S	20,556	101,140
Idaho	W	20,478	75,000	Oregon	W	24,393	88,300
Illinois	MW	28,202	130,261	Pennsylvania	NE	26,058	105,035
Indiana	MW	23,604	77,200	Rhode Island	NE	25,760	69,900
Iowa	MW	23,102	104,352	South Carolina	S	20,755	106,078
Kansas	MW	24,379	85,225	South Dakota	MW	21,447	82,271
Kentucky	S	20,657	95,526	Tennessee	S	23,018	85,000
Louisiana	S	20,680	95,000	Texas	S	23,656	115,345
Maine	NE	22,078	70,000	Utah	W	20,432	90,700
Maryland	S	28,969	120,000	Vermont	NE	23,401	105,402
Massachusetts	NE	31,524	90,000	Virginia	S	26,438	124,855
Michigan	MW	25,560	127,300	Washington	W	26,718	121,000
Minnesota	MW	26,797	114,506	West Virginia	S	18,957	90,000
Mississippi	S	18,272	83,160	Wisconsin	MW	24,475	115,899
Missouri	MW	24,001	112,755	Wyoming	W	22,648	95,000

This labeled scatterplot displays governor salary vs. average wage, using different symbols for the regions.

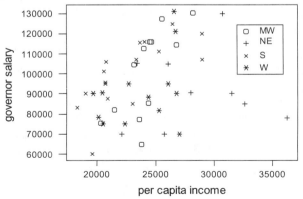

(a) Comment on the relationship between these two variables.

(b) List three pairs of states for which the state with the higher average pay has the lower governor salary.

(c) Explain in your own words how (b) reveals that these data represent a statistical *tendency.*

(d) Name a state that appears to have a governor salary much higher than would be expected for a state with its average pay.

(e) Identify a cluster of three states that seem to have much lower governor salaries than would be expected for states with their average pay. What region are they all from?

Activity 8-19: Comparison Shopping

The following data were collected by Cal Poly students over a two-day period at two grocery stores, Lucky's and Von's, in the same town (SHOPPING.83g).

Product	Luckys	Vons	Product	Luckys	Vons
Baker's angel flake coconut	1.60	1.50	Hershey's semi-sweet choc. chips (12 oz)	2.19	2.68
C&H granulated sugar (10 lb)	4.75	3.99	Iceberg lettuce	0.89	.99
Campbell's Tomato Soup	0.75	0.60	Jiffy Crunch Peanut Butter (18 oz)	2.49	2.42
Carnation evaporated milk (5 oz)	0.48	0.49	Kikkoman soy sauce	1.29	1.49
Carrots (2 lb bag)	1.86	2.09	Kraft Minute Rice (28 oz)	3.49	2.98
Crest regular toothpaste (6.4 oz)	1.99	2.85	Kraft Parmesan Cheese	3.99	3.86
Dannon yogurt, strawberry	0.85	0.84	Lender's raisin bagels (6 pack, fresh)	1.99	2.29
Ding Dongs (12 pack)	3.79	3.79	Milk (1/2 gallon)	2.07	2.15
Dr. Pepper (2 liter)	0.79	0.59	Minute Maid frozen OJ (8.54 oz)	1.89	1.99
Dryer's Vanilla Frozen Yogurt	4.99	4.99	Navel oranges (price/oz)	6.18	4.36
Excedrin (40 tablets)	6.99	6.99	Northern bath tissue (12 rolls)	4.01	3.99
Franco-American Spaghetti O's	1.29	1.39	Pepperidge Farm Brussel Cookies	2.00	1.98
Gladlock Zipper Sandwich bags	3.29	3.29	PopTarts - strawberry (8-pack)	1.67	2.03
Gold Medal flour (5 lb, regular)	1.62	1.69	Special K	5.29	5.39
Green Beans, cut (8 oz)	0.49	0.53	Tide (50 oz)	4.69	4.75
Ground round (22% fat, price/lb)	3.29	3.29	Wishbone Italian dressing (16 fl oz)	2.99	2.89
Heinz ketchup (28 oz)	2.49	2.18	Wonder bread, wheat	2.49	2.99

(a) Identify the observational units in this study.

(b) Examine a scatterplot of the prices at Lucky's versus the prices at Von's. Write a paragraph describing the relationship between the prices at the two stores. In particular, where do the points fall compared to the "y = x" line?

(c) Are there any products for which you are suspicious of the prices recorded? Explain.

Activity 8-20: Variables of Personal Interest (*cont.*)

Think of two *pairs* of quantitative variables whose relationship you might be interested in studying. (Remember that a quantitative variable is any characteristic of a person or object that can assume a range of numerical values.) Be very specific in describing these variables; identify the *observational units* as well as the variables.

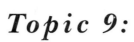

Topic 9:

CORRELATION COEFFICIENT

OVERVIEW

You have seen how scatterplots provide useful visual information about the relationship between two quantitative variables. Just as you made use of numerical summaries of various aspects of the distribution of a single variable, it would also be handy to have a numerical measure of the association between two variables. This topic introduces you to just such a measure and asks you to investigate some of its properties. This measure is one of the most famous in statistics: the **correlation coefficient**.

OBJECTIVES

- To explore basic properties of the **correlation coefficient** as a numerical measure of the degree of association between two variables.
- To discover some of the limitations of the correlation coefficient as a summary of the relationship between two variables.
- To recognize the important distinction between association and **causation**.
- To become familiar with judging correlation values from scatterplots.
- To learn to use scatterplots and correlations to look for and to describe relationships between variables when analyzing genuine data.

1. What is the cheapest property on the Monopoly game board? What is the most expensive?

    Cheapest: Most expensive:

2. Take a guess as to the number of people per television set in the United States in 1990; do the same for China and for Haiti.

    United States: 2:1

    China:

    Haiti: 1,000:1

3. Do you expect that countries with few people per television tend to have longer life expectancies, shorter life expectancies, or do you suspect no relationship between televisions and life expectancy?

IN-CLASS ACTIVITIES

Activity 9-1: Properties of Correlation

The *correlation coefficient*, denoted by r, is a measure of the degree to which two variables are associated. The calculation of r is very tedious to do by hand, so you will begin by letting your calculator compute correlations while you explore their properties.

(a) Look back on the nine scatterplots involving pairs of variables related to 1999 cars from Activity 8-2 on page 165. Reproduce in the "letter" row of the table below your ordering of these scatterplots from most strongly positive to most strongly negative association.

	strongly negative		mildly negative		virtually none		mildly positive		strongly positive
letter:									
r									

(b) Lists of variables related to 1999 cars are stored in a grouped file named CARS99.83g. Download this grouped file into your calculator and the program CORR.83p. The lists have been loaded as PNUM, CMPG, HMPG, FCAP, WGT, FWGT, ACC30, ACC60, MILE, TYPE, respectively.

Run the CORR program to calculate the correlation coefficient for each scatterplot. Record the value of the correlation in the above table underneath the appropriate letter designation.

(c) Based on these results, what do you suspect is the largest value that a correlation coefficient can assume? What do you suspect is the smallest value?

(d) Under what circumstances do you think the correlation assumes its largest or smallest value; i.e., what would have to be true of the observation pairs in that situation?

(e) How does the value of the correlation relate to the *direction* of the association?

(f) How does the value of the correlation relate to the *strength* of the association?

> These examples should convince you that a correlation coefficient has to be between +1 and −1. It can equal one of those values when the observations form a perfectly straight line. The sign of the correlation reflects the direction of the association. The magnitude of the correlation indicates the strength of the association, with values closer to +1 or −1 signifying stronger association.

Recall the data you analyzed in Activity 5-1 on page 89 of Raleigh's average monthly temperature. Consider the scatterplot of temperature vs. the number of the month (January = 1, February = 2, and so on):

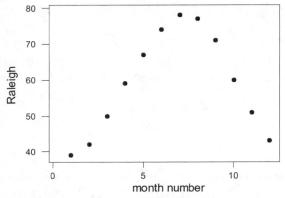

(g) Does there seem to be any relationship between temperature and month in Raleigh? If so, describe the relationship.

(h) Download CITYTEMPS.83g into your calculator and determine the correlation coefficient for these variables (MONTH and RALEI); record it below. Does its value seem to indicate a strong or a weak relationship? Is this consistent with your answer to (g)? Explain.

> The example above illustrates that the correlation coefficient measures only *linear* (straight-line) relationships between two variables. More complicated types of relationships (such as curvilinear ones) can go undetected by *r*. Thus, there might be a relationship, albeit not a linear one, even if the correlation is close to zero. One should be aware of such possibilities and examine a scatterplot as well as the value of *r*.

Consider the following scatterplot, which shows tuition fee vs. year of founding for 25 public four-year colleges in Pennsylvania (which is also discussed in Activity 3-10 on page 58 and Activity 6-15 on page 133):

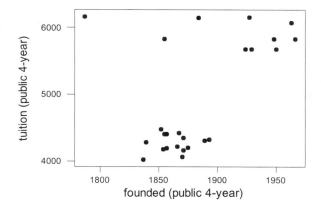

(i) Download COLL99.83g (the list names are PUBT and PUBF) into your calculator and determine the correlation coefficient for these variables; record it below. Does its value surprise you at all? Explain.

Activity 9-2: Monopoly Prices

The following table presents data on properties in the board game Monopoly. Listed are the property's position on the game board, purchase price, rental charge, rental charge with one house, and rental charge with one hotel:

	property	pos	price	rent	house	hotel		property	pos	price	rent	house	hotel
1	Mediterranean	1	60	2	10	250	12	Kentucky	21	220	18	90	1050
2	Baltic	3	60	4	20	450	13	Indiana	23	220	18	90	1050
3	Oriental	6	100	6	30	550	14	Illinois	24	240	20	100	1100
4	Vermont	8	100	6	30	550	15	Atlantic	26	260	22	110	1150
5	Connecticut	9	120	8	40	600	16	Ventnor	27	260	22	110	1150
6	States	11	140	10	50	750	17	Marvin Gardens	29	280	24	120	1200
7	St. Charles Place	13	140	10	50	750	18	Pacific	31	300	26	130	1275
8	Virginia	14	160	12	60	900	19	No. Carolina	32	300	26	130	1275
9	St. James Place	16	180	14	70	950	20	Pennsylvania	34	320	28	150	1400
10	Tennessee	18	180	14	70	950	21	Park Place	37	350	35	175	1500
11	New York	19	200	16	80	1000	22	Boardwalk	39	400	40	200	2000

(a) Suppose that you are interested in predicting the rent of a property based on its price. Download MONOP.83g and use your calculator to produce a scatterplot of rent vs. price. Based on the scatterplot, make a guess as to the value of the correlation coefficient between rent and price.

(b) Use your calculator to determine that correlation. Record it below and comment on the accuracy of your guess.

(c) Investigate the effect that one case can have on a correlation coefficient. Place the lists into the Stat List Editor and change Boardwalk's price and rent values to those listed in the table below. In each instance use your calculator to produce a scatterplot, make a guess for the value of the correlation from the scatterplot, and then use your calculator to compute the correlation. Record your results in the table below.

Boardwalk price	400	400	400	400	100	1	1
Boardwalk rent	40	100	1	1000	40	40	100
guess for correlation							
actual correlation							

(d) Based on your analyses of Boardwalk's effect, would you say that the correlation coefficient is a *resistant* measure of association? Explain.

One can gain some insight into how the correlation coefficient r measures association by examining the formula for its calculation:

$$r = \frac{1}{n-1} \sum_{i=1}^{n} \left(\frac{x_i - \bar{x}}{s_x} \right)\left(\frac{y_i - \bar{y}}{s_y} \right),$$

where x_i denotes the ith observation of one variable, y_i the ith observation of the other variable, $\bar{x}$ and $\bar{y}$ the respective sample means, s_x and s_y the respective sample standard deviations, and n the sample size. This formula says to standardize each x and y value into its z-score, to multiply these z-scores together for each pair, to add those results, and to divide the sum by one less than the sample size.

Activity 9-3: Cars' Fuel Efficiency (*cont.*)

For the fuel efficiency data that you analyzed in Activity 8-1 on page 163, the weights have a mean of 2997 and a standard deviation of 357.6 pounds. The mean of the miles per gallon ratings is 20.867, and their standard deviation is 3.044 miles per gallon. The table below begins the process of calculating the correlation between weight and miles per gallon by calculating the z-scores for the weights and miles per gallons and then multiplying the results.

model	weight	z-score	city MPG	z-score	product
BMW 318Ti	2790	–0.579	23	0.701	–0.406
BMW Z3	2960	–0.103	19	–0.613	0.063
Chevrolet Camaro	3545	1.532	17	–1.270	–1.946
Chevrolet Corvette	3295		17	–1.270	
Ford Mustang	3270	0.763	17	–1.270	–0.970

Honda Prelude	3040	0.120	22	0.372	0.045
Hyundai Tiburon	2705	–0.817	22	0.372	–0.304
Mazda MX-5 Miata	2365	–1.767	25	1.358	–2.400
Mercedes/Benz SLK	3020	0.064	22	0.372	0.024
Mercury Cougar	3140	0.400	20	–0.285	–0.114
Mitsubishi Eclipse	3235	0.666	23	0.701	0.466
Pontiac Firebird	3545	1.532	18	–0.942	–1.443
Porsche Boxster	2905	–0.257	19	–0.613	0.158
Saturn SC	2420	–1.613	27		
Toyota Celica	2720	–0.775	22	0.372	–0.288

(a) Calculate the z-score for the weight of a Chevrolet Corvette and for the miles per gallon rating of a Saturn SC. Then calculate the product of z-scores for these two cars. Show your calculations below and record the results in the table.

(b) Add the results of the "product" column and then divide the result by 14 (one less than the sample size of 15 cars) to determine the value of the correlation coefficient between weight and miles per gallon.

(c) What do you notice about the miles per gallon z-score of most of the cars with negative weight z-scores? Explain how this results from the strong negative association between weight and miles per gallon.

Activity 9-4: Televisions and Life Expectancy

The following table provides information on life expectancies for a sample of 22 countries. It also lists the number of people per television set in each country.

country	life exp	per TV	country	life exp	per TV
Angola	44	200	Mexico	72	6.6
Australia	76.5	2	Morocco	64.5	21
Cambodia	49.5	177	Pakistan	56.5	73
Canada	76.5	1.7	Russia	69	3.2
China	70	8	South Africa	64	11
Egypt	60.5	15	Sri Lanka	71.5	28
France	78	2.6	Uganda	51	191
Haiti	53.5	234	United Kingdom	76	3
Iraq	67	18	United States	75.5	1.3
Japan	79	1.8	Vietnam	65	29
Madagascar	52.5	92	Yemen	50	38

(a) Which of the countries listed has the fewest people per television set? Which has the most? What are those numbers?

(b) These data have been stored in a grouped file named TVLIFE.83g. Download this grouped file into your calculator. The lists are named LFEXP and PERTV. Use your calculator to produce a scatterplot of life expectancy vs. people per television set. Does there appear to be an association between the two variables? Elaborate briefly.

(c) Use your calculator to determine the value of the correlation coefficient between life expectancy and people per television.

(d) Since the association is so strongly negative, one might conclude that simply sending television sets to the countries with lower life expectancies would cause their inhabitants to live longer. Comment on this argument.

(e) If two variables have a correlation close to +1 or to −1, indicating a strong linear association between them, does it follow that there must be a cause-and-effect relationship between them?

> This example illustrates the very important distinction between **association** and **causation**. Two variables may be strongly associated (as measured by the correlation coefficient) without a cause-and-effect relationship existing between them. Often the explanation is that both variables are related to a third variable not being measured; this variable is often called a **lurking** or **confounding variable**.

(f) In the case of life expectancy and television sets, suggest a confounding variable that is associated both with a country's life expectancy and with the prevalence of televisions in the country.

Activity 9-5: Guess the Correlation

This activity will give you practice at judging the value of a correlation coefficient by examining the scatterplot.

(a) To generate random scatterplots:
* Download SCATSIM.83p into your calculator.
* Select and run the SCATSIM program.
* Indicate that you want 10 scatterplots and press ENTER.
* The program will generate some "pseudo-random" data and a scatterplot of these data. Based solely on the scatterplot, take a guess at the value of the correlation coefficient r, recording your guess in the table below, and press ENTER. Enter your guess into the calculator as well and press ENTER.
* The program will display the actual correlation coefficient in the lower right corner or your screen. Record this value in the table below.
* Repeat this for all ten of the "pseudo-random" data sets.

repetition	1	2	3	4	5	6	7	8	9	10
your guess										
actual										

(b) Before pressing ENTER to get your score (which is the correlation beween your guesses and actual values of r), make a guess as to what the value of the correlation coefficient between *your guesses* for r and the *actual values* of r would be.

(c) Press ENTER and record the correlation coefficient (score) below. Does the value surprise you?

(d) Examine a scatterplot of how your errors change over time (your guesses are in a list named GUESS and the actual values are in a list named ACTUA). Is there evidence that your guesses got better or worse as you went along? Explain.

(e) Examine a scatterplot of your errors (stored in ERROR) vs. the actual values. Is there evidence that you are better at guessing certain values of correlation than others?

(f) If all of your guesses were too high by exactly 0.1, what would the correlation between your guesses and the actual values be? [Think about what the scatterplot would look like.]

(g) Repeat (f) if your guesses were all too low by exactly 0.5.

(h) If the correlation between your guesses and the actual values is 1.0, does this mean that you guessed perfectly every time? What does this reveal about the utility of correlation as a measure of your guessing prowess? Explain.

WRAP-UP

In this topic you have discovered the very important **correlation coefficient** as a measure of the linear relationship between two variables. You have derived some of the properties of this measure, such as the values it can assume, how its sign and value relate to the direction and strength of the association, and its lack of resistance to outliers. You have also practiced judging the direction and strength of a relationship from looking at a scatterplot. In addition, you have discovered the distinction between correlation and **causation** and learned that one needs to be very careful about inferring causal relationships between variables based solely on a strong correlation.

The next topic will expand your understanding of relationships between variables by introducing you to **least squares regression**, a formal mathematical model that is often useful for describing such relationships.

HOMEWORK ACTIVITIES

Activity 9-6: Properties of Correlation (*cont.*)

Suppose that every student in a class scores ten points *higher* on the second exam than on the first exam.

(a) Produce (by hand) a rough sketch of what the scatterplot would look like.

(b) Enter some data that satisfy this condition into your calculator and calculate the value of the correlation coefficient between the two exam scores.

(c) Repeat (a) and (b) supposing that every student scored twenty points *lower* on the second exam than on the first.

(d) Repeat (a) and (b) supposing that every student scored *twice* as many points on the second exam as on the first.

(e) Repeat (a) and (b) supposing that every student scored *one-half* as many points on the second exam as on the first.

(f) Based on your investigation of these questions, does the degree of the slope evident in a scatterplot affect the correlation between the two variables?

Activity 9-7: Properties of Correlation (*cont.*)

Consider the following scatterplots of two hypothetical data sets (**HYPOCORR.83g**). Think of the context as scores on two exams in a class.

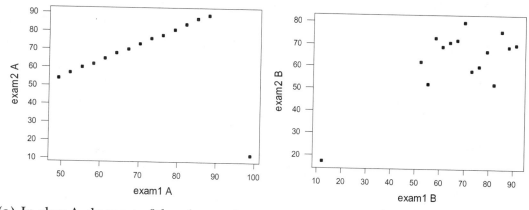

(a) In class A, do most of the observations seem to follow a linear pattern? Are there any exceptions?

(b) In class B, do most of the observations seem to be scattered haphazardly with no apparent pattern? Are there any exceptions?

(c) Use your calculator to determine the correlation coefficient for each of these classes; record them below. Are you surprised at either of the values? Explain.

(d) Describe how these scatterplots pertain to the issue of the resistance of the correlation coefficient.

(e) Consider the scatterplot of another hypothetical data set:

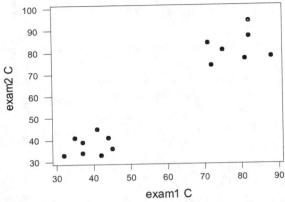

(f) Describe what the scatterplot reveals about the relationship between exam scores in class C.

(g) Use your calculator to determine the correlation coefficient between exam scores in class C. Is its value higher than you expected? Explain what this example reveals about correlation.

Activity 9-8: Properties of Correlation (*cont.*)

Consider the following four data sets, each with four pairs of observations:

Data Set A		Data Set B		Data Set C		Data Set D	
x	y	x	y	x	y	x	y
20	1	0	−5	2	10	2	10
25	2	18	20	3	9	12	17
27	3	11	7	4	6	22	9
52	4	−12	20	5	3	17	5

(a) Which data set clearly has a positive correlation coefficient? Explain based on the behavior of the variables without using your calculator.

(b) Which data set clearly has a negative correlation coefficient? Explain based on the behavior of the variables without using your calculator.

(c) Which data set has a correlation coefficient close to zero? Explain based on the behavior of the variables without using your calculator.

Activity 9-9: States' SAT Averages (*cont.*)

Reconsider the data from Activity 2-4 on page 32 concerning the percentage of high school seniors in a state who take the SAT test and the state's average SAT score (SAT98.83g).

(a) Use your calculator to determine the correlation between average SAT score vs. percentage of high school seniors who take the SAT. Is there a fairly strong relationship between these variables? Explain.

(b) Would you conclude that a cause-and-effect relationship exists between these two variables? Explain.

Activity 9-10: College Alumni Donations (*cont.*)

Reconsider the data on college alumni donations from Activity 8-15 on page 178 (HMCDONORS.83g). The scatterplots below are of average gift vs. percentage giving per class and of average gift vs. previous year's average gift per class.

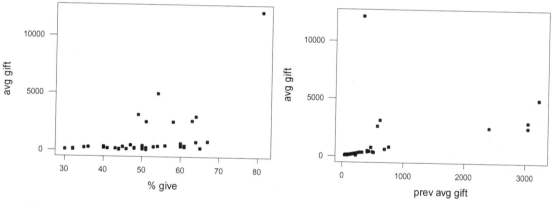

(a) Which graduating class is the outlier in its average gift?

(b) Use your calculator to determine the correlation coefficient for each pair of variables among YEAR, PERGV, AVGGV.

(c) Based on the scatterplot, do you suspect that the correlation between average gift and percentage giving will increase, decrease, or not change much when the outlier is removed? Explain.

(d) Based on the scatterplot, do you suspect that the correlation between average gift and previous average gift will increase, decrease, or not change much when the outlier is removed? Explain.

(e) Remove the outlier from each pair of variables, and your calculator to recalculate the correlation coefficients. How did they change?

Activity 9-11: Governors' Salaries (*cont.*)

Reconsider the data on governors' salaries and per capita income of the states from Activity 8-18 on page 180 (GOVSAL98.83g).

(a) Use your calculator to produce a scatterplot of governor's salary vs. per capita income. Does there appear to be an association between these variables? Would you describe it as positive or negative, as strong or moderate or weak?

(b) Based on the scatterplot, take a guess as to the value of the correlation coefficient.

(c) Use your calculator to determine the value of the correlation coefficient between governor's salary and per capita income. How good was your guess?

(d) Identify the three states with the highest per capita incomes but that have governor's salaries below the median.

(e) Consider removing these three states from the analysis. Make a prediction for the value of the correlation coefficient after this removal.

(f) Remove these three states from the analysis, and use your calculator to recalculate the correlation coefficient.

(g) Did the correlation go up or down? How accurate was your prediction?

Activity 9-12: Monopoly Prices (*cont.*)

Refer to the data from the board game Monopoly provided in Activity 9-2 on page 189 (MONOP.83g).

(a) Identify the observational units here.

(b) How many variables are recorded for each case?

(c) Use your calculator to determine the correlation coefficient for all ten pairs of variables among POS, PRICE, RENT, HOUSE, HOTEL. Report the values of these correlations.

(d) Would you expect most of these correlations to increase or decrease if Boardwalk were removed from the analysis? Explain.

(e) Remove Boardwalk from the analysis, and use your calculator to recalculate the correlation coefficients. How many increased, and how many decreased? How well do the results match your expectation in (e)?

Activity 9-13: Broadway Shows (*cont.*)

Reconsider Activity 8-7 on page 173, where you looked at scatterplots for variables pertaining to the data on Broadway shows (BROADWAY.83g with ATTEN, CAPAC, GROSS, TOPP).

(a) Use your calculator to compute the correlation coefficient between the following pairs of variables:
    A: gross and capacity
    B: top price and capacity

    C: gross and attendance
    D: gross and top price
    E: attendance and top price
    F: attendance and capacity

(b) Arrange these in order from the highest correlation to the lowest. Does this ordering agree with the one you based on scatterplots alone in Activity 8-7 on page 173?

(c) Which production has considerably higher attendance and gross than any other?

(d) See whether you can make the correlation between gross and attendance fall below .8 just by changing the *gross* value for this one production. Report the value to which you change its gross as well as the correlation.

(e) See whether you can make this correlation fall below .4 just by changing the *attendance* value for this one production. Report the value to which you change its attendance as well as the correlation.

(f) See whether you can make this correlation negative by changing the *gross* and *attendance* values for just this one production. Report the values to which you change its gross and attendance as well as the correlation.

Activity 9-14: Solitaire

The following table reports data on 25 winning games of computer solitaire. The "losses" column gives the number of losing games that preceded the win, the "time" column is the time in seconds to complete the game, and the "points" column is the number of points that the computer listed as the score of the game (SOLITAIRE.83g).

win #	losses	time	points	win #	losses	time	points	win #	losses	time	points
1	2	188	4349	10	6	241	3447	18	33	192	4267
2	3	193	4276	11	11	194	4257	19	7	185	4384
3	13	175	4636	12	4	150	5300	20	9	131	5979
4	11	218	3858	13	4	157	5110	21	11	142	5582
5	5	178	4541	14	11	183	4444	22	6	182	4444
6	4	254	3340	15	0	161	5008	23	4	152	5240
7	1	192	4292	16	7	175	4651	24	1	183	4404
8	9	181	4534	17	21	144	5502	25	4	160	5023
9	5	174	4666								

Use your calculator to examine scatterplots and to calculate correlation coefficients between all three pairs of these variables. Write a paragraph reporting your findings with regard to any relationships among these variables.

Activity 9-15: Ice Cream, Drownings, and Fire Damage

(a) Suppose that a community finds a strong positive correlation between the amount of ice cream sold in a given month and the number of drownings that occur in that month. Does this mean that ice cream *causes* drowning? If not, can you think of an alternative explanation for the strong association? Write a few sentences addressing these questions.

(b) Explain why one would expect to find a positive correlation between the number of fire engines that respond to a fire and the amount of damage done in the fire. Does this mean that the damage would be less extensive if fewer fire engines were dispatched? Explain.

Activity 9-16: Evaluation of Course Effectiveness

Suppose that a college professor has developed a new freshman course that he hopes will instill strong general learning skills in students. As a means of assessing the success of the course, he waits until a class of freshmen that have taken the course proceed to graduate from college. The professor then looks at two variables: score on the final exam for his freshman course and cumulative college grade point average. Suppose that he finds a very strong positive association between the two variables ($r = .92$, let's say). Suppose further that he concludes that his course must have had a positive effect on students' learning skills, for those who did well in his course proceeded to do well in college; those who did poorly in his course went on to do poorly in college. Comment on the validity of the professor's conclusion.

Activity 9-17: Climatic Conditions (*cont.*)

The following table lists a number of climatic variables for a sample of 25 American cities. The data are in CLIMATE.83g. These variables measure long-term averages of:
- January high temperature (in degrees Fahrenheit) (JANHI)
- January low temperature (JANLO)
- July high temperature (JLYHI)
- July low temperature (JLYLO)
- annual precipitation (in inches) (APCPT)
- days of measurable precipitation per year (DPCPT)
- annual snow accumulation (SNOW)
- percentage sunshine (PSUN)

city	Jan hi	Jan lo	July hi	July lo	precip	precday	snow	sun
Atlanta	50.4	31.5	88	69.5	50.77	115	2	61
Baltimore	40.2	23.4	87.2	66.8	40.76	113	21.3	57
Boston	35.7	21.6	81.8	65.1	41.51	126	40.7	58
Chicago	29	12.9	83.7	62.6	35.82	126	38.7	55
Cleveland	31.9	17.6	82.4	61.4	36.63	156	54.3	49
Dallas	54.1	32.7	96.5	74.1	33.7	78	2.9	64
Denver	43.2	16.1	88.2	58.6	15.4	89	59.8	70
Detroit	30.3	15.6	83.3	61.3	32.62	135	41.5	53
Houston	61	39.7	92.7	72.4	46.07	104	0.4	56
Kansas City	34.7	16.7	88.7	68.2	37.62	104	20	62
Los Angeles	65.7	47.8	75.3	62.8	12.01	35	0	73
Miami	75.2	59.2	89	76.2	55.91	129	0	73
Minneapolis	20.7	2.8	84	63.1	28.32	114	49.2	58
Nashville	45.9	26.5	89.5	68.9	47.3	119	10.6	56
New Orleans	60.8	41.8	90.6	73.1	61.88	114	0.2	60
New York	37.6	25.3	85.2	68.4	47.25	121	28.4	58
Philadelphia	37.9	22.8	82.6	67.2	41.41	117	21.3	56
Phoenix	65.9	41.2	105.9	81	7.66	36	0	86
Pittsburgh	33.7	18.5	82.6	61.6	36.85	154	42.8	46
St. Louis	37.7	20.8	89.3	70.4	37.51	111	19.9	57
Salt Lake City	36.4	19.3	92.2	63.7	16.18	90	57.8	66
San Diego	65.9	48.9	76.2	65.7	9.9	42	0	68
San Francisco	55.6	41.8	71.6	65.7	19.7	62	0	66
Seattle	45	35.2	75.2	55.2	37.19	156	12.3	46
Washington	42.3	26.8	88.5	71.4	38.63	112	17.1	56

(a) Determine the correlation coefficient between all pairs of these eight variables, recording your results in a table like the one below. (You need not record each value twice.)

	Jan hi	Jan lo	July hi	July lo	precip	precday	snow	sun
Jan hi	xxxx							
Jan lo		xxxx						
July hi			xxxx					
July lo				xxxx				
precip					xxxx			
precday						xxxx		
snow							xxxx	
sun								xxxx

(b) Which pair of variables has the *strongest* association? What is the correlation between them?

(c) Which pair of variables has the *weakest* association? What is the correlation between them?

(d) Suppose that you want to predict the annual snowfall for an American city and that you are allowed to look at that city's averages for these other variables. Which would be *most* useful to you? Which would be *least* useful?

(e) Suppose that you want to predict the average July high temperature for an American city and that you are allowed to look at that city's averages for these other variables. Which would be *most* useful to you? Which would be *least* useful?

(f) Use your calculator to explore the relationship between annual snowfall and annual precipitation more closely. Look at and comment on a scatterplot of these variables.

Activity 9-18: Space Shuttle O-Ring Failures (*cont.*)

Reconsider the data from Activity 8-4 on page 169 concerning space shuttle missions (CHALLENG.83g). Use your calculator to determine the value of the correlation between temperature and number of O-ring failures. Then exclude the flights in which no O-rings failed and recalculate the correlation. Explain why these correlations turn out to be so different.

Activity 9-19: Variables of Personal Interest (*cont.*)

Think of a situation in which you would expect two variables to be strongly correlated even though no cause-and-effect relationship exists between them. Describe these variables in a paragraph; also include an explanation for their strong association.

Topic 10:

LEAST SQUARES REGRESSION I

In previous topics you studied scatterplots as visual displays of the relationship between two quantitative variables and the correlation coefficient as a numerical measure of the linear association between them. With this topic you will begin to investigate **least squares regression** as a formal mathematical model often used to describe the relationship between quantitative variables.

OBJECTIVES

- To develop an awareness of **least squares regression** as a technique for modeling the relationship between two quantitative variables.
- To learn to use regression lines to make **predictions** and to recognize the limitations of those predictions.
- To understand some concepts associated with regression such as **fitted values**, **residuals**, and **proportion of variability explained**.
- To use your calculator to apply regression techniques with judgment and thoughtfulness to genuine data.

PRELIMINARIES

1. If one city is farther away than another, do you expect that it generally costs more to fly to the farther city?

2. Take a guess as to how much (on average) each additional mile adds to the cost of airfare.

3. Would you guess that distance explains about 50% of the variability in airfares, about 65% of this variability, or about 80% of this variability?

4. Do you think that an older private college is likely to charge a higher or lower tuition than a newer private college?

IN-CLASS ACTIVITIES

Activity 10-1: Airfares

The data below are airfares to various cities from Baltimore, Maryland (as of January 8, 1995):

178	138	94	278	158	258	198	188	98	179	138	98

Descriptive statistics for these airfares are:

n	mean	std. dev.	minimum	lower quartile	median	upper quartile	maximum
12	166.92	59.5	94	108	168	195.5	278

(a) If someone asks how much they can expect to pay for airfare from Baltimore, what prediction would you give? Explain.

(b) Suggest another variable (an *explanatory variable*) that might be useful for predicting the airfare to a certain destination (the *response variable*).

The following table reports the distance (in miles) as well as the airfare for twelve destinations. A scatterplot appears below.

destination	distance	airfare	destination	distance	airfare
Atlanta	576	178	Miami	946	198
Boston	370	138	New Orleans	998	188
Chicago	612	94	New York	189	98
Dallas	1216	278	Orlando	787	179
Detroit	409	158	Pittsburgh	210	138
Denver	1502	258	St. Louis	737	98

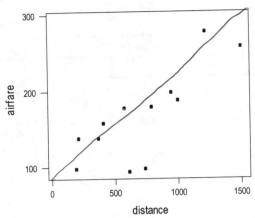

(c) Based on this scatterplot, does it seem that knowing the distance to a destination would be useful for predicting the airfare? Explain.

A natural goal is to try to use the distance of a destination to predict the airfare for flying there, and the simplest model for this prediction is to assume that a straight line summarizes the relationship between distance and airfare.

(d) Place a thread or straightedge over the scatterplot above so that it forms a straight line that roughly summarizes the relationship between distance and airfare. Then draw this line on the scatterplot.

(e) Roughly what airfare does your line predict for a destination that is $x_1 = 300$ miles away?

$y_1 =$

(f) Roughly what airfare does your line predict for a destination that is $x_2 = 1500$ miles away?

$y_2 =$

The equation of a line can be represented as $\hat{y} = a + bx$, where $\hat{y}$ denotes the (response) variable being predicted (which is plotted on the vertical axis), x denotes the (explanatory) variable being used for the prediction (which is plotted on the horizontal axis), a is the value of the **y-intercept** of the line, and b is the value of the **slope** of the line. In this case x represents distance and y airfare.

(g) Use your answers to (e) and (f) to find the *slope* of your line, remembering that since

$$\text{slope} = \frac{\text{rise}}{\text{run}} = \frac{\text{change in } y}{\text{change in } x}, \text{ the slope is } b = \frac{y_2 - y_1}{x_2 - x_1}$$

slope:

(h) Use your answers to (e) and (g) to determine the *intercept* of your line, remembering that $a = y_1 - bx_1$.

intercept:

(i) Put your answers to (g) and (h) together to produce the equation of your line. It is good form to replace the generic x and $\hat{y}$ symbols in the equation with the actual variable names, in this case *distance* and *airfare*, respectively.

$$\hat{airfare} = \underline{\hspace{2cm}} + \underline{\hspace{2cm}} distance$$

Naturally, we would like to have a better way of choosing the line to approximate a relationship than simply drawing one that seems about right. Since there are infinitely many lines that one could draw, however, we need some criterion to select which line is the "best" at describing the relationship. The most commonly used criterion is **least squares**, which says to choose the line that minimizes the sum of squared vertical distances from the points to the line. In other words, if you think of the line as predicting the y-value for a given x-value, then the least squares criterion says to choose whichever line produces the smallest sum of squared errors in those predictions.

We write the equation of the **least squares line** (also known as the **regression line**) as

$$\hat{y} = a + bx,$$

where the **slope coefficient** b and the **intercept coefficient** a are determined from the sample data. The most convenient expression for calculating these coefficients relates them to the means and standard deviations of the two variables and the correlation coefficient between them:

$$b = r\frac{s_y}{s_x}, \quad a = \bar{y} - b\bar{x},$$

where, you will recall, $\bar{x}$ and $\bar{y}$ represent the means of the variables, s_x and s_y their standard deviations, and r the correlation between them.

Activity 10-2: Airfares (*cont.*)

(a) Enter the airfare data into your calculator, naming the lists DIST and AIRF, respectively.

(b) Use the CORR program to compute the mean and standard deviation of distance and airfare, and the value of the correlation between the two. Record the results below:

	mean	std. dev.	correlation
airfare (y)			
distance (x)			

(c) Write the equation of the least squares line for predicting airfare from distance (using the variable names rather than the generic x and $\hat{y}$).

We will now use the calculator to produce the regression line. [*TI Hint*: In order for your calculator to display the correlation coefficient you will need to turn the diagnostic display mode to on. To do this, access the calculator's CATALOG menu (press ___2nd___ ___0___) and scroll down until you find "DiagnosticOn." Press your ENTER button twice.

(d) To use your calculator to find the equation of the least squares line, use the STAT CALC menu and select option LinReg(a+bx). You should have the following displayed on your home screen:

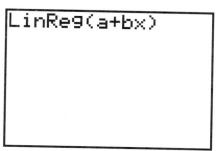

```
LinReg(a+bx)
```

(e) To find the least squares line for predicting airfare from distance, complete the command by entering the two appropriate lists so that you have the following (providing the X-list first):

```
LinReg(a+bx)  ∟DI
ST,∟AIRF
```

(f) Press ENTER and write the equation below. Compare this equation with the equation you found in (c). (Note that the calculator also gives the correlation coefficient.)

One of the primary uses of regression is for **prediction**, for one can use the regression line (another name for the least squares line) to predict the value of the y-variable for a given value of the x-variable simply by plugging that value of x into the equation of the regression line. This is, of course, equivalent to finding the y-value of the point on the regression line corresponding to the x-value of interest.

(g) What airfare does the least squares line predict for a destination that is 300 miles away?

(h) What airfare does the least squares line predict for a destination that is 1500 miles away?

(i) Draw the least squares line on the scatterplot below by plotting the two points that you found in (g) and (h) and connecting them with a straight line.

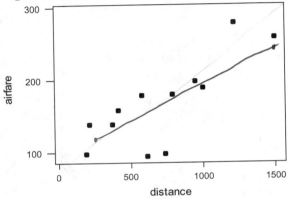

(j) To use your calculator to create a scatterplot of airfare vs. distance and then graph the least squares line:
- Select LinReg(a+bx) from the STAT CALC menu as before.
- Enter DIST, AIRF as before, but now type another comma. Press the │VARS│ button and then use the right arrow to see the Y-VARS menu. Press │ENTER│ to select Function and then │ENTER│ again to select Y1. You should have LinReg(a+bX) ʟDIST, ʟAIRF, Y1 on your home screen.
- Press │ENTER│. If you now press the blue │ Y= │ button at the top of your calculator, you will see the regression equation stored as the Y1 function.
- Press │GRAPH│.

(k) Just from looking at the regression line that you have drawn on the scatterplot, guess what value the regression line would predict for the airfare to a destination 900 miles away. [*TI Hint:* You can also use the │TRACE│ button.]

(l) Use the equation of the regression line to predict the airfare to a destination 900 miles away, and compare this prediction to your guess in (g). [*TI Hint:* You can substitute this value into the regression equation by obtaining Y_1 as before and entering $Y_1(900)$.]

(m) What airfare would the regression line predict for a flight to San Francisco, which is 2842 miles from Baltimore? Would you consider this prediction as reliable as the one for 900 miles? Explain.

The actual airfare to San Francisco at that time was 198 dollars. That the regression line's prediction is not very reasonable illustrates the danger of **extrapolation**, i.e., of trying to predict *y* for values of *x* beyond those contained in the data. Since we have no reason to believe that the relationship between distance and airfare remains roughly linear beyond the range of values contained in our data set, such extrapolation is not advisable.

(n) Use the equation of the regression line to predict the airfare if the distance is 900 miles. Record the prediction in the table below, and repeat for distances of 901, 902, and 903 miles.

distance	900	901	902	903
predicted airfare	188	189.02	189.14	189.24

(o) Do you notice a pattern in these predictions? By how many dollars is each prediction higher than the preceding one? Does this number look familiar (from your earlier calculations)? Explain.

> This exercise demonstrates that one can *interpret* the slope coefficient of the least squares line as the predicted change in the *y*-variable (airfare, in this case) for a one-unit change in the *x*-variable (distance).

(p) By how much does the regression line predict airfare to rise for each additional 100 miles that a destination is farther away?

Activity 10-3: Airfares (*cont.*)

(a) If you look back at the original listing of distances and airfares, you find that Atlanta is 576 miles from Baltimore. What airfare would the regression line have predicted for Atlanta?

(b) The actual airfare to Atlanta at that time was $178. By how much does the actual price exceed the prediction?

> A common theme in statistical modeling is to think of each data point as being composed of two parts: the part that is explained by the model (often called the **fit**) and the "leftover" part (often called the **residual**) that is either the result of chance variation or of variables not measured. In the context of least squares regression, the **fitted value** for an observation is simply the *y*-value that the regression line would predict for the *x*-value of that observation. The **residual** is the difference between the actual *y*-value and the fitted value $\hat{y}$ (residual = actual − fitted), so the residual measures the vertical distance from the observed *y*-value to the regression line.

(c) Record Atlanta's fitted value and residual (which you calculated above) in the table below. Then calculate Boston's residual and Chicago's fitted value using the definition of residual (i.e., *without* using the equation of the regression line), showing your calculations.

destination	distance	airfare	fitted value	residual	mean fare	deviation
Atlanta	576	178	(c)	(c)	166.92	11.08
Boston	370	138	126.70	(c)	166.92	−28.92
Chicago	612	94	(c)	−61.10	166.92	−72.92
Dallas	1216	278	226.00	52.00	166.92	(g)
Detroit	409	158	131.27	26.73	166.92	−8.92
Denver	1502	258	259.57	−1.56	166.92	91.08
Miami	946	198	194.30	3.70	166.92	31.08
New Orleans	998	188	200.41	−12.41	166.92	21.08
New York	189	98	105.45	−7.45	166.92	−68.92
Orlando	787	179	175.64	3.36	166.92	12.08
Pittsburgh	210	138	107.92	30.08	166.92	−28.92
St. Louis	737	98	169.77	−71.77	166.92	−68.92

(d) Which city has the largest (in absolute value) residual? What were its distance and airfare? By how much did the regression line err in predicting its airfare; was it an underestimate or an overestimate? Circle this observation on the scatterplot containing the regression line above.

(e) For observations with *positive* residual values, was their actual airfare greater or less than the predicted airfare?

(f) For observations with *negative* residual values, do their points on the scatterplot fall above or below the regression line?

(g) At the beginning of Activity 10-1 on page 206, you noted that the mean of these airfares is $166.92, with a standard deviation of $59.45. In the absence of information about distance, we could use the overall mean airfare as the prediction for each city's airfare. In this situation, the deviations from the mean would be the errors in those predictions. The last column of the table above reports these deviations. Determine the deviation from the mean airfare for Dallas, and record it in the table.

(h) Alternatively, the residuals represent the errors in predictions using the regression line, which takes the distance to a destination into account. For how many cities does the overall mean airfare result in a closer prediction to the actual airfare than the regression line does? In other words, how many cities have a deviation from the mean that is smaller than their residual from the regression line? Which cities are these?

(i) Do most cities have a smaller residual or a smaller deviation from the mean? Does this suggest that predictions from the regression line are generally better than the airfare mean? Explain.

Recall the idea of thinking of the data as consisting of a part explained by the statistical model and a "leftover" (residual) part due to chance variation or unmeasured variables. One can obtain a numerical measure of how much of the variability in the data is explained by the model and how much is "left over" by comparing the residuals from the regression line and the deviations from the mean. To do this comparison we will calculate the sums of squares of these residuals.

(j) Use your calculator to compute the sum of squared residuals from the regression line. The calculator automatically computes and enters the residuals into a list named RESID every time you use it to find the linear regression equation. You can use the sum(comand from the **LIST MATH** menu and also use the $\boxed{x^2}$ key to square all of the values in the list: $\texttt{sum(LRESID}^2)$.

   sum of squared residuals:
   (prediction errors using regression line)

(k) Calculate the sum of squared deviations from the overall mean airfare. [*Hints:* You found the overall mean airfare in Activity 10-2(b). Remember to square the differences before you apply the sum command.]

   sum of squared deviations in airfare from overall mean:
   (prediction errors using overall mean)

(l) To see how much the regression line has reduced prediction errors, divide the sum of squared residuals by the sum of squared deviations from the mean. Then subtract this result from 1. Record these values below.

This ratio of the sum of squared residuals and the sum of squared deviations is the proportion of the variability in the response variable that is left *unexplained* (residual) by the regression model. Subtracting this value from 1 gives the proportion of variability in the response variable that *is* explained by the regression model.

(m) Recall that you found the correlation between distance and airfare in part (a) of Activity 10-2 on page 210. Square this value and record the result below. Does this value look familiar? Explain.

> It turns out that the ***proportion of variability*** in the *y*-variable explained by the regression model with the *x*-variable is more efficiently calculated as the square of the correlation coefficient, written r^2. This proportion provides a measure of how closely the points fall to the least squares line and thus also provides an indication of how confident one can be of predictions made with the line.

Activity 10-4: College Tuitions (*cont.*)

The following scatterplots display the tuition charges and founding years for four-year colleges in Pennsylvania. The scatterplot on the left pertains to public institutions, the one on the right to private ones:

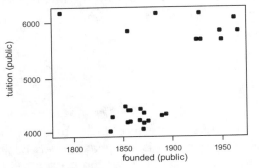 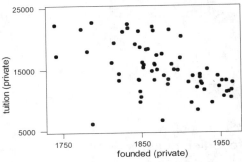

(a) Download COLLEGES99.83g into your calculator and determine the regression line for predicting tuition from year of founding for the public schools (the list names are PUBT and PUBF), storing the regression equation into Y1. Report the equation of the line and the r^2 value below. Reproduce the scatterplot with the regression line. Sketch the line on the scatterplot above.

public: r^2:

(b) Repeat the analysis in (a) for the private schools (the list names are PRIVF and PRIVT).

private: r^2:

(c) Are the equations and r^2 values similar for the public and private schools?

(d) Which line appears to do a better job of summarizing the relationship between tuition and founding year? Explain.

(e) What tuition charge would the line predict for a *public* school founded in 1900? What tuition charge would the line predict for a *private* school founded in 1900? Judging from the scatterplot, which of these is a more reasonable prediction? Explain.

public: private:

> This activity reveals that linear regression models are not appropriate for all sets of data. The correlation coefficient and r^2 values do not necessarily attest to how well a linear model describes the association. The importance of looking at visual displays of the data cannot be overstated.

WRAP-UP

This topic has led you to study a formal mathematical model for describing the relationship between two quantitative variables. In studying **least squares regression**, you have encountered a variety of related terms and concepts. These ideas include the use of regression in **prediction**, the danger of **extrapolation**, the interpretation of the **slope coefficient**, the concepts of **fitted values** and **residuals**, and the interpretation of r^2 as the **proportion of variability explained** by the regression line. Understanding all of these ideas is important to applying regression techniques thoughtfully and appropriately.

In the next topic you will continue your study of least squares regression. You will explore the distinction between outliers and influential observations, discover the utility of residual plots, and consider transformations of variables.

— $vs.$ —
Y X

HOMEWORK ACTIVITIES

Activity 10-5: Cars' Fuel Efficiency (*cont.*)

Refer back to the data in Activity 8-1 on page 163 concerning the relationship between sports cars' weight and fuel efficiency. The means and standard deviations of these variables and the correlation between them are reported below:

$b = r \dfrac{s_y}{s_x}$

$b \rightarrow$

$a = \bar{y} - b\bar{x}$

	mean	std. dev.	correlation
weight	2997	357.6	–0.816
mpg	20.867	3.044	

(a) Use this information to determine (by hand) the coefficients of the least squares line for predicting a car's miles per gallon rating from its weight. Report the equation of this line.

(b) Recall that the MPG rating for the Audi TT was not provided. Use the regression line to predict the city MPG rating for this car, whose weight is 2655 pounds.

(c) By how many miles per gallon does the least squares line predict a car's fuel efficiency to drop for each additional 100 pounds of weight? (Use the slope coefficient to answer this question.)

(d) What proportion of the variability in cars' miles per gallon ratings is explained by the least squares line with weight?

Activity 10-6: Governors' Salaries (*cont.*)

Reconsider the data from Activity 8-18 on page 180 concerning governor salaries and average pay in the state (GOVSAL98.83g).

(a) Use your calculator to determine the regression equation for predicting a state's governor salary from its average pay. Record the equation of the regression line.

(b) What proportion of the variability in governor salaries is explained by this regression line with average pay? $- r^2$

(c) Examine the RESID list to determine which state has the largest positive residual? Explain what this signifies about the state. [*TI Hint*: You could also use the TRACE feature to explore the scatterplot and determine the X-value of this state.]

(d) Which state has the largest (in absolute value) negative residual. Explain what this signifies about the state.

(e) Calculate the fitted value for each state. [*TI Hint*: You can pass the entire explanatory variable into the Y₁ function: Y₁(LINCM).] Examine the list to determine which state has the largest fitted value. Explain how you could determine this from the raw data without actually calculating any fitted values.

Activity 10-7: College Tuitions (*cont.*)

The following labeled scatterplot displays the relationship between tuition charges in 1997–98 and the year of founding for all two- and four-year colleges in Pennsylvania. The labels indicate the institution's classification as a two- or four-year school and as public or private

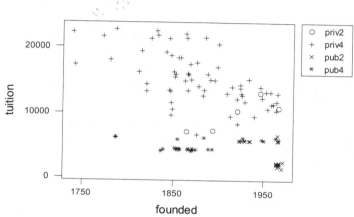

(a) Use your calculator (COLLEGES99.83g) to determine the regression line for predicting tuition from year of founding, storing the equation in Y1. The list names are PUBT, PRIVT, PUBF, and PRIVF. [*TI Hint*: You can use the augment(feature in the LIST OPS menu to concatenate two lists. Choose the augment(option and enter the two lists separated by a comma. Then use the STO▶ button to specify the name of the new list and press ENTER .]

(b) Is the slope of the line positive or negative? Interpret the value of the slope coefficient in the context of these tuition data.

(c) What tuition charge would the regression line predict for an institution founded in 1925?

(d) Draw a rough sketch of the line on the scatterplot. Does the line seem to summarize the relationship between tuition and founding year equally well for all four types of institutions? Explain.

Activity 10-8: College Tuitions (*cont.*)

Consider again the data on tuition charges and founding dates of Pennsylvania colleges from Activity 10-7 on page 221 (COLLEGES99.83g).

(a) Use your calculator to determine separate regression lines for the four types of institutions (private, four-year; public, four-year; private, two-year; public, two-year). [*TI Hints*: Use the SPRT program to separate the lists by TYPE. You can store the equations separately into Y1, Y2, Y3, and Y4. You have already determined two of these equations in Activity 10-4 on page 218.] Comment on the degree to which these lines differ.

(b) Use these four separate regression lines to determine the predicted tuition charge for an institution founded in 1925 for each of these four types.

Activity 10-9: Fast Food Sandwiches (*cont.*)

Reconsider the data from Activity 8-5 on page 171 about fast food sandwiches (ARBYS99.83g).

(a) Use your calculator to determine the least squares line for predicting calories from serving size.

(b) Use the LBLSCT program as you did in Activity 8-5 on page 171 to produce a labled scatterplot using the MEAT variable as the CATEGORICAL VARIABLE LIST. Do the points for roast beef sandwiches tend to fall above or below the line?

(c) Examine the RESID list to see how many of the residuals for roast beef sandwiches are positive and how many are negative? Comment on what this reveals about the relationship between calories and serving size for roast beef sandwiches in comparison with other types.

Activity 10-10: Electricity Bills

The following table lists the average temperature for a month and the amount of the electricity bill for that month (ELECBILL.83g):

month	temp	bill	month	temp	bill
Apr-91	51	$41.69	Jun-92	66	$40.89
May-91	61	$42.64	Jul-92	72	$40.89
Jun-91	74	$36.62	Aug-92	72	$41.39
Jul-91	77	$40.70	Sep-92	70	$38.31
Aug-91	78	$38.49	Oct-92	*	*
Sep-91	74	$37.88	Nov-92	45	$43.82
Oct-91	59	$35.94	Dec-92	39	$44.41
Nov-91	48	$39.34	Jan-93	35	$46.24
Dec-91	44	$49.66	Feb-93	*	*
Jan-92	34	$55.49	Mar-93	30	$50.80
Feb-92	32	$47.81	Apr-93	49	$47.64
Mar-92	41	$44.43	May-93	*	*
Apr-92	43	$48.87	Jun-93	68	$38.70
May-92	57	$39.48	Jul-93	78	$47.47

(a) Before you examine the relationship between average temperature and electric bill, examine the distribution of electric bill charges themselves. Create a dotplot of the electric bill charges, and write a few sentences describing the distribution of electric bill charges.

(b) Produce a scatterplot of electric bill vs. average temperature. Does the scatterplot reveal a positive association between these variables, a negative association, or not much association at all? If there is an association, how strong is it?

(c) These temperatures have a mean of 55.88 and a standard deviation of 16.21 degrees. The electric bills have a mean of $43.18 and a standard deviation of $4.99. The correlation between temperature and bill is –0.695. Use this information to determine the equation of the least squares (regression) line for predicting the electric bill from the average temperature.

(d) Interpret the slope coefficient in the context of these data.

(e) Use this equation to determine the fitted value and residual for March of 1992.

(f) Without doing any calculations, identify the month with the largest fitted value. Explain your answer.

(g) What proportion of the variability in electric bills is explained by the regression line with temperature?

Activity 10-11: Signature Measurements (*cont.*)

Consider again the data collected on students' signatures in Topic 1 (SIGS.83g).

(a) Produce a labeled scatterplot of signature length vs. number of letters, using gender as the categorical variable. [*TI Hint*: You can use the LBLSCT program as you did in Activity 8-5 on page 171.]

(b) Use your calculator to determine the regression equation for predicting a student's signature length from her or his number of letters, storing the equation in Y1. Report the equation and sketch it on the scatterplot.

(c) Use your calculator to determine the regression equation for predicting a male student's signature length from his number of letters. Record the value of the equation, and sketch it (clearly labeled) on the scatterplot. [*TI Hint*: You may want to use the SPRT program to separate the lists.]

(d) Use your calculator to determine the regression equation for predicting a *female* student's signature length from her number of letters. Record the value of the equation, and sketch it (clearly labeled) on the scatterplot.

(e) Compare these three regression lines, comments on any difference between them.

(f) What signature length would the regression equation predict for a male with 12 letters in his name?

(g) What signature length would the regression equation predict for a woman with 12 letters in her name?

(h) What signature length would the regression equation predict for a person of unknown gender with 12 letters in her or his name?

(i) Comment on how much these predictions differ.

Activity 10-12: Turnpike Tolls

If one enters the Pennsylvania Turnpike at the Ohio border and travels east to New Jersey, the mileages and tolls for the turnpike exits are as displayed in the scatterplot below. The regression line for predicting the toll from the mileage has been drawn on the scatterplot; its equation is: toll = $-0.123 + 0.0402$ mileage. The correlation between toll and mileage is 0.999.

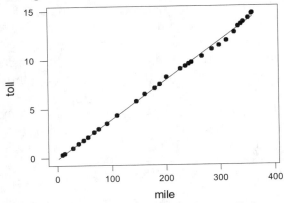

(a) What proportion of the variability in turnpike tolls is explained by the regression line with mileage?

(b) Use the regression equation to predict the toll for a person who needs to drive 150 miles on the turnpike.

(c) By how much does the regression equation predict the toll to rise for each additional mile that you drive on the turnpike?

(d) About how many miles do you have to drive in order for the toll to increase by one dollar?

Activity 10-13: Broadway Shows (*cont.*)

Consider again the data on Broadway shows from Activity 2-8 on page 36 (BROADWAY.83g).

(a) Use your calculator to produce a scatterplot of gross receipts vs. attendance and to calculate the least squares line for predicting gross receipts from attendance, storing this equation in Y_1.

(b) What would the least squares line predict for the receipts of a show with a weekly attendance of 7500?

(c) Calculate the fitted value and residual for *Death of a Salesman*.

(d) Which show has the largest fitted value?

Activity 10-14: Climatic Conditions (*cont.*)

Reconsider the climatic data presented in Activity 9-17 on page 202 (CLIMATE.83g).
(a) Choose any *pair* of variables (preferably a pair that is strongly correlated) and use your calculator to examine a scatterplot of these two variables and to determine the least squares line for predicting one variable from the other. Record the equation of this line, being sure to identify which variables you are considering.
(b) Select one particular value of the independent variable and use the least squares line to predict the value of the dependent variable.
(c) Examine the RESID list to determine which city has the largest (in absolute value) residual from your least squares line. What is the value of this residual? Interpret what this residual says about the city.
(d) What proportion of the variability in the dependent variable is explained by the least squares line?

Activity 10-15: Birth and Death Rates (*cont.*)

Reconsider the data from Activity 6-1 on page 118 on states' birth and death rates (BIRTHDEATH.83g).
(a) Use your calculator to create a scatterplot of the data with birth rate on the vertical axis.
(b) Determine the regression line for predicting a state's birth rate from its death rate.
(c) Examine the RESID list to identify the four states with the largest (in absolute value) residuals. Circle these states' values on the scatterplot.
(d) Which residuals are positive and which are negative? Comment on what the residuals signify about these states.

Activity 10-16: College Football Players (*cont.*)

Consider the weights and jersey numbers of college football players from Activity 12-16 on page 265, which you began to analyze in Activity 8-8 on page 175 (CPFOOTBALL.83g).
(a) Determine the regression line for predicting a player's weight from his jersey number.
(b) Use the regression equation to predict the weight of a player wearing number 50.
(c) Interpret the value of the slope coefficient in context.
(d) If Dan Loney, a 300-pound center wearing number 66, were to exchange his jersey in practice one day for number 16, previously worn by 185-pound quarterback Kevin Cooper, would you expect Loney to:
  • lose about 42.4 pounds (which is 50 times the slope coefficient)
  • fall to a weight of about 194.7 pounds (which is the fitted value for jersey number 16)

- neither gain nor lose weight

Explain your answer.

(e) Is knowing a player's jersey number of some use in predicting his weight? Explain.

Activity 10-17: Incorrect Conclusions

It can be shown that the sum of the residuals from a least squares line must equal zero.

(a) Does it follow that the mean of the residuals must equal zero? Explain.

(b) Does it follow that the median of the residuals must equal zero? Explain.

(c) Does the observation with the largest value of the predictor variable have to have the largest fitted value? Explain.

(d) Can an observation whose value of the predictor variable equals the mean of that variable have the largest residual? Explain.

Activity 10-18: Airfares (*cont.*)

Recall the least squares formulas for the slope b and intercept a of a regression line: $b = r\frac{s_y}{s_x}$ and $a = \bar{y} - b\bar{x}$. To investigate the reasonableness and behavior of these formulas, reconsider the airfare data from Activity 10-1 on page 206. For each of the following, use your calculator to change the variable as indicated. Then recalculate the regression equation and comment on how the slope and/or intercept change.

(a) Suppose that $500 is added to each fare.

(b) Suppose that each fare is doubled.

(c) Suppose that each distance is cut in half.

(d) Suppose that 1000 miles is added to each distance.

Topic 11:

LEAST SQUARES REGRESSION II

This topic extends your study of least squares regression. You will examine the impact that a single observation can have on a regression analysis, learn how to use **residual plots** to indicate when the linear relationship is not appropriate, and discover **transformation** of variables as a way to use regression even when the relationship between the variables is not linear.

OBJECTIVES

- To understand the distinction and importance of **outliers** and **influential observations** in the context of regression analysis.
- To learn to use **residual plots** to indicate when the linear relationship is not a satisfactory model for describing the relationship between two variables.
- To discover how to **transform** variables to create a linear relationship between variables.

PRELIMINARIES

1. Identify an animal that you suspect has an especially long lifetime.

2. If one animal tends to live longer than another, would you expect it to have a longer gestation period than the other?

3. Take a guess concerning the average lifetime and gestation period of a horse, rabbit, and human.

IN-CLASS ACTIVITIES

Activity 11-1: Gestation and Longevity

The following table lists the average gestation period (in days) and longevity (in years) for a sample of animals, as reported in *The 1993 World Almanac and Book of Facts*.

animal	gestation	longevity	animal	gestation	longevity
baboon	187	20	guinea pig	68	4
bear, black	219	18	hippopotamus	238	25
bear, grizzly	225	25	horse	330	20
bear, polar	240	20	kangaroo	42	7
beaver	122	5	leopard	98	12
buffalo	278	15	lion	100	15
camel	406	12	monkey	164	15
cat	63	12	moose	240	12
chimpanzee	231	20	mouse	21	3
chipmunk	31	6	opossum	15	1
cow	284	15	pig	112	10
deer	201	8	puma	90	12
dog	61	12	rabbit	31	5
donkey	365	12	rhinoceros	450	15
elephant	645	40	sea lion	350	12
elk	250	15	sheep	154	12
fox	52	7	squirrel	44	10
giraffe	425	10	tiger	105	16
goat	151	8	wolf	63	5
gorilla	257	20	zebra	365	15

(a) Load the grouped file ANIMALS.83g into your calculator. The list names are GEST and LONG. Use your calculator to determine the regression line for predicting an animal's gestation period from its longevity, storing the equation into Y_1. (The residuals are stored in a list named RESID.) Record the equation of the line below and sketch it on the following scatterplot. [*TI Hint:* You can reproduce the scatterplot with the regression line drawn over the points.]

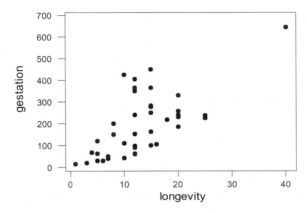

(b) Interpret the slope coefficient of the regression line. In other words, explain precisely what its value signifies about longevity and gestation period.

(c) What proportion of the variability in animals' gestation periods is explained by the regression line?

(d) Use your calculator to create a scatterplot of the animals' residual values vs. their longevities. Is there any relationship between residuals and longevities? If so, explain in a sentence or two what that signifies about the accuracy of predictions for animals with long vs. short lifetimes.

(e) Which of the animals is clearly an outlier both in longevity and in gestation period? Circle this animal on the scatterplot and record (look in the **RESID** list) its residual value. Does it seem to have the largest residual (in absolute value) of any animal?

> In the context of regression lines, ***outliers*** are observations with large (in absolute value) residuals. In other words, outliers fall far from the regression line, not following the pattern of the relationship apparent in the others. While the elephant is an outlier in both the longevity and gestation variables, it is not an outlier in the regression context.

(f) Which animal does have the largest (in absolute value) residual? Is its gestation period longer or shorter than expected for an animal with its longevity?

(g) Use the Stat List Editor to eliminate for the moment the giraffe's information from the analysis. Use your calculator to determine the equation of the regression line for predicting gestation period from longevity in this case. Record the equation and also the value of r^2 below.

regression line:

r^2:

(h) The following scatterplot displays the original data and both regression lines. Is this new regression line substantially different from the actual one?

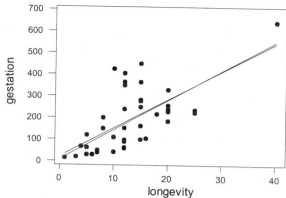

(i) Return the giraffe's values (GEST=425, LONG=10) to the lists (using [2nd] [DEL] or [INS]) but eliminate for now the elephant's. Use your calculator to determine the equation of the regression line in this case. Record this equation along with the value of r^2 below.

regression line:

r^2:

(j) The scatterplot below displays the original data with the actual regression line and the line found when the elephant had been removed. In which case (giraffe or elephant) did the removal of one animal affect the regression line more?

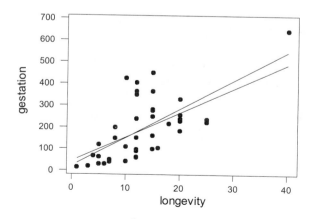

> In the context of least squares regression, an ***influential observation*** is one whose removal would substantially affect the regression line. Observations with extreme values of the predictor variable are potentially influential. Removing the elephant from the analysis has a considerable effect on the least squares line, whereas the removal of the giraffe does not. The elephant is an influential observation due to its exceptionally long lifetime.

(k) To appreciate even further the potential influence of the elephant, change its gestation period to 45 days instead of 645 days. Use your calculator to recalculate the equation of the regression line and of r^2, recording each below. The following scatterplot reveals the new data along with both the original and revised regression lines.

regression line:

r^2:

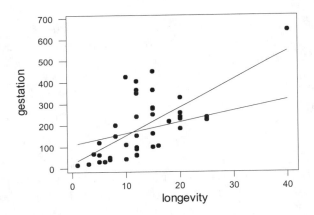

Activity 11-2: Residual Plots

Below are four scatterplots with regression lines drawn in:
1. MPG rating vs. weight for sports cars
2. distance from sun vs. position number (Mercury 1, Venus 2, etc.) for planets
3. rent vs. price for Monopoly properties
4. airfare vs. distance for selected destinations

1. city MPG = 41.7 − .00695 weight, r^2=66.6%

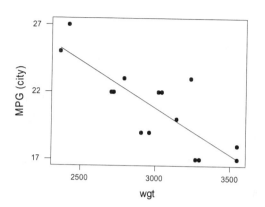

2. distance = −1126 + 446 position, r^2=82.8%

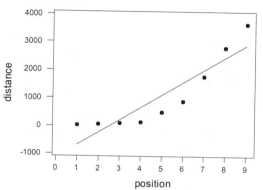

3. rent = −4.70 + 0.106 price, r^2=98.8%

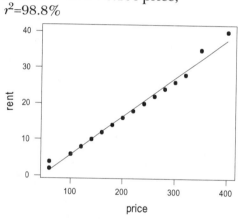

4. airfare = 83.3 + .117 distance, r^2=63.2%

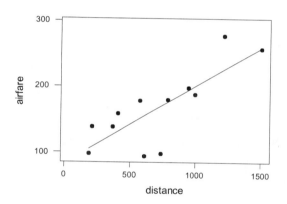

(a) For each of these data sets, a scatterplot of the residuals from the regression line vs. the explanatory or predictor variable is given below. Your task is to match up each residual plot with its regression scatterplot.

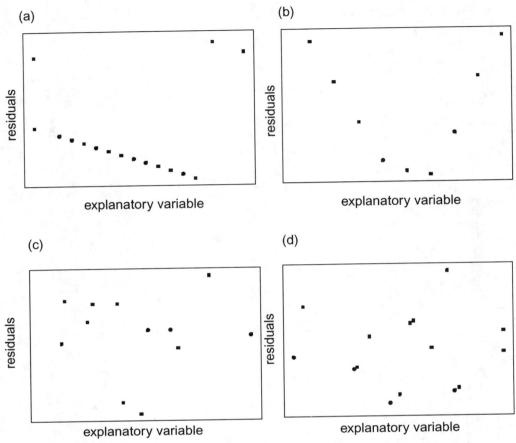

(a)
(b)
(c)
(d)

(b) Match up the data with the following descriptions of the residual plots:
- The residuals are randomly scattered.
- The residuals are largely randomly scattered except for two very large negative residuals.
- The residuals show a distinct curved pattern.
- The residuals show a clear linear pattern with three severe outliers.

(c) Look back at the original scatterplots with the regression lines drawn in. In which two plots do the lines summarize the relationship in the data about as well as possible? In which two plots do the lines fail to capture important aspects of the relationship? Explain.

(d) Do the scatterplots where the lines summarize the data about as well as possible correspond to the highest values of r^2? Explain.

> These examples show that residual plots can indicate when a linear model does not adequately describe the relationship in the data. When a straight line is a reasonable model, the residual plot should reveal a seemingly random scattering of points. When a nonlinear model would fit the data better, the residual plot reveals a pattern of some kind. Note that the value of r^2 alone is not sufficient for assessing the fit of the linear model.

Activity 11-3: Televisions and Life Expectancy (*cont.*)

When a straight line is not the best mathematical description of a relationship, one can **transform** one or both variables to make the association more linear.

Reconsider the data from Activity 9-4 on page 193 dealing with televisions and life expectancies (TVLIFE.83g).

(a) Use your calculator to look at a scatterplot of life expectancy (LFEXP) vs. people per television (PERTV) and to calculate the correlation between these two. Describe the relationship between the variables; in particular, does it appear to be linear?

(b) Use your calculator to create a new variable LPRTV: logarithm of the number of people per television. (You may use any base to take the logarithm, but base 10 is a natural choice.) You should see the following on your home screen:

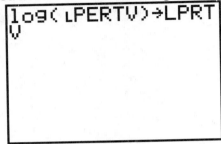

```
log( LPERTV)→LPRT
V
```

Now press ENTER .

(c) Use your calculator to determine the equation of the least squares line for predicting a country's life expectancy from the log of its number of people per television. Record the equation below.

(d) What proportion of the variability in countries' life expectancies is explained by the regression equation with the log of the people per television?

(e) What life expectancy would this regression line predict for a country with 10 people per television. To determine this, first take the log of this number of people per television and then plug that log value into the regression equation.

(f) Use the regression line to predict the life expectancy for a country with 100 people per television. Then calculate the difference between the prediction in (e) and this one. Does the value look familiar? Explain.

(g) Examine a scatterplot of the residuals of this regression equation vs. the log of the number of people per television. Does the scatterplot reveal any clear pattern?

(h) Is the linear regression model a better fit with the original data or with the transformed data?

WRAP-UP

This topic has extended your study of least squares regression. You have examined **outliers** and **influential observations**, noting the effects that they can have on regression analysis. You have also learned to use **residual plots** to judge whether a nonlinear model might better describe the relationship between two variables, and you have discovered how to **transform** variables when such a nonlinear model is called for.

This unit and the previous one have addressed exploratory analyses of data. In the next unit you will begin to study background ideas related to the general issue of drawing **inferences** from data. You will find that drawing meaningful inferences depends on having collected data well in the first place, and you will study again the ideas of random sampling and randomization that you encountered earlier. Your study of **randomness** will lay the foundation for procedures of **statistical inference**.

HOMEWORK ACTIVITIES

Activity 11-4: Planetary Measurements (*cont.*)

Reconsider the data from Activity 4-5 on page 80 on planetary measurements (PLAN-ETS.83g).

(a) Use your calculator to produce a scatterplot showing the relationship between a planet's distance (DISTI) from the sun and its position number (POSIT). Would a least squares line be a good fit for this relationship? Explain.

(b) Use your calculator to create two new variables: square root of distance and logarithm of distance. Look at scatterplots of each of these variables vs. position. Which transformation (square root or logarithm) seems to produce a more linear relationship?

(c) For whichever transformation you select in (b) as more appropriate, use your calculator to determine the regression equation for predicting that transformation of distance from position. Record this equation, indicating which transformation you work with.

(d) Report the value of r^2 for this regression equation.

(e) Look at a residual plot and comment on whether the residuals seem to be scattered randomly or to follow a pattern.

Activity 11-5: Planetary Measurements (*cont.*)

Refer again to the planetary data (PLANETS.83g), and now consider the problem of using a planet's distance (DISTI) from the Sun to predict the period of its revolution around the Sun (REV).

(a) Use your calculator to produce a scatterplot of period of revolution vs. distance. Does there appear to be a strong relationship between the two? Does it appear to be linear?

(b) Find a power of distance such that the relationship between period of revolution and this transformation of distance appears to be quite linear.

(c) Using this transformation of distance, find the regression equation for predicting period of revolution. (You should also store the residual values.)

(d) Examine a residual plot for this regression equation and comment on what it reveals about the fit of the model.

Activity 11-6: Car Data (*cont.*)

The following scatterplot shows the time to travel one-quarter mile vs. weight for all of the cars in *Consumer Reports 1999 New Car Buying Guide*:

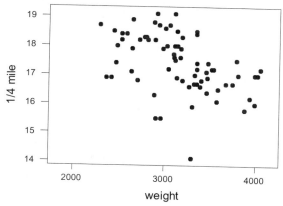

The Chevrolet Corvette took 14.1 seconds to travel one-quarter mile.

(a) Estimate as well as you can from the graph the Corvette's residual value.

(b) In this regression setting, would you say that the Corvette is more of an outlier or an influential observation or both? Explain.

(c) Use your calculator (CARS99.83g) to determine the equation of this regression line. Then calculate the residual value for the Corvette. How close was your answer in (a)?

(d) Remove the Corvette from the analysis, and use your calculator to recalculate the regression line. Has the slope coefficient changed much? How about the intercept?

(e) Based on your answers to (d), would you say that the Corvette is a very influential observation? Explain, and comment on your answer in (b).

Activity 11-7: Broadway Shows (*cont.*)

Consider again the data on Broadway shows from Activity 2-8 on page 36 (BROAD-WAY.83g).

(a) Use your calculator to produce a scatterplot of gross receipts vs. attendance and to calculate the regression line and store the residuals.

(b) What is the residual value for *The Lion King*? Does this show appear to be an outlier in this regression setting?

(c) What is the residual value for *Footloose*? Does this show appear to be an outlier in this regression setting?

(d) Based on the scatterplot, do you suspect that *The Lion King* is potentially influential? How about *Footloose*? Explain.

(e) Remove *The Lion King* from the analysis, and recalculate the regression line. Then do the same for *Footloose* (remember to put *The Lion King* back in before deleting *Footloose*). Which deletion caused a greater change in the regression line's slope? Which show is more influential?

(f) Explore how much the regression line can change if you change only the gross receipts value for *The Lion King*. Then explore how much the regression line can change if you change only the gross receipts value for *Footloose*. Write a paragraph describing your findings.

Activity 11-8: Solitaire (*cont.*)

Reconsider the solitaire data reported in Activity 9-14 on page 201 (SOLITAIRE.83g).

(a) Use your calculator to compute the correlation coefficient between the time of a game and the points earned in the game.

(b) Use this correlation coefficient to calculate r^2. Is it very close to 100%? Does this suggest that a linear regression would provide a good fit to the data? Explain.

(c) Use your calculator to create a scatterplot of points vs. time. Does the relationship appear to be linear? Explain.

(d) Use your calculator to determine the regression equation for predicting points from time. Draw this line on the scatterplot. Does it appear to summarize well the relationship between points and time?

(e) How many points would the line predict for a winning game that took three minutes? Would you expect this to be about right, an underestimate, or an overestimate? Explain.

(f) Repeat (e) for a winning game that takes two minutes. Then repeat again for a four-minute winning game.

Activity 11-9: College Football Players (*cont.*)

Consider the data on college football players' weights and jersey numbers from Activity 12-16 on page 265, which you analyzed in Activity 8-8 on page 175 and Activity 10-16 on page 225 (CPFOOTBALL.83g).

(a) Use your calculator to produce a scatterplot of weight (WGHT) vs. jersey number (JNUM) and to draw the least squares line on the scatterplot.

(b) Based on the appearance of this scatterplot, draw a sketch showing any patterns that you expect to see in a plot of residuals vs. jersey numbers. Write a sentence or two explaining your sketch.

(c) Use your calculator to determine the regression residuals and to create a scatterplot of residuals vs. jersey numbers. Does the pattern revealed in this plot generally match your sketch in (b)? Explain.

Activity 11-10: Monopoly Prices (*cont.*)

Refer again to the data on Monopoly properties (Activity 9-2 on page 189), and consider the issue of predicting a property's rent from its position on the board or from its purchase price (MONOPOLY.83g).

(a) Use your calculator to determine the regression equation for predicting rent from position. Report the equation and the value of r^2.

(b) Examine a scatterplot with the regression line drawn and a residual plot. Comment on what these reveal about the appropriateness of the linear model in this context.

(c) Repeat (a) and (b) for predicting rent from price.

(d) Despite the high values of r^2, are there reasons to doubt the adequacy of the linear model in both of these situations? Explain.

Activity 11-11: Box Office Blockbusters (*cont.*)

Refer back to the data in Activity 2-9 on page 37 concerning box office revenues for movies (MOVIES99.83g).

(a) Use your calculator to produce a scatterplot of the second week's box office revenue (REV2) vs. the first week's (REV1). Also report the correlation between these two. Describe any association that you find.

(b) Use your calculator to determine the regression equation for predicting a movie's second week box office revenue from its first week box office revenue; report this equation.

(c) Examine the RESID list to detetermine which movie has the largest (in absolute value) residual? What is the value of this residual? What specifically is unusual about this movie?

(d) Examine a dotplot of the residuals. Report the mean and median of the residuals. How many and what proportion of the residuals are negative? Does the regression line seem to be overestimating or underestimating the second week box office revenue for most movies?

(e) Which movie is most likely to be an influential observation? Remove this movie from the analysis and use your calculator to recalculate the regression equation and residuals. Report the regression equation. Does it seem to have changed much?

Activity 11-12: College Enrollments

The following data are the enrollments and faculty sizes of a sample of 34 American colleges and universities, taken from *The 1991 World Almanac and Book of Facts*.

college	enrollment	faculty	college	enrollment	faculty
1	12385	600	18	1671	114
2	1223	110	19	3778	234
3	2920	241	20	1250	71
4	1697	147	21	20110	733
5	890	113	22	16239	814
6	1131	87	23	766	63
7	1257	89	24	1875	115
8	2595	179	25	4170	390
9	52895	3796	26	1317	162
10	16500	950	27	3023	144
11	1080	45	28	15958	2362
12	1501	67	29	1220	44
13	1477	103	30	794	193
14	1369	114	31	2315	162
15	1385	49	32	1316	101
16	1966	160	33	800	63
17	2173	128	34	2381	144

(a) Use ENROLL.83g to produce a scatterplot of these data and to calculate the correlation coefficient between enrollment (ENROL) and faculty size (FACSZ). Write a few sentences commenting on whether there is an association between the two variables and, if so, on the direction and strength of the association.

(b) Suppose that you want to use enrollment information to predict the number of faculty members that a college has. Use your calculator to determine the least squares line, recording its equation (expressed in terms of the relevant variables).

(c) *Interpret* the slope coefficient of the least squares line.

(d) Comment on whether there seem to be any outliers or influential observations in the sample. Identify such colleges by number and explain your answers.

Activity 11-13: Gestation and Longevity (*cont.*)

Recall from the Activity 11-1 on page 228 the data that you analyzed concerning the relationship between an animal's longevity and its gestation period (ANIMALS.83g).

(a) Use the original regression equation that you found above to predict the gestation period of a human being. (Use your longevity guess from the "Preliminaries" section.) Show the details of your calculation.

(b) Do you accept this prediction as a reasonable one? If not, explain why the regression equation does not produce a reasonable prediction in this instance.

Activity 11-14: Gestation and Longevity (*cont.*)

Consider again the data from Activity 11-1 on page 228 concerning the relationship between an animal's longevity and its gestation period (ANIMALS.83g).

(a) Report the regression equation for predicting an animal's gestation period from its longevity.

(b) How long a gestation period does the line predict for an animal with longevity of 20 years?

(c) For what longevity does the regression line produce a prediction of a 365-day gestation period? [*Hint:* Use algebra on the regression equation to solve backwards.]

(d) Use your calculator to find the regression equation for predicting an animal's longevity from its gestation period. Report the equation.

(e) What longevity does this line predict for an animal with a gestation period of 365 days?

(f) Explain why your answers to (c) and (e) are not the same or even very close.

Activity 11-15: Turnpike Tolls (*cont.*)

Refer back to the regression analysis of Pennsylvania Turnpike tolls presented in Activity 10-12 on page 224. A scatterplot of the regression residuals vs. mileage is given below:

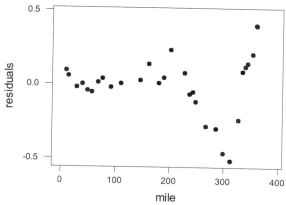

(a) Do the residuals seem to be randomly scattered or do some definite patterns emerge in this scatterplot?

(b) Despite the very high correlation of 0.999, does this residual plot give you reason to suspect that the relationship between toll and mileage is not completely linear? Explain. (You may also want to refer back to the scatterplot of toll vs. mileage in Activity 10-12.)

Unit III

Collecting Data

Topic 12:

SAMPLING

OVERVIEW

This topic explores methods for collecting your own data. You will learn the difference between a **sample** and a **population**. You will then learn how to select a sample that is **representative** of the population. You will begin to see the consequences of making conclusions from a sample that is **biased**.

OBJECTIVES

- To appreciate the fundamental distinctions between **population** and **sample** and between **parameter** and **statistic**.
- To recognize **biased** sampling methods and to be wary of conclusions drawn from studies that employ them.
- To discover the principle of **simple random sampling** and to be able to implement it using a **table of random digits** and your calculator.

PRELIMINARIES

1. In Topic 1, were all students in your school asked how many states they visited? In Topic 2, was an SAT average reported for every state?

2. What percentage of adult Americans do you think believe that Elvis Presley is still alive?

3. Who won the 1936 U.S. Presidential election? Who lost?

4. Do you know the names of the current U.S. Senators from your home state?

5. Have you sent an electronic mail message in the last week? Have you received and read an e-mail message in the last 24 hours? How many e-mail messages do you think you have received and read in the last 24 hours?

6. Collect from your classmates data on three variables:
  - whether the student has sent an e-mail message within the past week ~ 2 0
  - whether he or she has received and read an e-mail message in the last 24 hours
  - how many e-mail messages the person has received and read in the last 24 hours

IN-CLASS ACTIVITIES

Activity 12-1: States and SATs

Think again about the data on states' SAT scores you examined in Activity 2-4 on page 32, and the data we collected on students' travels in Activity 1-5 on page 11.

(a) Do these data tell you the SAT averages for all 50 states in 1998?

(b) Do you know how many students in your school have visited at least 25 states? If not, do you think you have a reasonable estimate of this number?

> In statistics, we use the term **population** to refer to the *entire* group of people or objects about which information is desired. A study that examines data on the entire population is called a **census**. However, conducting a census is rarely feasible. A **sample** is a (typically small) *part* of the population. If the sample is selected carefully, so that it is *representative* of the population, we still gain very useful information about the population. The number of **observational units** studied in a sample is the **sample size**. The essential idea of sampling is to learn about the whole by studying a part.

Activity 12-2: Elvis Presley and Alf Landon

On the twelfth anniversary of the death of Elvis Presley, a Dallas record company sponsored a national call-in survey. Listeners of over 1000 radio stations were asked to call a 1-900 number (at a charge of $2.50) to voice an opinion concerning whether or not Elvis was really dead. It turned out that 56% of the callers felt that Elvis was alive.

(a) Identify the population of interest and the sample actually used to study that population in this example.

population:

sample:

(b) Do you think that 56% is an accurate reflection of beliefs of *all* Americans on this issue? If not, identify some of the flaws in the sampling method.

In 1936, *Literary Digest* magazine conducted the most extensive (to that date) public opinion poll in history. They mailed out questionnaires to over 10 million people whose names and addresses they had obtained from telephone books and vehicle registration lists. More than 2.4 million people responded, with 57% indicating that they would vote for Republican Alf Landon in the upcoming Presidential election. (Incumbent Democrat Franklin Roosevelt won the election, carrying 63% of the popular vote.)

(c) Identify the population of interest and the sample actually used to study that population in this example.

population:

sample:

(d) Offer an explanation as to how *Literary Digest*'s prediction could have been so much in error. In particular, comment on why its sampling method made it vulnerable to overestimating support for the Republican candidate.

In both of these examples, the goal was to learn something about a very large population (all American adults, all American registered voters) by studying a sample. However, both illustrate a very poor job of sampling, i.e., of selecting the sample from the population. In neither case could one accurately infer anything about the population of interest from the sample results. This is because the sampling methods used were **biased**.

> A sampling procedure is said to be **biased** if it tends *systematically* to overrepresent certain segments of the population and systematically to underrepresent others.

These examples also indicate some common problems that produce biased samples. Both are **convenience samples** to some extent, since they both reached those people most readily accessible. Another problem is **voluntary response**, which refers to samples collected in such a way that members of the population decide for themselves whether or not to participate in the sample. The related problem of **nonresponse** can arise even if an unbiased sample of the population is contacted. Furthermore, the **sampling frame**, or list used to select the subjects (telephone books, vehicle registration records), was not representative of the population in 1936.

Activity 12-3: Sampling Senators

Consider the members of the 1999 U.S. Senate as the population of interest. These Senators are listed below, along with their gender, party, state, and years of service completed (as of 1999) in the Senate. Notice that each has been assigned a two-digit identification number.

ID#	Name	M/F	Party	State	Yrs	ID#	Name	M/F	Party	State	Yrs
01	Abraham	m	Rep	Michigan	4	51	Hutchinson, T	m	Rep	Arkansas	2
02	Akaka	m	Dem	Hawaii	9	52	Inhofe	m	Rep	Oklahoma	5
03	Allard	m	Rep	Colorado	2	53	Inouye	m	Dem	Hawaii	36
04	Ashcroft	m	Rep	Missouri	4	54	Jeffords	m	Rep	Vermont	10
05	Baucus	m	Dem	Montana	21	55	Johnson	m	Dem	South Dakota	2
06	Bayh	m	Dem	Indiana	0	56	Kennedy	m	Dem	Mass.	37
07	Bennett	m	Rep	Utah	6	57	Kerrey	m	Dem	Nebraska	10
08	Biden	m	Dem	Delaware	26	58	Kerry	m	Dem	Mass.	14
09	Bingaman	m	Dem	New Mexico	16	59	Kohl	m	Dem	Wisconsin	10
10	Bond	m	Rep	Missouri	12	60	Kyl	m	Rep	Arizona	4
11	Boxer	f	Dem	California	6	61	Landrieu	f	Dem	Louisiana	2
12	Breaux	m	Dem	Louisiana	12	62	Lautenberg	m	Dem	New Jersey	17
13	Brownback	m	Rep	Kansas	2	63	Leahy	m	Dem	Vermont	24
14	Bryan	m	Dem	Nevada	10	64	Levin	m	Dem	Michigan	20
15	Bunning	m	Rep	Kentucky	0	65	Lieberman	m	Dem	Connecticut	10
16	Burns	m	Rep	Montana	10	66	Lincoln	f	Dem	Arkansas	0
17	Byrd	m	Dem	West Virginia	40	67	Lott	m	Rep	Mississippi	10
18	Campbell	m	Rep	Colorado	6	68	Lugar	m	Rep	Indiana	22
19	Chafee	m	Rep	Rhode Island	13	69	Mack	m	Rep	Florida	10
20	Cleland	m	Dem	Georgia	2	70	McCain	m	Rep	Arizona	12
21	Cochran	m	Rep	Mississippi	21	71	McConnell	m	Rep	Kentucky	14
22	Collins	f	Rep	Maine	2	72	Mikulski	f	Dem	Maryland	12
23	Conrad	m	Dem	North Dakota	12	73	Moynihan	m	Dem	New York	22
24	Coverdell	m	Rep	Georgia	6	74	Murkowski	m	Rep	Alaska	18
25	Craig	m	Rep	Idaho	8	75	Murray	f	Dem	Washington	6
26	Crapo	m	Rep	Idaho	0	76	Nickles	m	Rep	Oklahoma	18
27	Daschle	m	Dem	South Dakota	12	77	Reed	m	Dem	Rhode Island	2
28	Dewine	m	Rep	Ohio	4	78	Reid	m	Dem	Nevada	12
29	Dodd	m	Dem	Connecticut	18	79	Robb	m	Dem	Virginia	10
30	Domenici	m	Rep	New Mexico	26	80	Roberts	m	Rep	Kansas	2
31	Dorgan	m	Dem	North Dakota	7	81	Rockefeller	m	Dem	West Virginia	14
32	Durbin	m	Dem	Illinois	2	82	Roth	m	Rep	Delaware	28
33	Edwards	m	Dem	N. Carolina	0	83	Santorum	m	Rep	Pennsylvania	4
34	Enzi	m	Rep	Wyoming	2	84	Sarbanes	m	Dem	Maryland	22
35	Feingold	m	Dem	Wisconsin	6	85	Schumer	m	Dem	New York	0
36	Feinstein	f	Dem	California	7	86	Sessions	m	Rep	Alabama	2
37	Fitzgerald	m	Rep	Illinois	0	87	Shelby	m	Rep	Alabama	12

38	Frist	m	Rep	Tennessee	4	88	Smith,B	m	Rep	New Hampshire	9
39	Gorton	m	Rep	Washington	10	89	Smith,G	m	Rep	Oregon	2
40	Graham	m	Dem	Florida	12	90	Snowe	f	Rep	Maine	4
41	Gramm	m	Rep	Texas	14	91	Specter	m	Rep	Pennsylvania	18
42	Grams	m	Rep	Minnesota	4	92	Stevens	m	Rep	Alaska	31
43	Grassley	m	Rep	Iowa	18	93	Thomas	m	Rep	Wyoming	4
44	Gregg	m	Rep	New Hampshire	6	94	Thompson	m	Rep	Tennessee	5
45	Hagel	m	Rep	Nebraska	2	95	Thurmond	m	Rep	South Carolina	43
46	Harkin	m	Dem	Iowa	14	96	Torricelli	m	Dem	New Jersey	2
47	Hatch	m	Rep	Utah	22	97	Voinovich	m	Rep	Ohio	0
48	Helms	m	Rep	North Carolina	26	98	Warner	m	Rep	Virginia	20
49	Hollings	m	Dem	South Carolina	33	99	Wellstone	m	Dem	Minnesota	8
50	Hutchinson, K	F	Rep	Texas	6	00	Wyden	m	Dem	Oregon	3

Some characteristics of this population are:

Males:	91
Females:	9

Democrats:	45
Republicans:	55

Years	Mean	Min	Max
of service	11.09	0	43

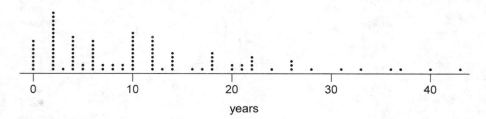

(a) For each of the four variables that are recorded about the Senators, identify whether it is a categorical or quantitative variable. If it is categorical, specify whether it is also binary.

Sex:

Party:

State:

Years of service:

(b) Choose five senators that you have heard of. Record the following information:

ID #	Name	Party	Years

(c) Do these five senators constitute a sample or a population?

(d) What is your home state? Is there a senator from your home state in the sample?

(e) How many Democrats and Republicans are in your sample?

No. of Democrats: No. of Republicans:

(f) Construct a dotplot of the years of service for the senators in your sample. Compute the average of the years of service in your sample. Also note the minimum and maximum values, and record these in the table below.

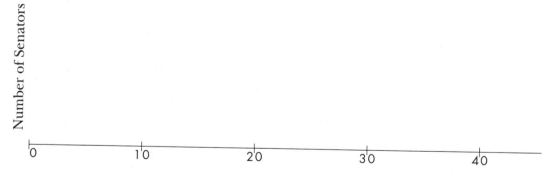

Years of service	Sample Mean	Min	Max

12

(g) Do you consider your sample of five senators *representative* of the population of senators? Explain.

(h) Combine your sample mean with those of the rest of the class and produce a dotplot below. (The population mean of 11.09 years has been marked on the axis.)

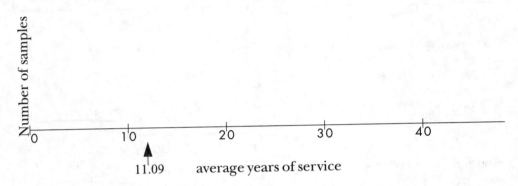

(i) How many students in your class obtained a sample average years of service that was higher than the population average of 11.09 years? What proportion of students?

(j) Explain why this sampling method (asking people to choose five senators with whom they are familiar) is biased and how this bias is exhibited in the above graph. Also identify the *direction* of the bias. In other words, does the sampling method tend to overestimate or underestimate the average years of service in the population?

Most people **tend** to be more familiar with senators who have served more years. If most people in your class estimate an average length of service that is higher than the population mean, this indicates that the sampling method is biased in the direction of overestimating the mean years of service.

(k) Would taking a sample of ten Senators in this manner be likely to produce more representative samples? Explain. [*Hint:* Think about the large sample taken by *Literary Digest* in 1936.]

Activity 12-4: Sampling Senators (*cont.*)

> In order to avoid biased samples, it seems fair and reasonable to give every member of the population the same chance of being selected for the sample. In other words, the selection method should ensure that every possible sample has an equal chance of being the sample ultimately selected. Such a sampling design is called **simple random sampling**.

While the principle of simple random sampling is probably clear, it is by no means simple to implement. One thought is to rely on physical mixing: write the names on pieces of paper, throw them into a hat, mix them thoroughly, and draw them out one at a time until the sample is full. Unfortunately, this method is fraught with the potential for hidden biases, such as different sizes of pieces of paper and insufficient mixing.

A better alternative for selecting a simple random sample (hereafter to be abbreviated **SRS**) is to use a computer-generated **table of random digits**. Such a table is constructed so that each position is equally likely to be occupied by any one of the digits 0, 1, 2, 3, 4, 5, 6, 7, 8, 9 and so that the value of any one position has no impact on the value of any other position. A table of random digits can be found in the back of the book (Table I).

The first column in the random number table gives a line number for you to refer to. The other columns give the random digits in groups of 5. It is often convenient to read across a line, but you can begin anywhere on a line and move in any direction. If you need more digits, just continue to the next line.

(a) Use the table of random digits to select a simple random sample of five U.S. Senators. Do this by entering the table at any point (does not have to be at the beginning of the line) and reading off the first 5 two-digit numbers that you happen across. (If you happen to get repeats, keep going until you have five different two-digit numbers.) Record the names and other information of the Senators corresponding to those ID numbers:

Line Number used:

ID #	Name	Party	Years	Your state?

(b) Is there a senator from your home state in the sample?

(c) Record the same information as before.

No. of Democrats: No. of Republicans:

Years of service	Sample Mean	Min	Max

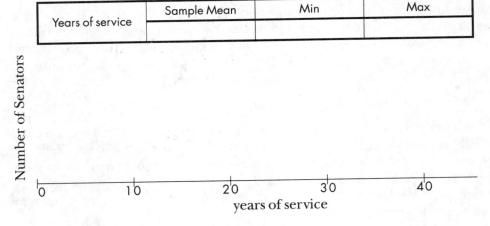

(d) Does the proportional breakdown of Democrats/Republicans in this sample equal that in the entire population of Senators? Does the mean years of service in your sample equal that of the population?

(e) If your answer to any part of question (d) is "no," does that mean that this sampling method is biased? Explain.

(f) To see how results vary from sample to sample, make a dotplot of the *mean* years of service from the samples of students in your class. Note that the population mean has been marked on the axis.

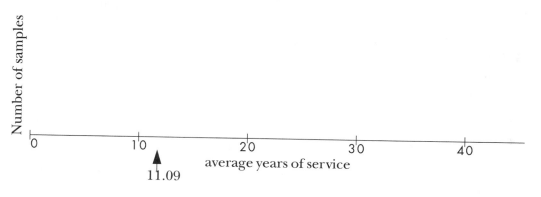

(g) How many students in your class obtained an average years of service that was higher than the population average of 11.09 years? How many were lower?

The ideas of population and sample are crucial in statistics. A closely related distinction concerns summaries called parameters and statistics. When we measure a variable about a group of objects, we summarize that information. A **parameter** is a numerical characteristic of the **population**, while a **statistic** is a numerical characteristic of a **sample**. (To help you keep this straight, notice that population and parameter start with the same letter, as do sample and statistic.)

We say that a statistic is an **unbiased** estimate of a parameter if the values of the statistic from different samples are centered at that parameter value.

(h) Based on your findings in (f) and (g), does the sample mean years of service appear to provide an unbiased estimate of the population mean? Explain.

(i) Identify each of the following as a parameter or a statistic:

  • The 56% of callers who believed that Elvis was alive ⌐

  • The 57% of voters who indicated they would vote for Alf Landon

  • The 63% of voters who voted for Franklin Roosevelt

  • The mean years of service among the 100 Senators

  • The mean years of service among your five Senators

  • The proportion of men in the entire 1999 Senate

Activity 12-5: Sampling Senators (*cont.*)

As you may have already guessed, the calculator can generate simple random samples more quickly than you can.

Download SENATORS.83g into your calculator. You will use your calculator to take a simple random sample of senators from this list as outlined below.

(a) Run the SENATORS program and select Activity 12-5. Indicate that you want 10 samples of size 20 by entering 10 and 20 at prompts, making sure to press ENTER after each entry. The program will generate a simple random sample of 20 ID numbers corresponding to 20 senators. The output that will appear on your screen are these 20 ID numbers, use the right arrow to view them. Press ENTER to view both the count of Democrats and the mean years of service for the senators in this sample. In the table below, record the sample proportion of Democrats in your sample and the sample mean years of service in your sample. Press ENTER again to generate the next sample. Record the values in the table for all ten samples.

sample	1	2	3	4	5	6	7	8	9	10
proportion Dem										
mean years										

(b) Did you get the same sample proportion of Democrats in each of your ten samples? Did you get the same sample mean years of service in each of your ten samples?

> This elementary question illustrates a very important statistical property known as **sampling variability**: The value of sample quantities varies from sample to sample. This simple idea is actually a very important and powerful one, which lays the foundation for most techniques of statistical inference.

(c) Create a dotplot of the sample mean years you found. Also include ten more sample means from your neighbor on the axis below:

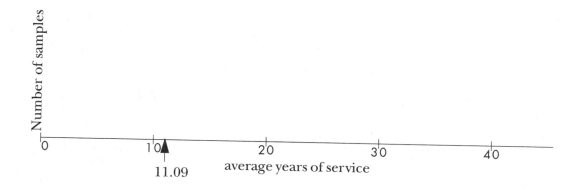

(d) Compare this distribution to that obtained in (f) of Activity 12-4. Do these sample means seem to be centered roughly at the population mean of 11.09 as they were with a sample size of five Senators?

(e) Which of the two distributions (sample size 5 or sample size 20) has less variability in the values of the sample mean years of service?

(f) In which case (sample size of 5 or sample size of 20) is the result of a *single* sample more likely to be close to the truth about the population?

> The ***precision*** of a sample statistic refers to its variability from sample to sample. Precision is related to ***sample size***: Sample means and sample proportions from larger samples are more precise/closer together than those from smaller samples. Since these statistics are also unbiased, those from larger samples provide a more *accurate* estimate of the corresponding population parameter.

Two caveats are in order, however. First, one still gets the occasional "unlucky" sample whose results are not close to the population even with large sample sizes. Second, the sample size means little if the sampling method is not random. Remember that the *Literary Digest* had a huge sample of 2.4 million people, yet their results were not close to the truth about the population.

Activity 12-6: Sampling Senators (*cont.*)

(a) Use the SENATORS program again, selecting Activity 12-6. Specify the sample size of 5, the population size as 100 to match the actual Senate, and 100 as the number of samples. The program will choose 100 samples, each containing 5 Senators, and will record the proportion of Democrats in each sample. The results are stored in a list named DEM. Construct a dotplot of the sample proportions of Democrats in these 100 samples. Comment on the features of this distribution.

(b) Copy the list DEM to a list named DEM1 and again use your SENATORS program to take 100 samples of size 5, but this time specify a population of 10,000 people. The calculator will sample from this second population, of whom 45% are Democrats (just as in the 1999 Senate), recording the proportion of Democrats in each sample. The results have been stored in a list named DEM. Construct a dotplot of the sample proportions in these 100 samples on the same scale as your data from (a). Do both distributions seem to be centered at the population value of .45? Would you say that the sample proportion is unbiased in both of these situations?

(c) Do these distributions seem to have similar variability? (You may want to use the DOTPLOT program to compare the two distributions.) Explain.

(d) In fact, did much change at all when you sampled from the much larger population?

> While the role of *sample* size is crucial to assessing how a sample statistic varies from sample to sample, the size of the *population* does not affect this sampling variability. As long as the population is large relative to the sample size (at least 10 times as large), the precision of a sample statistic depends on the sample size but not on the population size. Thus, a sample of size 1000 creates an equally precise estimate whether the population size is 100,000 or 260,000,000.

WRAP-UP

In this topic you began to consider issues related to how data are collected. You have learned four terms that are central to formal statistical inference: **population** and **sample, parameter** and **statistic.**

One of the key ideas to take away from this topic is that a poor method of collecting data can lead to misleading (if not completely meaningless) conclusions. Another fundamental idea is that of **simple random sampling** as a means of selecting a sample that will (most likely) be representative of the population that one is interested in.

In the next topic you will learn about different types of studies that can be conducted, whether you have a census or a sample from the population of interest.

HOMEWORK ACTIVITIES

Activity 12-7: Parameters Versus Statistics

(a) Suppose that you are interested in the population of all students at your school and that you are using the students enrolled in your course as a (nonrandom) sample. Identify each of the following as a parameter or a statistic.
- the proportion of your school's students who participate in school-sponsored athletics
- the proportion of students in your course who participate in school-sponsored athletics
- the mean grade point average for the students in your course
- the standard deviation of the grade point average for all students at your school

(b) Suppose that you are interested in the population of all high school or college students in the United States and that you are using students at your school as a (nonrandom) sample. Identify each of the following as a parameter or a statistic.
- the proportion of your school's students who have a car on campus
- the proportion of all U.S. high school or college students who have a car on campus
- the mean weekly income for all U.S. high school or college students
- the standard deviation of the weekly income for all students at your school

Activity 12-8: E-Mail Usage

Recall that in the "Preliminaries" section you gathered data concerning e-mail usage of your classmates.

(a) If your goal is to learn about e-mail usage among all students at your school, do your classmates form a population or a sample? Explain.

(b) What proportion of students in your class have sent at least one e-mail message in the past week? Is this number a statistic or a parameter? Explain.

(c) Do you and your classmates constitute a *random* sample from the students at your school? Explain.

(d) Do you think that you and your classmates form a *representative* sample of students at your school with regard to e-mail usage? Explain.

(e) If the population were all Americans in your age group, do you think that your class data would be representative in terms of e-mail usage? Explain.

Activity 12-9: E-Mail Usage (*cont.*)

Refer back to the data collected in the "Preliminaries" section on e-mail usage of students in your class.

(a) Produce a dotplot of the distribution of number of e-mail messages received in the last 24 hours.

(b) Based on the dotplot, make a guess as to the values of the mean and median of this distribution.

(c) Calculate the mean and median, and comment on how well you guessed in (b).

(d) If you were to remove the maximum and minimum from this distribution, what would you guess the effect to be on the mean and median? Explain.

(e) Remove the minimum and maximum, and recalculate the mean and median. What is the effect of removing those two values?

(f) What would you expect to be the effect of removing all zeros from the analysis? Explain.

(g) Remove all zeros, and recalculate the mean and median. What is the effect?

Activity 12-10: Web Addiction

The August 23, 1999 issue of the *Tampa Tribune* reported on a study involving data volunteered by 17,251 users of the `abcnews.com` web site. Users of the site were asked to respond to questions including whether they used the Internet to escape problems, tried unsuccessfully to cut back their Internet use, and found themselves preoccupied with the Internet even when not using it. Almost 6% of those responding confessed to some sort of addiction to the Internet.

(a) Identify the population and sample used in this study.

(b) The number 6% applies to the sample and so is a statistic. Identify in words the corresponding parameter of interest in this study.

(c) Do you think that 6% is a reasonable estimate of the parameter you defined in (b)? If not, indicate whether you think this estimate is too high or too low. Explain.

Activity 12-11: Emotional Support

In the mid-1980s Shere Hite undertook a study of women's attitudes toward relationships, love, and sex by distributing 100,000 questionnaires through women's groups. Of the 4500 women who returned the questionnaires, 96% said that they give more emotional support than they receive from their husbands or boyfriends.

(a) Comment on whether Hite's sampling method is likely to be biased in a particular direction. Specifically, do you think that the 96% figure overestimates or underestimates the truth about the population of all American women?

An ABC News/ *Washington Post* poll surveyed a random sample of 767 women, finding that 44% claimed to give more emotional support than they receive.

(b) Which poll surveyed the larger number of women?

(c) Which poll's results do you think are more representative of the truth about the population of all American women? Explain.

Activity 12-12: President's Popularity

Gallup conducted a poll on August 17, 1998, following a speech by President Clinton. Respondents were asked, "Now thinking about Bill Clinton as a person, do you have a favorable or unfavorable opinion of him?" The next day, Gallup asked another random sample the following question: "Now I'd like to get your opinion about some people in the news. As I read the name, please say if you have a favorable or unfavorable opinion of this person." In one of these polls, 40% of the respondents gave President Clinton a favorable rating. In the other poll, 55% of the responses were favorable. Give an argument for which poll received which favorability rating.

Activity 12-13: Alternative Medicine

In a spring 1994 issue, *Self* magazine reported that 84% of its readers who responded to a mail-in poll indicated that they had used a form of alternative medicine (e.g., acupuncture, homeopathy, herbal remedies). Comment on whether this sample result is representative of the truth concerning the population of all adult Americans. Do you suspect that the sampling method has biased the result? If so, is the sample result likely to overestimate or underestimate the proportion of all adult Americans who have used alternative medicine? Explain your answers.

Activity 12-14: Courtroom Cameras

An article appearing in the October 4, 1994 issue of *The Harrisburg Evening-News* reported that Judge Lance Ito (who was trying the O.J. Simpson murder case) had received 812 letters from around the country on the subject of whether to ban cameras from the courtroom. Of these 812 letters, 800 expressed the opinion that cameras should be banned.

(a) What proportion of this sample supports a ban on cameras in the courtroom? Is this number a parameter or a statistic?

(b) Do you think that this sample represents well the population of all American adults? Comment on the sampling method.

Activity 12-15: Sampling Representatives

Think again about how you sampled Senators in Activity 12-4 on page 255 using the random number table. Describe specifically how you would have to alter this procedure to use the table of random digits to select a simple random sample (SRS) of five from the population of 435 members of the U.S. House of Representatives.

Activity 12-16: College Football Players (*cont.*)

Below is the opening day roster for the 1999 Cal Poly football team:

#	Name	posit	weight	class	#	Name	posit	weight	class
1	Troy Henry	WR	175	SR	48	Joe Martinez	LB	225	FR
2	David Kellogg	WR	175	SR	50	Jesse Van Horn	LB	195	FR
3	Chris Jone	WR	150	JR	51	Harrison Stewart	LB	210	JR
4	Vaugn Jarrett	DB	185	FR	52	John Lloyd	C	270	FR
5	Kassim Osgood	DB	190	FR	53	Javier Ortiz	OL	270	SO
6	Raj Thompson	RB	215	FR	54	Tim Collins	OL	280	FR
7	Spencer Faddis	DB	190	FR	55	Juan Gonzalez	DL	270	SR
8	Kurt Rubin	DB	195	SO	57	Delon Craft	OL	290	SR
9	MikeDavies	DB	170	FR	58	Zak Repeka	DL	235	FR
10	Dave Woods	DB	205	FR	59	Eric Wicks	OL	265	JR
11	Andy Jepson	QB	190	SR	61	Travis Wheat	OL	270	JR
14	Jamar Isson	DB	180	FR	63	Matt Merritt	OL	265	FR
16	Kevin Cooper	QB	185	SO	64	Craig Knowles	OL	265	SO
17	Adam Herzing	WR	195	SO	66	Dan Loney	C	300	SR
18	Jason Brennan	WR	175	JR	68	Sal Rivas	K/P	225	JR
19	Robbie Bogdanovich	LB	215	FR	70	Kris Wicks	OL	280	SR
20	Craig Young	RB	190	SR	71	Dane Smith	DL	305	FR
22	David Richardson	DB	180	FR	72	Steve Prejean	DL	235	JR
23	Aaron Alston	DB	195	FR	73	Greg Mendonsa	OL	295	FR
24	Ameer Ross	DB	160	SO	74	Ryan Pittman	OL	310	SO
25	Mike Chavis	WR	175	FR	75	Dan Leahy	OL	285	JR
26	Vic Greco	LB	210	JR	76	Dustin Kroeker	OL	290	JR
28	Jeff Shaw	DB	235	JR	78	Chad McEwan	OL	305	SR
29	Dominic Washington	RB	205	FR	79	David Junod	OL	305	FR
30	Kevin Baga	DB	215	SR	80	Gary Parker	TE	225	JR

31	Kiko Griffin	CB	195	SR	81	Ryan McCarty	WR	180	FR
32	Brett Bakere	DB	195	SO	82	Chris Canoloes	WR	195	FR
33	Paki Bordon	RB	185	FR	83	Kyle Ray	WR	200	FR
34	Drew Ecklund	RB	175	SO	84	Russ Havens	TE	235	SO
35	Steve Sarubi	K/P	175	FR	86	Eric Martineau	TE	200	FR
36	Isaac Dixon	LB	230	FR	87	Mike O'Gorman	WR	220	FR
37	Jeese Wilson	DB	185	FR	88	Sam Carlson	TE	235	FR
39	Brian Villa	LB	200	SR	90	Jeff Zinn	DL	230	FR
40	Jeff Dennis	RB	210	JR	91	Nick Tignor	DL	250	FR
41	Mike Draper	DB	185	FR	92	Jason Roberts	DL	255	FR
42	Christian Broce	LB	235	SR	94	Ron Nelthorpe	DL	245	FR
43	Billy Stubblefield	LB	190	SR	95	Scott Brenton	DL	275	SO
44	Ryan Bianchi	FB	250	SO	96	Wade Green	DL	245	FR
45	Osbaldo Orozco	LB	225	SR	97	Emanuel Pasternak	DL	265	FR
46	Ron Cooper	LB	220	SO	98	Billy Beltz	DL	175	JR
47	Mike Toussaint	DB	170	SO	99	Garret Ford	DL	235	FR

(a) Identify each of the following variables as quantitative or categorical: number, position, weight, class.

(b) Identify the sampling frame (See page 250 for the definition).

(c) Use the random number table to take a simple random sample of 15 players. (Please be sure to list the line number you used in the table.) Calculate the average weight of these players. How do you think this sample mean weight compares to the average weight of all 82 players?

Activity 12-17: Phone Book Gender

The authors of your text were curious about the proportion of women living in San Luis Obispo County. They randomly selected a page from the 1998–99 SLO phone book (p. 40) and then randomly selected columns 1 and 4 on that page (This is an example of a multistage sampling design). They found the following results: 36 listings had both male and female names, 77 had male names, 14 had female names, 34 had initials only, 5 had pairs of initials.

(a) Identify the parameter and the statistic in this study.

(b) Do you believe that this sampling technique will give an unbiased estimate for the proportion of women living in San Luis Obispo? If not, explain whether you think the statistic will be an overestimate or an underestimate of the population parameter.

Activity 12-18: Sampling Senators (*cont.*)

(a) Use your calculator (SENATORS.83p) to take twenty random samples of size $n = 5$ Senators each. Examine a dotplot of the sample proportions of Democrats in these 20 samples, and comment on key features (shape, center, spread) of the distribution.

(b) Repeat (a) by generating twenty random samples of size $n = 20$ Senators each. Again comment on key features of the distribution of sample proportions of Democrats.

(c) Are the centers of the distributions of sample proportions similar for the two sample sizes?

(d) How do the spreads of the two distributions compare? Explain why this makes sense.

Activity 12-19: Sampling Senators (*cont.*)

Suppose that you were to repeatedly take simple random samples from a population and compute the proportion of Democrats, and then examine how much those proportions varied from sample to sample. Arrange the following four situations from the one that would produce the most variability to the one that would produce the least variability in these sample proportions, and explain your reasoning.

(a) a sample of size 20 from the population of U.S. Senators

(b) a sample of size 1000 from the population of New York residents

(c) a sample of size 100 from the population of New York residents

(d) a sample of size 500 from the population of Rhode Island residents

Activity 12-20: Voter Turnout

In the 1998 General Social Survey, a random sample of 2613 adult Americans were asked whether they had voted in the 1996 Presidential election, and 1783 said yes.

(a) What proportion of these people said that they had voted?

(b) Is this number a parameter or a statistic? Explain.

(c) Create a bar graph to display the proportions who claimed to have voted and not. The Federal Election Commission reported that 49.0% of those eligible to vote in the 1996 election had actually voted.

(d) Is this number a parameter or a statistic? Explain.

(e) Does the sample result seem to be consistent with the truth about the election?

(f) Do you suspect that the difference between the sample and population values here is just the result of random sampling? Explain

(g) Suggest a concern with doing reliable survey work that this example illustrates.

Activity 12-21: Nonsampling Sources of Bias

(a) Suppose that simple random samples of adult Americans are asked to complete a survey describing their attitudes toward the death penalty. Suppose that one group is asked, "Do you believe that the U.S. judicial system should have the right to call for executions?" while another group is asked, "Do you believe that the death penalty should be an option in cases of horrific murder?" Would you anticipate that the proportions of "yes" responses might differ between these two groups? Explain.

(b) Suppose that simple random samples of students on this campus are questioned about a proposed policy to ban smoking in all campus buildings. If one group is interviewed by a person wearing a T-shirt and jeans and smoking a cigarette while another group is interviewed by a nonsmoker wearing a business suit, would you expect that the proportions declaring agreement with the policy might differ between these two groups? Explain.

(c) Suppose that an interviewer knocks on doors in a suburban community and asks the person who answers whether he or she is married. If the person is married, the interviewer proceeds to ask, "Have you ever engaged in extramarital sex?" Would you expect the proportion of "yes" responses to be close to the actual proportion of married people in the community who have engaged in extramarital sex? Explain.

(d) Suppose that simple random samples of adult Americans are asked whether or not they approve of the President's handling of foreign policy. If one group is questioned prior to a nationally televised speech by the President on his or her foreign policy and another is questioned immediately after the speech, would you be surprised if the proportions of people expressing approval differed between these two groups? Explain.

(e) List four sources of bias that can affect sample survey results even if the sampling procedure used is indeed a random one. Base your list on the preceding four questions.

Activity 12-22: (Hypothetical) Service Times

(a) Suppose that you want to estimate the average time that a bank customer spends making transactions during a visit to the bank. Is this value a parameter or a statistic? Explain.

(b) If you enter the bank at a given time and use the service times of the customers in the bank at that particular time as your sample, would the average length of their service times provide an unbiased estimate? Explain your answer.

You can investigate this question more concretely by considering a hypothetical and greatly simplified population of 30 customers. Suppose these 30 customers enter the bank between 9:00 and 9:30 on a certain day and obtain the following service times:

ID#	start	time	ID#	start	time	ID#	start	Time
1	9:01	10	11	9:11	7	21	9:21	8
2	9:02	5	12	9:12	4	22	9:22	6
3	9:03	3	13	9:13	2	23	9:23	4
4	9:04	2	14	9:14	3	24	9:24	5
5	9:05	6	15	9:15	4	25	9:25	9
6	9:06	4	16	9:16	12	26	9:26	3
7	9:07	5	17	9:17	4	27	9:27	4
8	9:08	3	18	9:18	3	28	9:28	6
9	9:09	8	19	9:19	6	29	9:29	2
10	9:10	2	20	9:20	5	30	9:30	5

(c) Calculate the average service time for this population of 30 customers.

(d) If you enter the bank at 9:12:30 to select your sample, which customers (by ID#) would be in the bank and therefore in your sample?

(e) Calculate the average service time for the customers in this sample. Is it greater or less than the population average you calculated in (d)?

(f) Repeat (d) and (e) for the following times 9:21:30, 9:27:30, and two other times that you select yourself.

(g) How many of these five samples produced an overestimate of the population average?

(h) In light of your analysis, explain why this sampling method is biased in the direction of overestimating the average service time.

Activity 12-23: (Hypothetical) Service Times (*cont.*)

Refer to the hypothetical service times for the population of 30 bank customers given in Activity 12-22.

(a) Use the table of random digits to take a sample of five bank customers. Report the line number of the table that you use and the ID numbers for the sample.

(b) Calculate the average service time for this sample. Is it greater or less than the population mean service time?

(c) Repeat (a) and (b) for a total of five samples.

(d) Does this sampling procedure seem to be biased in one direction or the other? Explain.

Activity 12-24: Prison Terms and Car Trips

Explain how the principle that you discovered in Activity 12-22 would apply to estimating the average prison term by taking a sample of inmates at any one time and to estimating the average length of a car trip by taking a sample of cars on the road at any one time. Then describe a similar situation to which the same principle would apply.

Activity 12-25: Other Probability Samples of U.S. Senators

The random number table allows us to take a sample in such a way that every member of the population has the same chance of being selected. In fact, any sample of size 10 has the same chance of being selected. We may get an "unlucky" sample, but we will not systematically over- or underestimate the parameters of the population.

Still, it is not realistic for an organization like Gallup to take a simple random sample of all United States citizens. This has sometimes led statisticians to use other methods of random sampling. We will now examine two of these other techniques.

(a) Break the Senators into groups by their ID #: 0–9, 10–19, ..., 90–99. Use the random number table to select a number between 0 and 9. Let this number correspond to the ID# of a Senator in the first group. Now repeatedly add 10 to that ID# to select a Senator from each group. Record the Senators in your sample.

(b) Do you think that this is an unbiased sampling method? Explain.

The above technique is called **systematic sampling** and can be a quick alternative to simple random sampling.

(c) Suppose an insurance company wants to conduct a survey on flood awareness of apartment building residents in your town. The apartment buildings are all 3 stories high, so he randomly selects a number between 1 and 3 and then takes that floor in each building. Is this a systematic sample? Is this an unbiased sampling method?

While systematic sampling can be convenient, you still need to make sure that you are not introducing any bias.

(d) Since the population of 1999 senators is roughly 10% female, it is very possible that we could obtain a completely male sample. Suggest a reasonable method, still using randomness, that guarantees that a sample of ten senators will include exactly one female.

We can break the senators into two **strata**, men and women, and then take a simple random sample from each stratum. This technique, **stratified sampling**, can be used to further ensure that your sample is representative of the population if you can identify an important, obvious way of dividing the population. After dividing the sampling frame into groups, select a separate simple random sample from each group.

(e) Create a sample of 20 senators that contains 10 Democrats and 10 Republicans. Label the Democrats from 01 to 45 and select ten. Then label the Republicans from 01 to 55 and select 10. Make sure you randomly select a different set of ID#s for the two groups.

Activity 12-26: Sampling Employees

Suppose that a company employs 1000 males and 200 females. A management task force officer polls a stratified random sample of 100 males and 20 females. [*Hint*: See Activity 12-25 for more on stratification.]
(a) What is the chance that a particular female employee will be polled?
(b) What is the chance that a particular male employee will be polled?
(c) Does this sample meet the requirements of a simple random sample? Explain carefully.
(d) Can the sample be representative of the population even if it is not a simple random sample? Explain.

Activity 12-27: College Football Players (*cont.*)

Refer to the opening day roster for the 1999 Cal Poly football team on page 265.
(a) Suggest a variable you could use to break the team into several strata that you suspect is related to the players' weights. Explain how you would carry out a stratified random sample. [*Hint*: See Activity 12-25 for more on stratification.]
(b) Take a systematic sample of 10 *nonkickers* by splitting the list into ten sections of eight players each. Show the details of your sampling technique. [*Hint*: See Activity 12-25 for more on systematic samples.]

Topic 13:

DESIGNING STUDIES

You have studied the idea of **random sampling** as a fundamental principle by which to gather information about a **population**. Sampling is often used to conduct opinion **surveys** to describe how the population feels about an issue or to examine people's habits and practices. Sometimes our goal is not to describe a population but to investigate the effects that one variable has on other variables. This topic introduces you to the design of controlled **experiments** for this purpose and contrasts them with **observational studies**. You will discover some principles for designing studies, and you will learn that the scope of one's conclusions depends critically on how the data were collected.

- To learn to distinguish between **response variables** and **explanatory variables**.
- To understand the distinction between **observational studies**, **surveys**, and **controlled experiments** and the different conclusions that one can draw from each.
- To recognize the need for carefully designed **experiments** in order to detect a cause-and-effect relationship between variables.
- To appreciate the principle of **control** as a fundamental idea of experimental design.
- To identify some **principles** through which experiments achieve control.
- To investigate the concept of **confounding** and the effects of **randomization**.

273

PRELIMINARIES

1. If high school students who study a foreign language tend to perform better on the verbal portion of the SAT than students who do not study a foreign language, does that establish that studying a foreign language improves one's verbal skills?

2. If 95% of the participants in a large SAT coaching program improve their SAT scores after attending the program, would you conclude that the coaching was responsible for their improvement?

3. Do you think that eating SmartFood™ popcorn really makes a person smarter?

4. If recovering heart attack patients who own a pet tend to survive longer than recovering heart attack patients who do not own a pet, would you conclude that owning a pet has therapeutic benefits for heart attack patients?

5. Your instructor will pass out a sequence of letters and give you twenty seconds to memorize as many as you can. Record how many you correctly list in order.

6. Record these memory scores for your classmates. Your score is the number of letters you remember correctly in order before the first mistake. Also keep track of which version of the letters each person had.

IN-CLASS ACTIVITIES

Activity 13-1: An Apple a Day

Suppose that you want to collect data to study whether the expression "an apple a day keeps the doctor away" has any validity. In other words, you want to investigate whether eating apples has any health benefit. Consider four different designs of such a study:

   (i) You take a random sample of individuals, identify which do and do not eat apples regularly, and then follow them for six months to see who requires a visit to a doctor and who does not.
   (ii) You take a random sample of physicians and ask them whether they have noticed any health benefits of eating apples.
   (iii) You take a random sample of individuals, randomly assign half to eat an apple a day for the next six months and the other half not to, and then see who requires a visit to a doctor and who does not.
   (iv) You recall that your Uncle Joe loved apples and was never sick a day in his life, while your Uncle Tom despised apples and was often ill.

> These four examples illustrate four types of studies:
>
> • **anecdotes**, with which the investigator merely recounts instances known to him or her.
>
> • **surveys**, with which the investigator asks people to answer questions about their opinions or practices.
>
> • **observational studies**, with which the investigator passively observes and records information on observational units, such as people's practices.
>
> • **experiments**, with which the investigator deliberately imposes some condition on the subjects or **experimental units**, observing and recording the results.

(a) Identify the type of study represented by (i)–(iv) above.

(i)   (ii)

(iii)   (iv)

(b) Do you think study (iv) is of much value? Explain.

Anecdotal evidence results from situations that come easily to mind. This type of evidence is of little value in scientific research. Much of the practice of statistics involves designing studies and collecting data so that people will not have to rely on anecdotal evidence.

(c) What type of information is gained from study (ii)?

(d) For studies (i) and (iii), identify the observational units (cases) and the two variables involved. Also indicate whether the variables are categorical or quantitative.

   observational/experimental unit:

   variable: type:

   variable: type:

> The variable whose effect one wants to study is called the ***explanatory*** (or independent) ***variable***. The variable that one suspects is affected by the other is known as the ***response*** (or dependent) ***variable***.

(e) Identify which is the explanatory and which is the response variable in studies (i) and (iii).
   explanatory:

   response:

We can denote the structure of such a study through a simple graphic:

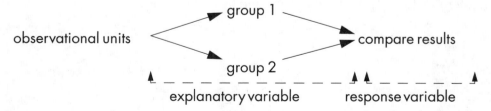

where we can list any number of groups for the explanatory variable.

(f) Produce such a graphic that describes studies (i) and (iii), replacing the generic terms with a description of the variables in this context.

(g) If the findings of study (i) are that those who eat apples regularly tend to visit a doctor significantly less often than those who do not eat apples regularly, would that enable you to conclude that apples are responsible for the fewer doctor visits? Explain.

(h) Explain why study (iii) is a more effective design than (i) for establishing whether or not there is a cause-and-effect relationship between apples and health.

An ***experiment*** is necessary to establish a cause-and-effect relationship between variables. In contrast to an observational study, one of the key characteristics of an experiment is that the experimenter *actively imposes* the treatment on the subjects. The experimenter then hopes to see the direct effect of this explanatory variable on the response.

(i) What would be lacking in an experiment that took a random sample of individuals, asked them to eat an apple a day for the next few months, and recorded the number of visits to a physician by each person?

> A well-designed experiment exerts several forms of **control** on the subjects to help minimize the effects of extraneous variables, so that any effects on the response variable can be directly attributed to the explanatory variable. One fundamental form of control is that of **comparison**.

Activity 13-2: Foreign Language and SATs

Suppose that a high school is considering adopting a foreign language requirement, partly on the grounds that students who study a foreign language tend to do better on the verbal portion of the SAT exam than students who do not. You decide to investigate this argument by examining records of random samples of students taking and not taking a foreign language. You find the following data and construct the dotplots below:

SAT Verbal scores:

| foreign language study | 570 | 570 | 600 | 610 | 660 | 660 | 670 | 700 | 700 | 780 |
| no foreign language study | 450 | 480 | 490 | 490 | 550 | 570 | 570 | 610 | 610 | 690 |

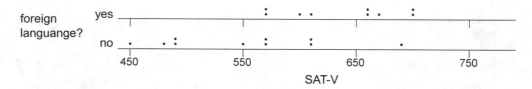

(a) Identify the explanatory and response variables in this study and produce a graphic summarizing the structure of the study. Also report for each variable whether it is categorical or quantitative.

explanatory: type:

response: type:

graphic:

(b) Is this an observational study or a controlled experiment? Explain.

(c) Do these sample data indicate that those studying a foreign language perform better on the verbal portion of the SAT than those who do not study a foreign language? Explain.

(d) Can you suggest an alternative explanation for why this difference might occur even if foreign language study does not improve students' verbal skills?

Suppose that you look back at the academic records of these 20 students and find a measure of their verbal aptitude taken from a standardized test prior to their foreign language study. Suppose that the data are as follows:

No foreign language study			Foreign language study		
name	SAT-V	Verbal aptitude	name	SAT-V	Verbal aptitude
Alice	690	41	Karla	570	82
Bob	570	33	Larry	600	67
Carol	490	41	Max	610	65
Dennis	480	34	Nancy	700	89
Ellen	610	38	Oscar	660	74
Frank	570	52	Peter	780	78
Greta	610	32	Qian	570	84
Harry	490	34	Randy	670	67
Isaac	550	36	Sally	660	82
Julie	450	37	Tara	700	75

The following scatterplot presents SAT vs. verbal aptitude score:

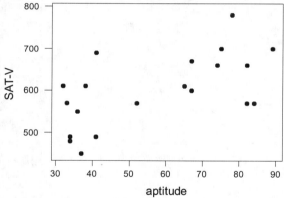

(e) Does the scatterplot reveal an association between verbal SAT score and prior verbal aptitude score? Explain.

(f) Circle the points on the scatterplot corresponding to the students who did not take a foreign language. What do you notice about the verbal aptitude scores of these students?

(g) From the data given, can you rule out the possibility that foreign language study has no effect on students' verbal SAT scores but those students do better because of a higher verbal aptitude that was present even before the foreign language study began? In fact, do the data provide you with any way of distinguishing between the effects of foreign language study and the effects of previous verbal aptitude?

An observational study does not control for possible effects of variables that are not considered in the study but could have an effect on the response variable. These unmonitored variables are often called *lurking variables*. These lurking variables can have effects on the response variable that are **confounded** with those of the explanatory variable. A **confounding variable** is one whose effects on the response are indistinguishable from those of the explanatory variable. This prevents the experimenter from isolating the effects of each variable.

In this study, those who studied a foreign language were the same students who had high verbal aptitude. Thus, when these students did better on the SAT-verbal, we have no way of knowing whether it was due to their higher initial verbal ability or their foreign language study. In this sense, verbal aptitude and foreign language study are confounded.

Now consider another lurking variable in this study: gender.

(h) Is there an association between gender and studying a foreign language as there is between verbal aptitude and taking a language? Investigate this by determining the gender breakdown in the group that studies a language and in the group that does not.

(i) Is the gender variable *confounded* with studying a language? [*Hint:* Ask yourself whether one gender studied a language and the other one did not, in which case the effects of gender and language study would be inseparable.]

These last questions reveal that a lurking variable need not be confounded with the explanatory variable.

(j) Suggest a potential confounding variable for observational study (i) in Activity 13-1 on page 275. Be sure you explain how this variable behaves differently for those who eat apples regularly and those who do not.

Activity 13-3: Foreign Language and SATs (*cont.*)

(a) If you could design an *experiment* to assess the possible effect of foreign language study on SAT score, would you want all of the high verbal aptitude students to be in one group and all of the low aptitude students in the other group? How would you want these students divided?

(b) Even if you did not have access to measures of verbal aptitude, how might you assign students to treatment groups in an effort to balance out the aptitudes between the two groups?

> Experimenters try to assign subjects to groups in such a way that lurking and potentially confounding variables tend to balance out between the two groups. ***Randomization*** provides a simple but effective way to achieve this balance. By randomly assigning subjects to treatment groups, experimenters ensure that hidden confounding variables will balance out in the long run between the treatment groups.

In this case, your graphic should indicate this randomization:

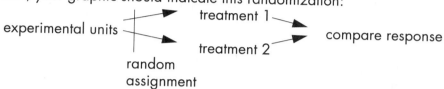

(c) Randomly assign the twenty students to either study a foreign language or not. You could flip a coin for each student, with heads designating one group and tails the other, until ten students are in one of the groups. You might also use the random number table, with even digits signifying one group and odd digits the other. Record the name and the verbal aptitude score in the assigned group below.

Subjects:
Alice, Bob, Carol, Dennis, Ellen, Frank, Greta, Harry, Isaac, Julie, Karla, Larry, Max, Nancy, Oscar, Peter, Qian, Randy, Sally, Tara

	No foreign language		Foreign language	
	Name	Aptitude	Name	Aptitude
1				
2				
3				
4				
5				
6				
7				
8				
9				
10				

(d) Use your calculator to create visual displays and to calculate numerical summaries of the distributions of the aptitude scores between the two groups. Record the summaries in the table below, and comment on how well randomization balanced out the aptitude scores between the two groups.

	Minimum	Mean	Maximum
Foreign language study			
No foreign language study			

(e) Consider another lurking variable: degree of motivation to study hard. Even though we have not measured this variable for each student, can we expect randomization to have reasonably balanced out this variable between the two groups? Explain.

(f) Now if there turned out to be a significant difference in the SAT scores between the two groups, would you be able to reasonably attribute the difference to the foreign language study alone? Explain.

> Concluding that a *causal* relationship exists between an explanatory and a response variable is appropriate only when the data come from a well-designed, controlled experiment. Randomization plays a crucial role in such experiments by aiming to balance out potential effects of lurking variables.

Activity 13-4: Parkinson's Disease and Embryo Treatment

Current drug therapies or neuroprotective methods do not provide sufficient help to patients with Parkinson's disease. A promising new treatment for Parkinson's disease involves injecting fetal tissue into the brain of the patient to replace damaged brain cells and compensate for the loss of essential nerve cell groups. A clinical trial was recently set up in Colorado to determine the effectiveness of this treatment. The trial involved 40 patients with Parkinson's disease. Twenty were given the fetal tissue and 20 were not.

(a) Identify the explanatory variable in this study and sketch a graphic of the experimental design.

(b) How would you suggest dividing the forty patients into the surgery group and the control group?

(c) Suppose the group who undergo surgery begin to feel more "positive" and "healthier" on subsequent patient follow-up questionnaires. Would you be willing to conclude that this response could be attributed solely to the injection of the cells? Explain.

13

> The very fact that subjects in the treatment group realize that they are being given something that researchers expect will improve their condition may affect their responses differently than those of subjects who are not given any treatment. This phenomenon has been detected in many circumstances, especially medical studies, and is known as the ***placebo effect***. Experimenters control for this confounding variable by administering a placebo (referring to the "sugar pill" or empty drug given in medical studies) to those subjects in the control group. This method of control is called ***blindness***, since subjects are not told whether or not they are receiving the real treatment or the placebo.

Medical researchers know from long experience that roughly 1 in every 3 patients feels better with only a sugar pill (*Time*, February 22, 1999). The placebo effect appears to be especially strong with Parkinson's patients.

(d) In the Colorado study, the twenty patients in the control group were given a surgical procedure that only involved cutting two holes in their skull without the injection of fetal tissue. Were the patients in this study blind to whether they were in the control group or the treatment group? Explain the benefits of this approach.

(e) The Colorado study caused a bit of controversy in the medical world. George Annas, director of the law, medicine, and ethics program at the Boston University School of Medicine, stated, "Not only is it ethical to do it this way, it's probably unethical to do it any other way" (*Boston Globe*, February 4, 1999). Do you believe that the knowledge gained from this type of comparative experiment is worth the potential dangers of fake surgery?

(f) Explain why it would be important for the person evaluating the health of the patient also to be blind to which treatment group the patient is in.

> When possible, experiments should be **double blind** in that the person responsible for evaluating the subjects should also be unaware of which subjects receive which treatment. In this way the evaluator's judgment is not influenced (consciously or subconsciously) by any hidden biases.

Activity 13-5: Pregnancy, AZT, and HIV (*cont.*)

In an experiment reported in the March 7, 1994 issue of *Newsweek*, 164 pregnant, HIV-positive women were randomly assigned to receive the drug AZT during pregnancy, and 160 such women were randomly assigned to a control group that received a placebo. It turned out that 40 of the mothers in the control group gave birth to babies who were HIV-positive, compared to only 13 in the AZT group.

(a) Identify the explanatory and response variables in this study and provide a graphic of the experimental design. Are these quantitative or categorical variables? If they are categorical, are they also binary?

(b) Explain how the study makes use of the principle of *comparison*.

(c) Explain how the study incorporates the principle of *randomization*.

(d) Explain how the study takes into account the principle of *blindness*.

Activity 13-6: Effectiveness of Gasoline Additive

While the goal of randomization is to create groups that are as similar as possible prior to administering treatment, if we can identify an obvious variable that we know has an effect on the response variable, we may wish to impose an additional level of control.

Suppose that an automotive engineer wants to see whether a newly developed gasoline additive successfully increases the mileage of small and luxury cars. The table below reports the city MPG of small and luxury cars from *Consumer Reports 1999 New Car Buyer's Guide*.

Small Cars	MPG	Luxury Cars	MPG
Acura Integra	25	Acura RL	18
Chevrolet Cavalier	23	Audi A8	17
Chevrolet Metro	30	BMW 5-Series	18
Chevrolet Prism	28	BMW 740I	17

Daewoo Lanos	23	Cadillac Deville	17
Daewoo Nubria	22	Cadillac Seville	17
Dodge Neon	25	Infiniti Q45	17
Ford Escort	25	Jaguar XJ8	17
Honda Civic	28	Lexus GS300	20
Hyundai Accent	26	Lexus LS 400	18
Hyundai Elantra	22	Lincoln Town Car	17
Kia Sephia	23	Mercedes/Benz E/Class	21
Mazda Protege	24		
Mitsubishi Mirage	26		
Nissan Sentra	27		
Plymouth Neon	25		
Pontiac Sunfire	23		
Saturn	25		
Subaru Impreza	23		
Suzuki Swift	30		
Toyota Corolla	28		
Volkswagon Golf	22		
Volkswagon Jetta	22		
Volkswagon New Beetle	24		

13

The researcher decides to divide the cars into two groups, giving one group the treatment and nothing to the other group of cars. The goal is for these two groups to be as similar as possible prior to assigning the treatment.

(a) Flip a coin for each car. If the coin is heads, place that car in the treatment group; if tails, place the car in the control group. Continue until you have 18 cars in one of the groups, and put the rest of the cars in the other group. List the mileages for the cars in the treatment group:

Treatment:

(b) Calculate the average gas mileage for the treatment group. Use the rest of the cars as the control group. Calculate the average gas mileage for the control group. Calculate the difference between these two numbers.

Treatment average:

Control average:

Difference in averages (treatment – control):

(c) What number should this difference be close to if the two groups have similar performance prior to treatment?

(d) Below is a dotplot of the result of 1000 such random assignments:

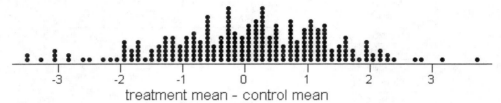

Does the randomization typically create similar groups? Explain.

(e) Identify an obvious variable, other than the treatment, that you think may also have an effect on the response variable.

> In a manner similar to that of a stratified sample (see Activity 12-25 on page 270), we can also separate our subjects into similar groups before administering the treatment. This technique, called **blocking**, allows us to further ensure the equivalence of the treatment groups.

Graphic:

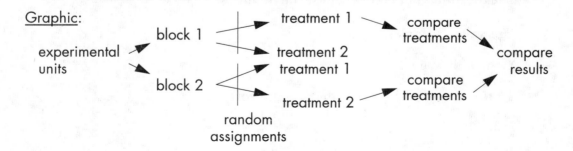

(f) If we allow the type of car, small or luxury, to be the blocking variable, complete a graphic like the one above using the context of this example.

(g) Carry out this design by randomly putting half of the the luxury cars into a treatment group and half into a control group. Then randomly place half of the small cars into a treatment group and half into a control group. Combine the two treatment groups and the two control groups. Find the average gas mileage of all cars now in the treatment group and the average gas mileage of all cars now in the control group. Find the difference of these two averages.

Treatment group average:

Control group average:

Difference (treatment – control):

(h) Pool your results for the difference with the rest of your class and construct a dotplot:

treatment mean – control mean differences

(i) How do the distributions of the differences in group averages plotted in (d) and (h) compare? Would you consider the treatment group and the control group even more similar to each other before the treatment is imposed?

13

One of the key advantages to blocking is a reduction in variation. By ensuring that the groups are balanced with respect to car type, we have helped to make the treatment and control groups even more similar to each other before we administer the treatment. This will make it easier for us to identify any effect the treatment might have.

WRAP-UP

This topic has introduced you to the three main types of studies: *surveys*, *observational studies*, and *experiments*. You have learned the limitations of observational studies with regard to establishing causal relationships between variables. You have also learned that *control* is the guiding principle of *experimental design* and discovered the principles of *comparison, randomization,* and *blindness* as specific techniques for achieving control. You have investigated the concept of *confounding* and the effects of *randomization*. We also discussed *blocking* as an extra step we sometimes take to control the most important unavoidable sources of variability, allowing randomization to average out effects of the remaining variation.

HOMEWORK ACTIVITIES

Activity 13-7: Pet Therapy

Suppose that you want to study whether pets provide a therapeutic benefit for their owners. Specifically, you decide to investigate whether heart attack patients who own a pet tend to recover more often than those who do not. You randomly select a sample of heart attack patients from a large hospital and follow them for one year. You then compare the sample proportions who have survived and find that 92% of those with pets are still alive, while only 64% of those without pets have survived.

(a) Identify the explanatory and response variables in this study.
(b) Is this study a controlled experiment or an observational study? Explain.
(c) Does this study make use of a comparison group? Explain.
(d) Does this study make use of randomization? Explain.
(e) Does the design of the study enable you to conclude that owning a pet does indeed have a therapeutic benefit for heart attack survivors? Explain.
(f) Describe in principle how you could design a controlled experiment to address this issue.
(g) Is such an experiment feasible in practice? Explain.

Activity 13-8: Foreign Language and SAT (*cont.*)

(a) Recall the question of whether foreign language study *improves* students' scores on the verbal section of the SAT, examined in Activity 13-2 on page 278. Describe the design of a controlled experiment to address this question.
(b) Would it be ethical to carry out the experiment? Explain.

Activity 13-9: Winter Heart Attacks

An article in the October 12, 1999 issue of *USA Today* reported on a study that recorded the number of deaths by heart attack in each month over a 12-year period in Los Angeles. December and January were found to have substantially higher numbers of deaths than the other months. Researchers conjectured that the stress and overindulgence associated with the holiday season might explain this finding.
(a) What type of study is this? Explain.
(b) Similar studies had previously been conducted in Northeastern cities such as New York and Boston. Identify a confounding variable and alternative explanation in these studies that is not relevant with the Los Angeles study.
(c) Does the Los Angeles study prove that holiday stress causes an increase in heart attack deaths during December and January? If not, identify a confounding variable in this study.

Activity 13-10: Friendly Observers

In a study published in the *Journal of Personality and Social Psychology* (Butler and Baumeister, 1998), researchers investigated a conjecture that having an observer with a vested interest would decrease subjects' performance on a skill-based task. Subjects were given time to practice playing a video game that required them to navigate an obstacle course as quickly as possible. They were then told to play the game one final time with an observer present. Subjects were randomly assigned to one of two groups. One group (A) was told that the participant and observer would each win $3 if the participant beat a certain threshold time, and the other group (B) was told only that the participant would win the prize if the threshold were beaten. The threshold was chosen to be a time that they beat in 30% of their practice turns. It turned out that 3 of the 12 subjects in group A beat the threshold, while 8 of 11 subjects in group B achieved success.
(a) Is this an observational study or an experiment?
(b) Explain how the study makes use of the principles of comparison, randomization, and blindness.
(c) Consider the following variables that were recorded for each subject:
   • Time to complete the game
   • Whether or not the threshold was beaten
   • Whether the observer was to share in the prize or not

Identify each variable as quantitative or categorical. If it is categorical, indicate whether it is binary.

(d) Identify each variable as explanatory or response.

(e) Calculate the proportion of each group that beat the threshold.

(f) Do these proportions differ in the direction conjectured by the researchers? Explain.

(g) Suppose that the 23 subjects had been labeled with letters from A through W. Explain how you could use a table of random digits to randomly assign 12 of them to group A and 11 to group B.

(h) What would be a disadvantage of flipping a coin to determine the group assignment of each subject (heads to A and tails to B, say)?

Activity 13-11: Children's Television Viewing (*cont.*)

Consider the Stanford study described in Activity 1-13 on page 16 to see whether reducing children's television viewing might help to prevent obesity. Third and fourth grade students at two public elementary schools in San Jose were the subjects. One of the schools incorporated a curriculum designed to reduce watching television and playing video games, while the other school made no changes to its curriculum. At the beginning and end of the study a variety of variables were measured on each child. These included body mass index, triceps skinfold thickness, waist circumference, waist-to-hip ratio, weekly time spent watching television, and weekly time spent playing video games. The researchers found significant decreases in these variables for the treatment group but not for the control group.

(a) Explain why this is an experiment and not an observational study.

(b) Identify the explanatory variable in this study. What type of variable is it?

(c) Explain why the findings are more conclusive than if the study had been an observational study.

Activity 13-12: Baldness and Heart Disease (*cont.*)

Consider the study mentioned in Activity 7-12 on page 152, where researchers took a sample of male heart attack patients and a sample of men who have not suffered from a heart attack and compared baldness ratings between the two groups. Their goal was to determine whether baldness has an effect on one's likelihood of experiencing a heart attack.

(a) Identify the explanatory and response variables.

(b) Is this study a controlled experiment or an observational study? Explain.

(c) Even if a strong relationship exists between heart disease and baldness, does that necessarily mean that heart disease is *caused* by baldness? Explain your answer.

Activity 13-13: Gender and Lung Cancer (*cont.*)

Activity 7-14 on page 153 describes a study of gender and incidence of lung cancer among smokers. Researchers screened 1000 people who were smokers of age 60 or higher. They found 459 women and 541 men, with 19 of the women and 10 of the men suffering from lung cancer.
(a) What type of study is this? Explain.
(b) Identify the observational/experimental units in this study.
(c) Identify the explanatory and response variables.
(d) What proportion of the men had lung cancer? What proportion of the women had lung cancer?
(e) Identify some potential confounding variables in this study. Make sure you indicate how this variable affects the two "treatment groups" differently.

Activity 13-14: Reducing Cold Durations

To study the effectiveness of a zinc nasal spray for reducing the duration of a common cold, researchers recruited 104 subjects who agreed to report to their lab within 24 hours of getting cold symptoms. Each subject was randomly assigned to one of three groups: One received full dosage of the zinc spray, another received a low dosage, and a third received a placebo spray. The cold symptoms lasted an average of 1.5 days for the full dosage group, 3.5 days for the low dosage group, and 10 days for the placebo group.
(a) Is this an observational study or an experiment? Explain.
(b) Identify the observational/experimental units in this study.
(c) Identify the explanatory and response variables.
(d) Why did the researchers use a placebo spray, as opposed to just providing no treatment for that group of subjects?

Activity 13-15: Religious Lifetimes

A national study of 3617 adult Americans concluded that people who attend religious services at least once per month live substantially longer than those who do not.
(a) Is this an experiment or an observational study? Explain.
(b) Identify the explanatory and response variables.
(c) Can one conclude from the study that attendance at religious services causes people to live longer? If not, identify a confounding variable and a plausible alternative explanation.
(d) Since the United States contains over 260 million people, is the sample size of 3617 woefully inadequate for describing the population? Explain. [*Hint:* Remember what you learned about the effect of population size in Activity 12-6 on page 260.]

Activity 13-16: Natural Light and Achievement (*cont.*)

Recall the study by the Heschong Mohone group, based near Sacramento, discussed in Activity 1-12 on page 16, which found that students who took their lessons in classrooms with more natural light scored as much as 25 percent higher on standardized tests than other students in the same school district.

(a) Describe the design of a controlled experiment that could determine whether natural lighting in classrooms improves test scores.

(b) Explain briefly why it would be difficult to carry out such an experiment.

(c) John B. Lyons, an Educational Department official, commented that this was "one of the first studies that shows a clear correlation" between daylight and achievement (*The Tribune*, Dec. 21, 1999). How could he reword his conclusion if this study was conducted as a controlled experiment?

Activity 13-17: Memory Experiment

Recall the data collected in the "Preliminaries" section about how many letters you could memorize in 20 seconds. Everyone was to have the same sequence of letters, but they were to be presented in two different formats. One group was to receive JFK-CIA-FBI-USA-SAT-GPA-GRE-IBM-NBA-CPR, and the other was to receive JFKC-IAF-BIU-SASA-TGP-AGR-EIB-MN-BAC-PR. Similar experiments have shown that those receiving the letters already organized in convenient chunks are able to memorize more than those with the less memorable chunks.

(a) Explain why this study is an experiment and not an observational study.

(b) Explain the method of randomization used and why it was important in this study.

(c) Calculate five-number summaries of the memory scores for each group, and draw boxplots to compare the distributions.

(d) Comment on how the centers of the two distributions compare. Does the conjecture about which group would do better seem to be supported?

(e) Comment on how the spreads of the two distributions compare.

(f) Comment on how the shapes of the two distributions compare.

(g) Do you notice any particular granularity in either distribution? If so, describe it and suggest an explanation.

Activity 13-18: SAT Coaching

Suppose that you want to study whether an SAT coaching program actually helps students to score higher on the SATs, so you gather data on random students who have attended the program. Suppose that you find that 95% of the sample scored higher on the SATs after attending the program than before attending the program. Moreover, suppose you calculate that the sample mean of the improvements in SAT scores was a substantial 120 points.

(a) Identify the explanatory and response variables in this study.

(b) Is the SAT coaching study as described a controlled experiment or an observational study?

(c) Explain why you cannot legitimately conclude that the SAT coaching program caused these students to improve on the test. Suggest some other explanation for their improvement.

Activity 13-19: Popcorn Smarts

Suppose that you want to study whether eating SmartFood popcorn actually makes students smarter. You instruct the students in your statistics class to eat two bags of SmartFood popcorn per week between the midterm and final examinations.

(a) Is this an observational study or an experiment? Explain.

(b) Suppose that all of the students do significantly better on the final than on the midterm exam. Would you be able to conclude that the SmartFood is responsible? If so, explain. If not, identify a potentially confounding variable.

(c) Indicate how one could improve upon the design of the SmartFood experiment.

(d) Explain why it would weaken the experiment to allow students to decide for themselves whether to join the popcorn group or the control group.

(e) Identify a better way to assign students to groups in order to balance out potentially confounding variables between the groups.

(f) Would you recommend that all students be told about the purpose of the experiment? Also, should those in the control group be given nothing or another type of popcorn? Explain.

Activity 13-20: Capital Punishment

Suppose that you want to study whether the death penalty acts as a deterrent against homicide, so you compare the homicide rates between states that have the death penalty and states that do not.

(a) Is this a controlled experiment? Explain.

(b) If you find a large difference in the homicide rates between these two types of states, can you attribute that difference to the deterrent effect of the death penalty? Explain.

(c) If you find no difference in the homicide rates between the two types of states, can you conclude that the death penalty has no deterrent effect? Explain.

Activity 13-21: Literature for Parolees

In a recent study 32 convicts were given a course in great works of literature. To be accepted for the program the convicts had to be literate and to convince a judge of their intention to reform. After thirty months of parole only six of these 32 had committed another crime. This group's performance was compared against a similar group of 40 parolees who were not given the literature course; 18 of these 40 had committed a new crime after thirty months.

(a) What proportion of the literature group committed a crime within thirty months of release? What proportion of this group did not commit a crime?

(b) What proportion of the control group committed a crime within thirty months of release? What proportion of this group did not commit a crime?

(c) Which fundamental principles of control does this experiment lack? Comment on how this lack hinders the conclusion of a cause-and-effect relationship in this case.

Activity 13-22: Parking Meter Reliability

In 1998, for her sixth-grade science project, Ellie Lammer randomly selected 50 parking meters along Solano and Shattuck avenues in Berkeley, California. The *Tri-Valley Times* reported that she put in 1 hour's worth of coins in each meter and used three stopwatches to see how long the meters actually lasted. She found that only three meters provided the correct amount of time.

(a) Was this a survey, observational study, or experiment? Explain.

(b) Would you be willing to generalize Ellie's results to all Berkeley parking meters? To all California parking meters?

Activity 13-23: Therapeutic Touch

Practitioners of a controversial medical practice known as "therapeutic touch" claim that they can manipulate a person's "human energy field" in order to provide healing powers, without actually touching the patient's body. Emily Rosa, an 11-year-old in Colorado, recruited 21 practitioners of therapeutic touch in a study she conducted in 1996. Her study consisted of placing a screen between a subject's eyes and hands. For several trials, she hovered her hand over one of theirs, flipping a coin each time to decide which hand to hold hers over. She then asked the practitioner to decide which of their hands Emily's hand was near to see whether the practitioner could detect Emily's energy field.

(a) Was this a survey, observational study, or experiment? Explain.

(b) Explain how you think the principle of randomization was employed in this study.

(c) Was the study double blind?

(d) Would you be willing to generalize the results Emily obtained for these 21 practitioners to all practitioners? Explain.

Activity 13-24: Effectiveness of Gasoline Additive (*cont.*)

In Activity 13-6 on page 288 you flipped a coin to decide which group the car was assigned to.

(a) Explain how we could use the random number table, instead of the coin flips, to assign cars to control and treatment groups.

(b) We suggested blocking on type of car to control that source of variation. Suggest another experimental design that would control for the fact that these cars are of different ages and have different drivers. [*Hint:* Explain how we could further isolate the with and without additive effects on a car's performance.]

Activity 13-25: Prayers and Cell Phones

Select one of the following issues, and describe in detail the design of an experiment to address it. Also explain in detail why observational studies investigating the issue could not settle it.

(i)   Is the use of cell phones by automobile drivers a hazard?

(ii) Does prayer help to reduce suffering?

(iii) Does requiring students to wear uniforms in school lead to better academic achievement?

13

Activity 13-26: Survey of Student Opinion

(a) Find 20 students and ask each person *one* of the following questions:

- Do you think our school should allow public speeches on campus that might incite violence?
- Do you think our school should forbid public speeches on campus that might incite violence?

You should decide which version you will ask based on a coin toss. You should get 10 responses to each question.

(b) Do you think that people responded differently to the two versions? If so, why do you think this happened?

(c) How do you think these results will appear if we combine your results with those obtained by your classmates?

(d) Was this study primarily a survey, an observational study, or a controlled experiment?

Activity 13-27: Survey of Student Opinion (*cont.*)

(a) Find 20 people and ask each person *one* of the following questions:

- Social science research shows that the old saying is true, "Out of sight, out of mind." Do you find this surprising?
- Social science research shows that the old saying is true, "Absence makes the heart grow fonder." Do you find this surprising?

You should decide which version you will ask based on a coin toss. You should get 10 responses to each question.

(b) Do you think that people responded differently to the two versions? If so, why do you think this happened?

(c) Discuss how these differing responses relate to the "placebo effect" discussed in this topic.

(d) Was this study primarily a survey, an observational study, or a controlled experiment?

Unit IV:

Randomness in Data

Topic 14:

PROBABILITY

You have been studying methods for analyzing data, from displaying them graphically to describing them verbally and numerically. Earlier, you saw how to create your own samples of data by sampling from a larger population. Now we turn our attention to drawing inferences about the population based on a sample. As you learned earlier, this inference process is feasible only if you have *randomly* selected the sample from the population. At first glance it might seem that introducing randomness into the process would make it more difficult to draw reliable conclusions. Instead, you will find that randomness produces patterns that allow us to quantify how close the sample will come to the population result. This topic introduces you to the idea of probability and asks you to explore some of its properties.

OBJECTIVES

- To develop an intuitive sense for the notion of **probability** as a long-term property of repeatable phenomena.
- To understand the use of **simulation** for acquiring empirical estimates of probabilities.
- To acquire a sense for whether an outcome of a random process is rare, unlikely, or not uncommon.
- To understand the idea of **equally likely** events and to develop a sense for when that assumption is and is not warranted.
- To continue to investigate the role that **sample size** plays in random phenomena.
- To be able to conduct simulation studies through physical devices, tables of random digits, and your calculator.

303

PRELIMINARIES

1. What does it mean to say that there is a 30% chance of rain tomorrow?

2. If four executives get their cell phones mixed up and decide that each of them will just take one at random, would it be more likely for everyone to get the right phone or for nobody to get the right phone?

3. In the above scenario, do you think it would be more likely for exactly one person to get the right phone or for exactly two people to get the right phone?

4. Suppose that a couple has four children. Which do you think is more likely, that the gender breakdown will be 2-2 or 3-1?

5. Which family would be more likely to have an exact 50/50 gender breakdown, one with four children or one with ten children?

6. Which would you expect to have more days on which more than 60% of its births are girls, a hospital that has 10 births per day or a hospital that has 50 births per day?

7. How many games of solitaire would you expect to play in order to win once?

In-Class Activities

Activity 14-1: Random Babies

Suppose that on one night at a certain hospital, four mothers (named Johnson, Miller, Smith, and Williams) give birth to baby boys. Each mother gives her child a first name alliterative to his last: Jerry Johnson, Marvin Miller, Sam Smith, and Willy Williams. As a very sick joke, the hospital staff decides to return babies to their mothers completely at random.

We want to investigate questions such as, How often will at least one mother get the right baby? How often will every mother get the right baby? What is the most likely outcome? On average, how many mothers will get the right baby?

Since it is clearly not feasible to actually carry out this exercise over and over to investigate what would happen in the long run, we will use simulation instead. **Simulation** is an artificial representation of a random process used to study its long-term properties. We will represent the process of distributing babies to mothers at random by shuffling and dealing cards (representing the babies) to regions on a sheet of paper (representing the mothers).

(a) Take four index cards and one sheet of scratch paper. Write a baby's first name on each index card, and divide the sheet of paper into four areas with a mother's last name written in each area. Shuffle the four index cards well, and then deal them out randomly with one going to each area of the sheet. Finally, turn over the cards to reveal which babies were randomly assigned to which mothers. Record the requested information below:

Did Mrs. Johnson get the right baby?

Did all mothers get the wrong baby?

How many mothers got the right baby?

(b) Repeat the random "dealing" of babies a total of five times, recording in each case the information requested:

repetition #	1	2	3	4	5
Johnson match?					
All wrong?					
# of matches					

(c) Combine your yes/no results sequentially with those of your classmates until you obtain a total of 100 repetitions. For each classmate, record in the table how often in their five repetitions Johnson received the right baby and how often all mothers received the wrong baby. Then calculate the cumulative totals and cumulative proportions:

Student	Cum reps	Johnson matches?			All wrong?		
		Of these 5	Cum tot	Cum prop	Of these 5	Cum tot	Cum prop
1	5						
2	10						
3	15						
4	20						
5	25						
6	30						
7	35						
8	40						
9	45						
10	50						
11	55						
12	60						
13	65						
14	70						
15	75						
16	80						
17	85						
18	90						
19	95						
20	100						

(d) Plot the cumulative relative frequencies as a function of repetitions for both of these variables below:

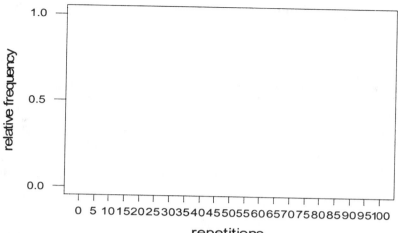

(e) Do the relative frequencies appear to be "settling down" and approaching one particular value? What might you guess that one value is?

The ***probability*** of a random event is the long-run proportion (or relative frequency) of times the event would occur if the random process were repeated over and over under identical conditions. One can *approximate* a probability by ***simulating*** the process a large number of times. Simulation leads to an *empirical* estimate of the probability.

(f) Now combine your results on number of matches with the rest of the class, obtaining a tally of how often each outcome occurred. Record the counts and proportions in the table below:

# of matches	0	1	2	3	4	Total
Count						
Proportion						1.00

(g) In what proportion of these simulated cases did at least one mother get the correct baby?

(h) Based on the simulation results of the class, what is your empirical estimate of the probability of no matches?

(i) Based on the simulation results of the class, what is your empirical estimate of the probability of at least one match?

(j) Explain why an outcome of exactly three matches is impossible.

(k) Is it impossible to get four matches? Would you call it rare? Unlikely?

(l) Would you consider a result of 0 matches, or of 1 match, or of 2 matches, to be unlikely?

Activity 14-2: Random Babies (*cont.*)

> In situations where the outcomes of a random process are **equally**
> **likely**, exact probabilities can be calculated by listing all of the possible
> outcomes and counting the proportion that correspond to the event of
> interest. The listing of all possible outcomes is called the **sample space**.

The sample space for the "random babies" consists of all possible ways to distribute the
four babies to the four mothers. Let 1234 mean that the first baby went to the first
mother, the second baby to the second mother, the third baby to the third mother, and
the fourth baby to the fourth mother. In this scenario all four mothers get the correct
baby. As another example, 1243 would mean that the first two mothers got the right
baby, but the third and fourth mothers had their babies switched. All of the possibilities are listed here:

1234	1243	1324	1342	1423	1432
2134	2143	2314	2341	2413	2431
3124	3142	3214	3241	3412	3421
4123	4132	4213	4231	4312	4321

(a) How many different arrangements are there for returning the four babies to their
mothers?

(b) For each of these arrangements, indicate how many mothers get the correct baby.

1234: 4 *matches*	1243: 2 *matches*	1324 ~ 2	1342 ⟩	1423 ⟩	1432 ∿
2134 2	2143 ○	2314 ⟩	2341 ○	2413 ○	2431 ⟋
3124 ⟋	3142 ○	3214 ∿	3241 ○	3412 ○	3421 ○
4123 ○ ～	4132 ⟋	4213 ⟋	4231 ⟋	4312 ○	4321 ○

(c) In how many arrangements is the number of "matches" equal to exactly:

    4:           3:   2:   1:   0:

(d) Calculate the (exact) probabilities by dividing your answers to (c) by your answer
to (a). Comment on how closely the exact probabilities correspond to the empirical estimates from the simulation recorded in (f) in Activity 14-1.

    4:           3:   2:   1:   0:

14

> An empirical estimate from a simulation generally gets closer to the actual probability as the number of repetitions increases.

Below you will find histograms of the number of matches resulting from simulating this process 100 times, 1000 times, and 10,000 times:

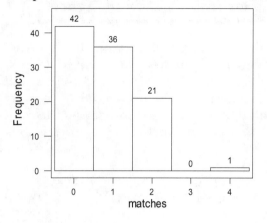

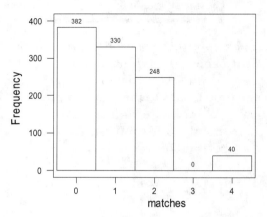

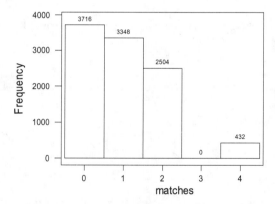

(e) Generally speaking, which of these three simulations produces empirical estimates closest to the actual probabilities?

(f) For your class simulation results summarized in (f) of Activity 14-1, calculate the average (mean) number of matches per repetition of the process by multiplying each outcome by the number of occurrences, summing the products, and then dividing by the total number of repetitions.

> The long-run average value achieved by a numerical random process is called its **expected value**. To calculate this expected value from the (exact) probability distribution, multiply each outcome by its probability, and then add these up over all of the possible outcomes.

(g) Calculate the *expected* number of matches from the (exact) probability distribution, and compare that to the average number of matches from the simulated data.

Activity 14-3: Weighted Coins

Suppose that six coins are weighted so that the probability of landing heads is not necessarily equal to one-half. Specifically, suppose that the probabilities of landing heads for the six coins are

coin A: 1/4   coin B: 1/3   coin C: 1/2
coin D: 3/4   coin E: 4/5   coin F: 99/100.

Suppose that in an effort to determine which coin is which, you flip each coin five times, obtaining the following results:

repetition	1st coin	2nd coin	3rd coin	4th coin	5th coin	6th coin
1	H	H	T	H	H	T
2	H	H	T	H	H	H
3	T	H	H	H	H	T
4	H	H	T	T	H	T
5	H	H	T	H	H	T
relative frequency						
coin guess (letter)						

(a) Fill in the "relative frequency" row with the proportion of heads obtained in these five tosses for each coin. Then fill in the bottom row of the table with guesses for which outcomes go with which coins. Use each of the six probabilities given above once and only once.

(b) Are you confident that all of your guesses are correct? Explain.

(c) Now suppose that you flip each coin five more times, obtaining the relative frequencies given in the table below. With this additional information, again make guesses (in the bottom row of the table) for which probabilities go with which coins.

n = 10	1st coin	2nd coin	3rd coin	4th coin	5th coin	6th coin
relative frequency	0.70	0.90	0.20	0.80	1.00	0.20
coin guess						

(d) Now suppose that you flip each coin 15 more times and then another 25 times, obtaining relative frequencies as shown in the tables below. In each instance supply your guess for which probabilities go with which coins.

n = 25	1st coin	2nd coin	3rd coin	4th coin	5th coin	6th coin
rel freq	0.56	0.88	0.28	0.88	1.00	0.20
coin guess						

n = 50	1st coin	2nd coin	3rd coin	4th coin	5th coin	6th coin
rel freq	0.58	0.92	0.26	0.78	1.00	0.32
coin guess						

(e) After a total of 50 flips, are you reasonably confident that your guesses are correct? Explain.

(f) The graph below shows the relative frequencies for the six coins changing as they are flipped more and more often:

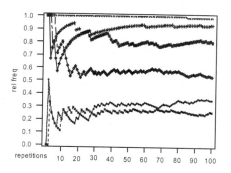

Comment on what this graph reveals about probability as a concept about the long-term and not the short-term behavior of random processes.

Activity 14-4: Boy and Girl Births

Suppose that a couple has four children. Approximate the probabilities of the various gender breakdown possibilities by conducting a simulation. While about 51% of all children born are male (according to the *National Vital Statistics Report*), for simplicity we will assume that the probability of a boy is .50, independently from child to child.

(a) Use a table of random digits, letting an even digit represent a girl and an odd represent a boy, to simulate the genders of one family with four children. Record the random digits and the corresponding genders in the table:

Random Number Table Line Number:

Family 1					
Child number	1	2	3	4	# girls
Random digit					
Gender					

(b) Repeat this for a total of five families, recording below the number of girls in each family:

Family #	1	2	3	4	5
# girls					

(c) Combine your results with the rest of the class, recording below the number of simulated families having the indicated number of girls in the table below. Then divide by the total number of simulated families to obtain empirical probabilities of the possible outcomes:

# of girls in family	0	1	2	3	4
# of simulated families					
Empirical probability					

In situations like this where each trial has only two possible outcomes and the probabilities remain the same on each trial independently from trial to trial, exact probabilities can be calculated from the **binomial distribution**.

The exact probabilities in this situation are:

# of girls in family	0	1	2	3	4
(Exact) probability	.0625	.25	.375	.25	.0625

(d) Comment on how closely your empirical estimates of the probabilities in (c) match the exact probabilities.

(e) According to the exact probabilities, is it more likely for a family to have two children of the same gender or 3 children of the same gender? Explain why this makes sense.

(f) Do you expect the likelihood of an exact 50/50 gender split to increase or to decrease with larger families? Explain briefly.

(g) Use your calculator to simulate the gender breakdown for 500 families with *four* children and then for 500 families with *ten* children (*Note:* This activity will go faster if you choose a partner and have one calculator run the simulation with four children and the other calculator run the simulation with 10 children. WARNING:

14

This simulation requires a significant amount of memory. You may want to remove unnecessary lists from your calculator before running this simulation):

- Go to the MATH PRB directory and select randBin(.
- The first entry is the sample size (4 or 10 depending on which simulation you are running).
- Press comma, and then the second entry is the probability of success (here, .5).
- Press comma, and then the last entry is the number of observations you want to simulate (here, 500 families). You should see `randBin(4, .5, 500)` in your homescreen.
- Store these data into a named list so that you can view it easily within the Stat List Editor. Press ENTER.

What proportion of the 4-children families have 2 boys and 2 girls? What proportion have of the 10-children families have 5 boys and 5 girls? (You may want to create an appropriate histogram to help answer these questions.) Was your expectation in (f) confirmed or refuted?

Activity 14-5: Hospital Births

Suppose that a region has two hospitals. Hospital A has 10 births per day, while Hospital B has 50 births per day. The following histograms display the results of a simulated year of births, with the variable being the number of girls born per day:

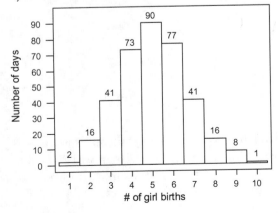
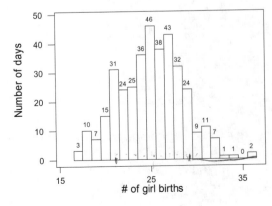

(a) In about what proportion of the 365 days did hospital A observe an equal count of girls and boys (a 5/5 split)? In about what proportion of the 365 days did hospital B have a 25/25 split of girls and boys? Does the larger or the smaller hospital have more days with an exact 50/50 gender split?

(b) Which of the two hospitals has more days on which 60% or more of the births are girls? Is your prediction from the "Preliminaries" section supported?

(c) Which of the two hospitals has more days on which between 41% and 59% of the births are girls?

(d) Explain what the previous questions reveal about the effect of sample size on the distribution of sample proportions.

> This activity reveals that while a larger sample size makes it less likely to get an exact 50/50 split in the observed counts, the probability of getting a sample proportion close to 1/2 increases with a larger sample. Consequently, we are less likely to obtain a sample proportion far away from the long-term probability of 1/2. Also note that a larger sample produces a probability distribution that is quite symmetric and mound-shaped.

WRAP-UP

This topic has initiated your study of randomness by introducing you to the concept of probability. You have learned that probability is a long-run property of events, and you have studied probability by conducting simulations. You have carried out these simulations using physical devices such as index cards, using a table of random digits, and using your calculator. You have also studied probability more theoretically, through the notion of equally likely events, sample space, and expected value.

HOMEWORK ACTIVITIES

Activity 14-6: Random Cell Phones

Suppose that three executives bump into each other on an elevator and drop their identical cellular phones as the doors are closing, leaving them with no alternative but to pick up a phone at random.

(a) Describe in detail how you could conduct a simulation of this situation to produce empirical estimates of the probabilities involved.

(b) List all of the possible outcomes in the sample space for this situation. [*Hint:* Compare it to the "random babies" scenario.]

(c) Use the sample space to determine the probability that:
- Nobody gets the correct phone
- Exactly one person gets the correct phone
- Exactly two people get the correct phone
- All three people get the correct phone
- At least one person gets the correct phone

(d) If this situation were to befall a group of five people, one could calculate the probability that nobody gets the correct phone as 11/30. Explain in your own words what this probability means about the likelihood of nobody getting the correct phone. Be sure to include the long-run interpretation of probability in your answer and to relate your response to the context.

Activity 14-7: Equally Likely Events

Indicate which of the following outcomes are equally likely. Provide a brief justification in each case.
(a) Whether a fair die lands on 1, 2, 3, 4, 5, or 6
(b) The sum of two fair dice landing on 2, 3, 4, 5, 6, 7, 8, 9, 10, 11, or 12
(c) A fair coin landing Heads or Tails when tossed
(d) A fair coin landing Heads or Tails when spun on its side
(e) A tennis racquet landing with the label "up" or "down" when spun on its end
(f) Your grade in this course being A, B, C, D, or F
(g) Whether or not California experiences a catastrophic earthquake within the next year
(h) Whether or not your waitress correctly brings you the meal you ordered in a restaurant
(i) Whether or not there is intelligent life on Mars
(j) Whether or not a woman will be elected President in the next U.S. election
(k) Whether or not a woman will be elected President before the year 2100
(l) Colors of Reese's Pieces candies: orange, yellow, and brown

Activity 14-8: Interpreting Probabilities

Explain in your own words what is meant by the following statements. Be sure to include the long-run interpretation of probability in your answer and to relate your response to the context.
(a) There is a .3 probability of rain tomorrow.
(b) Your probability of winning at this lottery game is 1/1000.
(c) The probability that a five-card poker hand contains "four of a kind" is .00024.
(d) The probability of obtaining a red M&M candy is .2.
(e) There is a 70% probability of having a white Christmas in Minneapolis.
(f) There is approximately a 50% probability that a brand new company will fail within the first 3 years of business.

Activity 14-9: Racquet Spinning

Tennis players often spin a racquet on its end and observe whether it lands with the handle's label facing "up" or "down" as a random mechanism for determining who serves first. To estimate the probability that a particular racquet would land "up" and to see whether it seems to be about 50%, Author A spun his racquet 100 times. A graph of the relative frequency of "up" results after each spin follows:

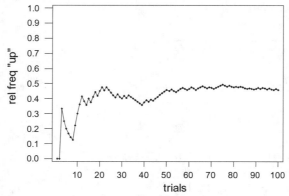

(a) Explain in your own words what the phrase "probability of the racquet landing up" means and how this graph relates to that.

(b) As well as you can from the graph, indicate the empirical estimate of this probability after 10 spins, after 20 spins, after 40 spins, and after 100 spins.

(c) Does this sample of spins appear to provide evidence that the racquet is not equally likely to land "up" or "down"? Explain.

(d) Spin your own racquet 100 times. How does your estimate of the probability of this racquet landing up compare?

Activity 14-10: Committee Assignments

A college professor found herself assigned to a committee of six people, composed of four men and two women. The committee had to select two officers to carry out the majority of its administrative work, and it turned out that both of the women were selected. The professor wondered whether this constituted evidence of subtle discrimination, so she considered how unlikely such an event would be if the two officers had been chosen at random from the six committee members.

(a) Suppose that the six committee members are named Alice, Bonnie, Carl, Danny, Evan, and Frank. Describe how you might use a fair, six-sided die to simulate the random selection of two people to be officers for this committee. [*Hint:* Be careful to consider that the officers have to be different people.]

(b) Use a die to carry out this simulation for a total of 25 repetitions. In how many of these 25 repetitions did the officers end up being the two women? What is your empirical estimate of that probability?

(c) To pursue a theoretical analysis, begin by listing all possible pairs of officers that could be chosen from these six people.

(d) How many pairs are possible? How many of them consist of two women?

(e) What is the theoretical probability of obtaining two women if one randomly chooses two people from these committee members? Would you say that this outcome is impossible? Rare? Uncommon? Likely?

(f) If the process of randomly selecting two people from these six were repeated over and over, in the long run what percentage of the time would two men be selected? Would such an outcome be a surprising result from a random selection?

(g) Repeat (f) for the outcome of one man and one woman.

(h) If the two officers were chosen at random, what would be the most likely gender breakdown, two men, two women, or one of each?

(i) Return to the results of your simulation, and calculate the mean number of men in those 25 simulated pairs of officers.

(j) Use your listing of possible outcomes in (c) to calculate the theoretical expected number of men among the two officers. How close does your simulated estimate come?

Activity 14-11: Simulating World Series

Suppose that two baseball teams, the Domestic Shorthairs and the Cache Cows, are to play each other in the World Series and that the Shorthairs have a .6 probability of winning any one game, regardless of the outcomes of preceding games. The winner of the best-of-seven series is the first team to win four games. Use simulation to estimate the probability that the Shorthairs will win the Series through the following steps:

(a) Enter the table of random digits. Let 1–6 represent a win for the Shorthairs and 7–0 a win for the Cows. Read digits until one team has won four games; this team is the winner of your first simulated series. (Please report the line numbers that you use.)

(b) Repeat this for a total of 40 simulated series. In each case, record the team that won and also how many games the series entailed.

(c) What proportion of the 40 simulated series did the Shorthairs win?

(d) Construct a visual display of the distribution of games played per series. Comment on the distribution. Also calculate the mean number of games per simulated series.

(e) Repeat (a)–(d) for a best-of-three series for which the winner is the first to win two games.

(f) Which length of series gives the greater advantage to the stronger team (Shorthairs, in this case)? Explain in your own words why this makes sense.

Activity 14-12: Hospital Births (*cont.*)

About 25% of all babies born in Texas are Hispanic (*1999 World Almanac and Book of Facts*). Again consider a city with two hospitals, and suppose that Hospital A has 10 births per day, while Hospital B has 50 births per day.

(a) Use your calculator to simulate 365 days of births at each of these two hospitals. Let "number of Hispanic children" be the variable, and simulate the binomial distribution with a "success" probability of .25. [*TI hint*: Use the randBin (command.]

(b) Which hospital has more days on which more than 40% of the births were Hispanic?

(c) Which hospital has more days on which between 15% and 35% of the births were Hispanic?

(d) Which hospital has more days on which fewer than 40% of the births were Hispanic?

(e) Explain how you could have answered (b)–(d) without the benefit of the simulation, based on what you learned in this topic about randomness and sample size.

Activity 14-13: Runs and "Hot" Streaks

Reconsider the setting of Activity 14-9, involving the spinning of a tennis racquet. Notice from the graph that among the first ten spins occurred a string (or "run") of five consecutive "down" results.

(a) In ten tosses of a fair coin, would you be very surprised to observe a string of five or more consecutive heads or tails? Take a guess as to how often such a string would appear in, say, 1000 repetitions of ten tosses.

(b) Use your calculator (randBin) to simulate 100 repetitions of ten tosses of a fair coin from the binomial distribution. For each of the 100 repetitions, record the length of the longest streak (run) of either heads or tails.

(c) Create a histogram of the distribution of the longest streaks in these 100 simulations.

(d) In how many of these 100 simulations of ten coin tosses did a streak of five or more appear? Would you characterize this event as very surprising?

(e) What was the most common length of the longest streak?

(f) Calculate and report the mean and median of the distribution of these 100 longest streaks.

(g) Would it be very surprising to obtain a series of ten coin tosses for which the longest streak is one? Explain, based both on your simulation results and on the ways in which such an event could happen.

Activity 14-14: Simulating Solitaire

In order to simulate wins and losses at the card game solitaire, you must first make an assumption about (or an estimate of) the probability of winning a single game. Enamored of the game's program on his new computer, author A set out to estimate this probability and won 25 games while losing 192.

(a) What proportion of games played did he win?

For ease of simulation, let us simplify this a bit by assuming that the probability of this person winning a game of solitaire is 1/9.

(b) Describe how you might use a table of random digits to simulate repetitive plays of games of solitaire with this success probability. [*Hint:* Use only one digit per simulated game.]

(c) Use the table of random digits to simulate the playing of games of solitaire until the first win is achieved. Record the number of games needed (including the win) as your first observation. Then repeat this process a total of twenty-five times.

(d) Create a dotplot of the distribution of the number of games needed to obtain the first victory for these twenty-five repetitions.

(e) Calculate the mean and median of these values.

(f) Based on your simulation, about how many games of solitaire can this person expect to play before winning for the first time?

(g) Based on your simulation, about how many games must this person play to have at least a 50% chance of winning at least once?

Activity 14-15: Simulating Solitaire (*cont.*)

Refer to the simulation study of Activity 14-14 on page 322. Anxious to outdo author A, author B played 444 games of solitaire and won 74.

(a) What proportion of games played did she win?

(b) Assuming the winning probability to be 1/6, simulate the number of games needed to obtain a win for the first time. Then repeat this for a total of 25 wins. [You may use your calculator, a random digit table, or a physical device such as a die to do the simulation.]

(c) Create a dotplot of the distribution of the number of games needed to obtain the first victory for these 25 repetitions.

(d) Calculate the mean and median of these values.

(e) Based on your simulation, about how many games of solitaire can this person expect to play before winning for the first time?

(f) Based on your simulation, about how many games must this person play to have at least a 50% chance of winning at least once?

(g) Describe how your findings differ when the win probability is 1/6 as opposed to 1/9.

NORMAL DISTRIBUTIONS

You began studying randomness and probability in the previous topic. Toward the end of that topic you saw that a mound-shaped distribution described the outcomes of a particular variable, such as counts and proportions. Such a pattern arises often enough that it has been very extensively studied mathematically. In this topic you will investigate mathematical models known as **normal distributions**, which describe this pattern of variation very accurately. You will learn how to use normal distributions to calculate **probabilities** of interest in a variety of contexts.

- To become familiar with the idea of using **normal curves** as mathematical models for approximating certain distributions.
- To discover how to use a **table of standard normal probabilities** to perform calculations pertaining to any normal distribution.
- To learn how to use your calculator to calculate probabilities from a normal distribution.
- To assess the usefulness of a normal model by comparing its predictions to observed data.
- To develop an intuitive sense for how normal distributions relate to the questions of statistical inference.
- To acquire familiarity with judging what type of population a sample may have come from.

15

1. How much do you think a typical baby weighs at birth?

2. About what proportion of babies would you guess weigh less than 6 pounds? More than 10 pounds?

3. Suppose that Professors Fisher and Savage assign A's to students scoring above 90 and F's to those scoring below 60 on the final exam. Suppose further that the distribution of scores on Professor Fisher's final has mean 74 and standard deviation 7, while the distribution of scores on Professor Savage's final is approximately normal with mean 78 and standard deviation 18. Which professor would you expect to assign more A's?

4. Which of Professors Fisher and Savage would you expect to assign more F's?

5. From which of Professors Fisher and Savage would you personally rather take a course (all other factors being equal)?

6. How many days do you think a human pregnancy lasts on the average?

7. About what proportion of human pregnancies would you guess last less than 8 months? more than 10 months?

8. What was Forrest Gump's IQ?

IN-CLASS ACTIVITIES

Activity 15-1: Placement Scores and Hospital Births

The following histograms display the distributions of the placement exam scores from Activity 3-8 on page 57 and the simulated hospital births from Activity 14-5 on page 316:

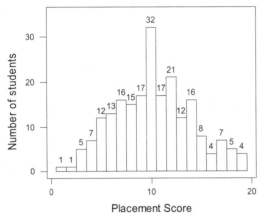 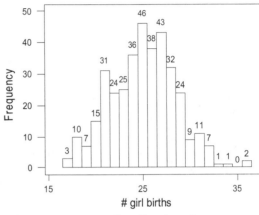

(a) What similarities do you notice about the shapes of these distributions?

(b) Draw a sketch of a smooth curve ("smooth" meaning having no breaks or jagged edges) that seems to approximate the general shape apparent in the two histograms above.

Data that display the general shape seen in the examples above occur very frequently. Theoretical mathematical models used to approximate such distributions are called **normal distributions**. Every normal distribution shares three distinguishing characteristics. All are symmetric, have a single peak at their center, and follow a bell-shaped curve. Two things distinguish one normal distribution from another: the **mean** and its **standard deviation**. The mean μ determines where its center is; the peak of a normal curve occurs at its mean, which is also its point of symmetry. The standard deviation σ indicates how spread out the distribution is.

[Note: We reserve the symbols $\bar{x}$ and s to refer to the mean and standard deviation computed from sample data.] The distance between the mean μ and the points where the curvature changes is equal to the standard deviation σ. The following sketch displays this relationship:

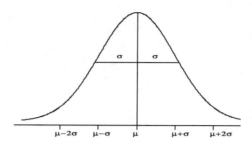

(c) The sketch below contains three normal curves; think of them as approximating the distribution of exam scores for three different classes. One (call it A) has a mean of 70 and a standard deviation of 5; another (call it B) has a mean of 70 and a standard deviation of 10; the third (call it C) has a mean of 50 and a standard deviation of 10. Identify which is which by labeling each curve with its appropriate letter.

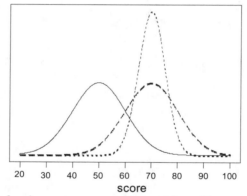

As a consequence of sharing a common shape, the placement scores and simulated births share other features such as the empirical rule discussed in Topic 5. The following table reports the proportions of observations falling within one, two, and three standard deviations of the mean for each data set.

	Placement exam scores	# of girl births	Empirical rule
Mean, standard deviation	10.221, 3.859	25.060, 3.472	
Within one std. dev. of mean	146/213 (68.5%)	244/365 (68.8%)	68%
Within two std. devs. of mean	202/213 (94.8%)	348/365 (95.3%)	95%
Within three std. devs. of mean	213/213 (100.0%)	363/365 (99.5%)	99.7%

(d) Are these proportions quite close to each other and to the predictions of the empirical rule that you studied in Topic 5?

Recall from Topic 5 the idea of standardization to produce a *z*-score, which indicates a value's relative position in the data set.

$$z\text{-score} = \frac{x - \mu}{\sigma} = \frac{\text{observation} - \text{mean}}{\text{standard deviation}}.$$

(e) Find the *z*-score for the value 8.5 in the distribution of placement exam scores ($\mu = 10.221, \sigma = 3.859$). You can round the *z*-score to the nearest hundredth.

(f) Find the *z*-score for the value 23.5 in the distribution of the number of girl births ($\mu = 25.060, \sigma = 3.472$).

(g) What is true about these two *z*-scores?

(h) How would you interpret these *z*-scores?

(i) Use the histogram to determine the proportion of observations falling below 8.5 in the placement exam data.

(j) Use the other histogram to find the proportion of observations falling below 23.5 in the simulated births data. Are these proportions fairly close?

The closeness of these percentages indicates that to find the proportion of data falling in a given region for normal distributions, all we need to determine is the *z*-score. Values with the same *z*-score will have the same percentage of observations lying below them, for any normal distribution. Thus, instead of finding percentages for all normal distributions, we need only the percentages corresponding to these *z*-scores.

(k) If we randomly select one of the 213 students who took this placement exam, what is the probability that his or her placement score will be below 8.5? [*Hint*: Refer back to your answer to (i).]

> The probability of a randomly selected observation falling in a certain interval is equivalent to the proportion of the population's observations falling in that interval. Since the total area under the curve of a normal distribution is 1, this probability can be calculated by finding the area under the normal curve for that interval.

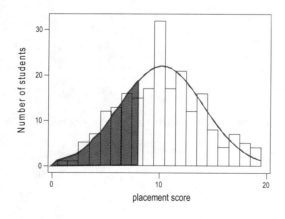

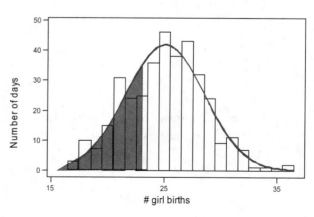

To find areas under the normal curve, you can use either your calculator or tables. Table II: on page 589 (the **standard normal probability table**) reports the area to the left of a given z-score under the normal curve. Enter the table at the row corresponding to the first two digits in the z-score. Then move to the column corresponding to the hundredths digit.

(l) Use Table II to look up the area to the left of $z = -0.45$ under the normal curve. [*Hint:* Find −0.4 along the left and the .05 along the top.] Record this area below, corresponding to the probability of a placement score falling below 8.5.

(m) Is this value reasonably close to your answers for the proportions of the placement and birth data less than their z-scores of −0.45?

We use Z to denote the standard normal distribution. The notation $\Pr(a < Z < b)$ denotes the probability lying between the values a and b, calculated as the area under the standard normal curve in that region. The notation $\Pr(Z \leq z)$ denotes the area to the left of a particular value z, while $\Pr(Z \geq z)$ refers to the area above a particular z value.

Activity 15-2: Birth Weights

Birth weights of babies in the United States can be modeled by a normal distribution with mean 3250 grams and standard deviation 550 grams. Those weighing less than 2500 grams are considered to be of low birth weight.

(a) A sketch of this normal distribution appears below. Shade in the region whose area corresponds to the probability that a baby will have a low birth weight.

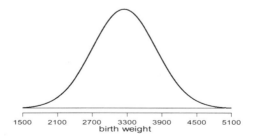

(b) Based on this shaded region (remembering that the total area under the normal curve is 1), make an educated guess as to the proportion of babies born with a low birth weight.

(c) Calculate the *z*-score for a birth weight of 2500 grams.

(d) Look this *z*-score up in Table II to determine the proportion of babies born with a low birth weight. In other words, find Pr($Z<z$), where *z* represents the *z*-score calculated in (c).

(e) You can use your calculator to find the proportion of observations below a particular value in a normal model:
   - Open the DISTR (short for distributions) menu by pressing 2nd   VARS . Select normalcdf(and press ENTER . This command allows you to calculate Pr($a<Z<b$).
   - The first entry is the lower limit of the interval of interest, *a*. In this example, there is no lower limit, so you will enter a very large negative number by typing (-)1 EE 99.
   - Press the comma key. The second entry is the upper limit of the interval of interest, *b*. In this case, *b*=2500, so type 2500.
   - Press the comma key, and then enter the mean and standard deviation of the normal distribution you are working with:

$$\texttt{normalcdf(-1EE99, 2500, 3250, 550)}$$

Note: If you do not specify a mean and standard deviation, they are assumed to be 0 and 1, respectively (the standard normal distribution).

Verify (to within rounding discrepancies) your answer to (d).

(f) What proportion of babies would the normal distribution predict as weighing more than 10 pounds (4536 grams) at birth? [*Hints:* Always start with a sketch of the normal curve and the area you are looking for. Then use your calculator or calculate the *z*-score and use Table II. To use the calculator, enter 4536 as the lower endpoint and a very large positive number (1EE 99) as the upper endpoint.]

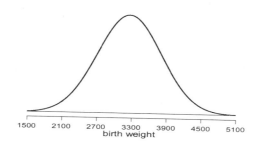

(g) Describe two different ways that you could have used your calculator/Table II to answer (f).

(h) Determine the probability that a randomly selected baby weighs between 3000 and 4000 grams at birth. [*Hint:* Start with a sketch, and then use your calculator or subtract the proportions obtained from Table II for the two relevant z-scores.]

(i) Data from the *National Vital Statistics Report* indicate that there were 3,880,894 births in the United States in 1997. A total of 291,154 babies were of low birth weight, while 2,552,852 babies weighed between 3000 and 4000 grams. Calculate the observed proportions in each of these two groups, and comment on how well the normal calculations in (d) and (h) approximate these values.

(j) How little would a baby have to weigh to be among the lightest 2.5% of all newborns? [*Hints*: Start with a sketch. To use your calculator, select invNorm (from the DISTR menu. Specify the probability below, the mean, and the standard devaition of the normal distribution of interest (or do not specify a mean or standard deviation to assume the standard normal distribution):

$$\texttt{invNorm(.025, 3250, 550)}$$

To use Table II, you will need to do so "in reverse": Look up the probability in the middle of the table and read backwards to find the relevant z-score. Then you will have to unconvert the z-score back to the birth weight scale.]

(k) How much would a baby have to weigh to be among the heaviest 10% of all newborns?

Activity 15-3: Matching Samples to Density Curves

While normal distributions are the most common, they are not the only kind of theoretical probability model. Any curve under which the total area is one and for which areas correspond to probabilities represents a probability model. Such curves are called **density curves**.

The following graphs show four very different density curves that could model a population of exam scores:

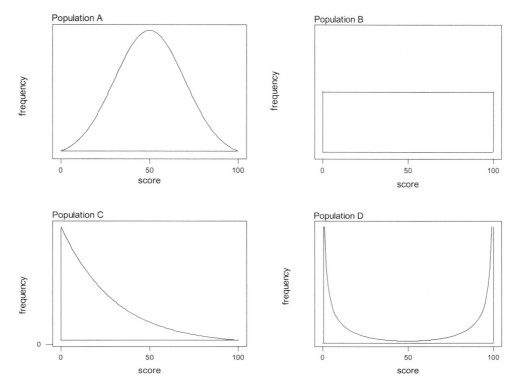

(a) The following four dotplots display random samples of size 100 from these populations. Your task is to match up each sample with the population from which it was drawn.

- Sample 1:

- Sample 2:

- Sample 3:

- Sample 4:

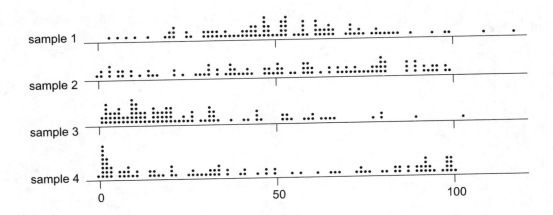

(b) The following four dotplots display random samples of size 10 from these populations. Your task again is to match up each sample with the population from which
it was drawn.

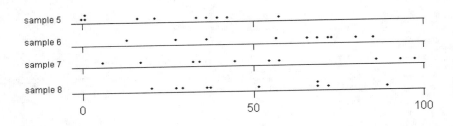

(c) With which sample size is it easier to discern the shape of the population from
which the sample was drawn?

Especially with small sample sizes, sample data from normal populations may not
look very normal and may be hard to distinguish from sample data from other
shapes of populations.

WRAP-UP

This topic has introduced you to the most important mathematical model in all of statistics: the **normal distribution**. You have discovered how z-scores provide the key to using a **table of standard normal probabilities** to perform calculations related to normal distributions. You have also compared predictions from a normal model to observed data as a means of assessing the usefulness of the model in a given situation. You have seen that the model predictions are not exact, but are close, especially with larger samples.

The next two topics will reveal how the normal distributions describe the pattern of variation that arises when one repeatedly takes samples from a population. In the next topic you will explore how sample proportions vary, while the variation of sample means will occupy your attention in Topic 17. These topics point toward the key role of normal distributions in the most important theoretical result in all of statistics: the **Central Limit Theorem**.

HOMEWORK ACTIVITIES

Activity 15-4: Normal Curves

For each of the following normal curves, identify (as accurately as you can from the graph) the mean μ and standard deviation σ of the distribution.

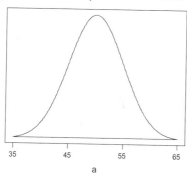

a

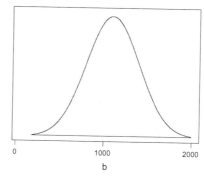

b

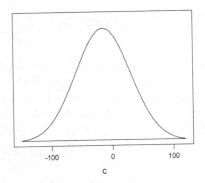

c

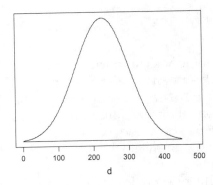

d

Activity 15-5: Pregnancy Durations

Data from the *National Vital Statistics Report* reveal that the distribution of the duration of human pregnancies (i.e., the number of days between conception and birth) is approximately normal with mean $\mu = 270$ and standard deviation $\sigma = 15$. Use this normal model to determine the proportion of all pregnancies that come to term in:

(a) less than 244 days (which is about 8 months).

(b) more than 275 days (which is about 9 months).

(c) over 300 days.

(d) between 260 and 280 days.

(e) Data from the *National Vital Statistics Report* reveal that of 3,880,894 births in the United States in 1997, the number of pregnancies that resulted in a preterm delivery, defined as 36 or fewer weeks since conception, was 436,600. Compare this observed proportion to the prediction from the normal model.

Activity 15-6: Professors' Grades

Suppose that you are deciding whether to take Professor Fisher's class or Professor Savage's next semester. You happen to know that each professor gives A's to those scoring above 90 on the final exam and F's to those scoring below 60. You also happen to know that the distribution of scores on Professor Fisher's final is approximately normal with mean 74 and standard deviation 7 and that the distribution of scores on Professor Savage's final is approximately normal with mean 78 and standard deviation 18.

(a) Produce a sketch of both teachers' grade distributions (on the same scale, as in Activity 15-1 on page 327, part (c)).

(b) Which professor gives the higher proportion of A's? Show the appropriate calculations to support your answer.

(c) Which professor gives the higher proportion of F's? Show the appropriate calculations to support your answer.

(d) Suppose that Professor DeGroot has a policy of giving A's to the top 10% of the scores on his final, regardless of the actual scores. If the distribution of scores on his final turns out to be normal with mean 69 and standard deviation 9, how high does your score have to be to earn an A?

Activity 15-7: Professors' Grades (*cont.*)

Suppose that Professors Wells and Zeddes have final exam scores that are approximately normally distributed with mean 75. The standard deviation of Wells' scores is 10, and that of Zeddes' scores is 5.

(a) With which professor is a score of 90 more impressive? Support your answer with appropriate probability calculations and with a sketch.

(b) With which professor is a score of 60 more discouraging? Again support your answer with appropriate probability calculations and with a sketch.

Activity 15-8: IQ Scores

Suppose that the IQ scores of students at a certain college follow a normal distribution with mean 115 and standard deviation 12.

(a) Draw a sketch of this distribution. Be sure to put some labels along the horizontal and vertical axes.

(b) Shade in the area corresponding to the proportion of students with an IQ below 100. Based on this shaded region, make an educated guess as to this proportion.

(c) Use the normal model to determine the proportion of students with an IQ score below 100.

(d) Find the proportion of these undergraduates having IQs greater than 130.

(e) Find the proportion of these undergraduates having IQs between 110 and 130.

(f) With his IQ of 75, Forrest Gump would have a higher IQ than what percentage of these undergraduates?

(g) Determine how high must one's IQ must be to be in the top 1% of all IQs at this college.

Activity 15-9: Candy Bar Weights

Suppose that the wrapper of a certain candy bar lists its weight as 2.13 ounces. Naturally, the weights of individual bars vary somewhat. Suppose that the weights of these candy bars vary according to a normal distribution with mean $\mu = 2.2$ ounces and standard deviation $\sigma = .04$ ounces.

(a) What proportion of candy bars weigh less than the advertised weight?

(b) What proportion of candy bars weigh more than 2.25 ounces?

(c) What proportion of candy bars weigh between 2.2 and 2.3 ounces?

(d) If the manufacturer wants to adjust the production process so that only 1 candy bar in 1000 weighs less than the advertised weight, what should the mean of the actual weights be (assuming that the standard deviation of the weights remains 0.04 ounces)?

(e) If the manufacturer wants to adjust the production process so that the mean remains at 2.2 ounces but only 1 candy bar in 1000 weighs less than the advertised weight, how small does the standard deviation of the weights need to be?

(f) If the manufacturer wants to adjust the production process so that the mean is reduced to 2.15 ounces but only 1 candy bar in 1000 weighs less than the advertised weight, how small does the standard deviation of the weights need to be?

Activity 15-10: SATs and ACTs (*cont.*)

Refer to the information presented in Activity 5-6 on page 99 about SAT and ACT tests. Recall that among the college's applicants who take the SAT, scores have a mean of 896 and a standard deviation of 174. Further recall that among the college's applicants who take the ACT, scores have a mean of 20.6 and a standard deviation of 5.2. Consider again applicant Bobby, who scored 1080 on the SAT, and applicant Kathy, who scored 28 on the ACT.

(a) Assuming that SAT scores of the college's applicants are normally distributed, what proportion of applicants score higher than Bobby on the SATs?

(b) Assuming that ACT scores of the college's applicants are normally distributed, what proportion of applicants score higher than Kathy on the ACTs?

(c) Which applicant seems to be the stronger in terms of standardized test performance? Compare your answer to that of question (g) of Activity 5-6.

Activity 15-11: Heights

Heights of American men aged 20 to 29 approximately follow a normal distribution with mean 70 inches and standard deviation 3 inches. The same shape and spread hold for heights of women in the same age group, but with a mean of 65 inches.

(a) Use this normal model to determine the probability that an American man in this age group will be shorter than 5'6" tall.

(b) According to this normal model, what proportion of American men aged 20–29 are 6' or taller?

(c) How tall must a man be to be among the tallest 10%?

(d) Answer (a)–(c) for American women in this age group.

(e) Sample data from the National Center for Health Statistics indicate that 11.7% of a sample of men and 74.0% of a sample of women are shorter than 5'6". Are these results generally consistent with your calculations?

(f) Sample data from the National Center for Health Statistics indicate that 29.9% of a sample of men and 0.5% of a sample of women are 6' or taller. Are these results generally consistent with your calculations?

Activity 15-12: Weights

Sample data from the National Center for Health Statistics reveal that weights of American men aged 20–29 have a mean of about 175 pounds and a standard deviation of about 35 pounds. For women the mean is about 140 pounds and the standard deviation is about 30 pounds.

(a) If these distributions are roughly normal, what percentage of men would you expect to weigh less than 150 pounds? Less than 200 pounds? Less than 250 pounds?

(b) Answer (a) for women.

(c) Sample data from the National Center for Health Statistics reveal that the observed percentages in these ranges are 29.0%, 82.1%, and 96.2% for men, compared to 70.4%, 92.5%, and 99.0% for women. How well does the normal model predict these percentages?

Activity 15-13: Coin Ages

A person with too much time on his hands collected 1000 pennies that came into his possession in 1999 and calculated the age (as of 1999) of each. The distribution has mean 12.264 years and standard deviation 9.613 years. Knowing these summary statistics but without seeing the distribution, can you comment on whether the normal distribution is likely to provide a reasonable model for these penny ages? Explain.

Activity 15-14: Empirical Rule

(a) Use the table of standard normal probabilities (Table II) or your calculator to find the proportion of the normal curve that falls within *one* standard deviation of its mean, i.e., between z-scores of -1 and $+1$.

(b) Repeat (a) for *two* standard deviations.

(c) Repeat (a) for *three* standard deviations. These calculations provide the theoretical basis for the empirical rule.

(d) Determine the z-score such that the area under a normal curve between it and its negative is .5000. Then subtract these two z-scores to find the interquartile range of a normal distribution.

(e) Recall that the outlier rule of Topic 6 identified as an outlier any observation falling more than $1.5 \times IQR$ away from its nearer quartile. Use Table II or your calculator and your answer to (d) to determine the z-scores for outliers and then the probability that an observation from a normal distribution will be classified as an outlier.

15

Activity 15-15: Critical Values

(a) Use the table of standard normal probabilities (Table II) or your calculator to find the z-score such that the area to its right under a normal curve equals .10 (or as close as possible from the table). In other words, find the value z^* such that $\Pr(Z > z^*) = .10$

(b) Repeat (a) for an area of .05.

(c) Repeat (a) for an area of .025.

(d) Repeat (a) for an area of .01.

(e) Repeat (a) for an area of .005. These values are called the *critical values* of the normal distribution.

Activity 15-16: Random Normal Data

The following dotplots display distributions of samples of hypothetical exam scores. Two of these samples were drawn from populations that are normally distributed. Identify the *three* samples that do *not* come from normal populations, and explain in each case why the sample is clearly nonnormal.

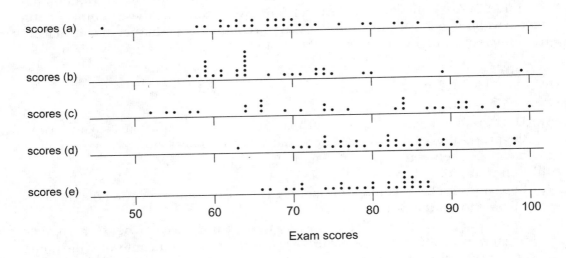

Activity 15-17: Miscellaneous Normal Distributions

(a) Identify at least three variables that you have studied in this course (either data collected in class or supplied in the text) that have a roughly normal distribution.

(b) List at least three variables that you have studied in this course that clearly do not have a normal distribution.

Topic 16:

SAMPLING DISTRIBUTIONS I: PROPORTIONS

OVERVIEW

You have been studying **probability** and **normal** distributions. These ideas arise in the practice of statistics because sound data collection strategies involve the deliberate introduction of randomness into the process. Therefore, drawing meaningful conclusions from sample data requires an understanding of properties of **randomness**. In this topic you will study how sample proportions vary from sample to sample, not in a haphazard way but in a predictable manner that enables us to draw conclusions about the underlying population proportion. This topic will also mark your introduction to the two essential concepts of statistical inference: **confidence** and **significance**.

OBJECTIVES

- To continue to practice distinguishing between **parameters** and **statistics**.
- To gain an understanding of the fundamental concept of **sampling variability**.
- To discover and understand the concept of **sampling distributions** as representing the long-term pattern of variation of a statistic under repeated sampling.
- To explore the sampling distribution of a sample proportion through actual experiments and calculator simulations.
- To discover the effect of sample size on the sampling distribution of a sample proportion.
- To begin to understand the concepts of statistical **confidence** and **significance** as they relate to estimating a population parameter based on a sample statistic.

16

PRELIMINARIES

1. Which color of Reese's Pieces candies do you think is the most common: orange, brown, or yellow?

2. Take a guess concerning the proportion of Reese's Pieces candies that have that color.

3. If each student in this class takes a random sample of 25 Reese's Pieces candies, would you expect every student to obtain the same number of orange candies?

4. If an ESP (extrasensory perception) test involves predicting which of four different shapes appears on the back of a card, is it possible for a person to get 20 or more correct out of 40 if he or she is just guessing?

5. Would you be very surprised to see a person get 20 or more correct out of 40 on this ESP test if he or she is just guessing?

6. What percentage of the popular vote in the 1996 U.S. Presidential election went to Bill Clinton? to Bob Dole? to Ross Perot?

7. Did you go trick-or-treating door to door during the Halloween season when you were a child?

8. Make a guess for the proportion of American families that hand out treats from the door of their home during Halloween.

9. Does it sound plausible to you that one-quarter of all adult Americans believe in witches?

10. If someone asks you to call the outcome of a coin flip, would you call "heads" or "tails"?

11. Record for your class the numbers of "heads" and "tails" responses to the previous question.

12. In Activity 16-3 you will be asked to study distributions of proportions of orange candies from random samples of Reese's Pieces. To complete Activity 16-3 you need the calculator to simulate the taking of 500 random samples of size 25 and of size 75 of Reese's Pieces. Download the SIMSAMP.83p program into your calculator.

Choose a partner and run SIMSAMP.83p. One of you should enter 25 for sample size, and the other should enter 75. Both of you should enter 500 for the number of samples and .45 entered for the true population proportion (we are assuming that 45% of the population is orange). The program places the simulated sample proportions into a list named PROP.

The simulation for the samples of size 25 will take approximately 8 minutes to run, while the simulation for the samples of size 75 will take approximately 22 minutes. The results of these simulations will be used in Activity 16-3.

WARNING: This program requires a significant amount of memory. You should delete any lists from your calculator that are no longer necessary.

Activity 16-1: Parameters and Statistics

Recall from Topic 12 that a *population* consists of the entire group of people or objects of interest to an investigator, while a *sample* refers to the part of the population that the investigator actually studies. Also remember that a *parameter* is a numerical characteristic of a population and that a *statistic* is a numerical characteristic of a sample.

In certain contexts a population can also refer to a *process* (such as flipping a coin or manufacturing a candy bar) that in principle can be repeated indefinitely. With this interpretation of population, a sample is a specific collection of process outcomes.

We will be very careful to use different symbols to denote parameters and statistics. For example, we use the following symbols to denote proportions, means, and stan-

16

dard deviations (note that we consistently use Greek letters for parameters):

	(population) parameter		(sample) statistic	
proportion	θ	"theta"	$\hat{p}$	"p-hat"
mean	μ	"mu"	$\bar{x}$	"x-bar"
standard deviation	σ	"sigma"	s	

Identify each of the following as a parameter or a statistic, indicate the symbol used to represent it, and specify its value.

- the proportion of men in the entire 1999 U.S. Senate

- the proportion of Democrats among the following five Senators:

ID#	Name	Sex	Party	State	Years
01	Abraham	m	Rep	Michigan	4
21	Cochran	m	Rep	Mississippi	21
41	Gramm	m	Rep	Texas	14
61	Landrieu	f	Dem	Louisiana	2
81	Rockefeller	m	Dem	West Virginia	14

- the mean years of service among these five senators

- the standard deviation of the years of service in the 1999 U.S. Senate

Activity 16-2: Colors of Reese's Pieces Candies

Consider the *population* of the Reese's Pieces candies manufactured by Hershey. Suppose that you want to learn about the distribution of colors of these candies but that you can afford to take a *sample* of only 25 candies.

(a) Take a random sample of 25 candies and record the count and proportion of each color in your sample.

	orange	yellow	brown
count			
proportion			

(b) Is the proportion of orange candies among the 25 that you selected a *parameter* or a *statistic?* What symbol is used to denote it?

(c) Is the proportion of orange candies manufactured by Hershey's process a parameter or a statistic? What symbol represents it?

(d) Do you *know* the value of the proportion of orange candies manufactured by Hershey?

(e) Do you know the value of the proportion of orange candies among the 25 that you selected?

> These simple questions point out the important fact that one typically knows (or can easily calculate) the value of a sample statistic, but only in very rare cases does one know the value of a population parameter. Indeed, a primary goal of sampling is to *estimate* the value of the parameter based on the statistic.

(f) Do you suspect that every student in the class obtained the same proportion of orange candies in his or her sample?

(g) Use the axis below to construct a dotplot of the sample proportions of orange candies obtained by the students in the class.

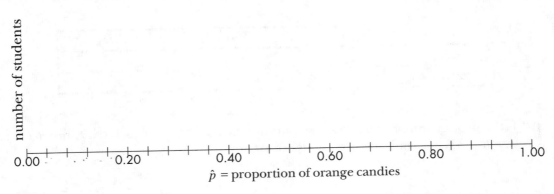

(h) *Did* everyone obtain the same number of orange candies in their samples?

(i) Identify the observational units in this display and the variable being measured from unit to unit.

(j) Comment on the shape, center, and spread of these sample proportions.

These simple, perhaps obvious, questions illustrate a very important statistical property known as ***sampling variability***: Values of sample statistics vary from sample to sample.

(k) If every student were to estimate the population proportion of orange candies by the proportion of orange candies in his or her sample, would everyone arrive at the same estimate?

(l) Based on what you learned in Topic 12 about *unbiased* statistics, and having the benefit of seeing the sample results of the entire class, take a guess concerning the population proportion of orange candies.

(m) Again assuming that each student had access only to her or his sample, would most estimates be reasonably close to the true parameter value? Would some estimates be way off? Explain.

(n) Remembering what you learned about the effect of sample size in Topic 14, in what way would the dotplot have looked different if each student had taken a sample of *ten* candies instead of 25?

(o) In what way would the dotplot have looked different if each student had taken a sample of *75* candies instead of 25?

Activity 16-3: Simulating Reese's Pieces

Our class results suggest that even though sample values vary depending on which sample you happen to pick, there seems to be a *pattern* to this variation. We need more samples to investigate this pattern more thoroughly, however. Since it is time-consuming (and possibly fattening) to *literally* sample candies, we will use your calculator to *simulate* the process (which we started in the preliminaries section of this topic).

To perform these simulations we need to suppose that we know the actual value of the parameter. Recall that we assumed that 45% of the population was orange in the Preliminaries.

(a) Examine the PROP list for the sample size of 25 to see the 500 sample proportions that were generated by the calculator simulation. (Pretend that this is really 500 students, each taking 25 candies and counting the number of orange ones.) Look at a visual display of the sample *proportions* of orange obtained.

(b) Reproduce a rough sketch of your visual display below, making sure to label both axes appropriately.

(c) Do you notice any *pattern* in the way that the resulting 500 sample proportions vary? Explain.

The pattern displayed by the variation of the sample proportions from sample to sample is called the **sampling distribution** of the sample proportion. Even though the sample proportion of orange candies varies from sample to sample, there is a recognizable long-term pattern to that variation. These simulated sample proportions approximate the theoretical sampling distribution derived from all possible samples.

(d) Write down the mean and standard deviation of these sample proportions below:

mean of $\hat{p}$ values:

standard deviation of $\hat{p}$ values:

(e) Roughly speaking, are there more sample proportions *close* to the population proportion (which, you will recall, is .45) than there are *far* from it?

(f) Let us quantify the previous question. Use the TRACE button of your calculator to count how many of the 500 sample proportions are within ± .10 of .45 (i.e., between .35 and .55). Note: .10 should be close to the standard deviation you calculated in (d), so that we are going about one standard deviation on each side of θ. Then repeat for within ± .20 and for within ± .30. Record the results below:

	how many of the 500 sample proportions	percentage of the 500 sample proportions
within ± .10 of .45		
within ± .20 of .45		
within ± .30 of .45		

(g) Forget for the moment that you have designated the population proportion of orange candies to be .45. Suppose that each of these 500 imaginary students was to estimate the population proportion of orange candies by going a distance of .20 on either side of her or his sample proportion. What percentage of the 500 students would capture the actual population proportion (.45) within this interval?

(h) Still forgetting that you actually know the population proportion of orange candies to be .45, suppose that you were one of those 500 imaginary students. Would you have any way of knowing *definitively* whether your sample proportion was within .20 of the population proportion? Would you be reasonably "confident" that your sample proportion was within .20 of the population proportion? Explain.

16

While one cannot use a sample proportion to determine a population proportion *exactly*, one can be reasonably **confident** that the population proportion is within a certain distance of the sample proportion. This "distance" depends primarily on how confident one wants to be and on the size of the sample. You will study this notion extensively when you encounter **confidence intervals** in Topics 19 and 20.

(i) Examine the PROP list for the sample size of 75 to see the 500 sample proportions of orange candies that were generated by the calculator simulation. (These samples are three times larger than the ones you gathered in class and simulated earlier.) Look at a display of the sample *proportions* and write down their mean and standard deviation.

   mean of $\hat{p}$ values:

   standard deviation of $\hat{p}$ values:

(j) How has the sampling distribution changed from when the sample size was only 25 candies?

(k) Use the TRACE button of your calculator to count how many of these 500 sample proportions are within ± .10 of .45. Record this number and the percentage below.

(l) How do the percentages of sample proportions falling within ± .10 of .45 compare between sample sizes of 25 and 75?

(m) In general, is a sample proportion more likely to be close to the population proportion with a larger sample size or with a smaller sample size?

Since these sample proportions follow approximately a normal distribution, the **empirical rule** establishes that about 95% of the sample proportions fall within two standard deviations of the mean of these sample proportions.

(n) Remembering that you found the mean and standard deviation of these sample proportions in (i) for a sample size of 75, double the standard deviation. Then subtract this value from the mean and also add this value to the mean. Record the results.

(o) Use your calculator to count how many of the sample proportions fall within this interval. What percentage of the 500 sample proportions is this? Is this percentage close to the 95% predicted by the empirical rule?

(p) If each of the 500 imaginary students would subtract this value (twice the standard deviation) from her or his sample proportion and also add this value to her or his sample proportion, about what percentage of the students' intervals would contain the actual population proportion of .45?

This activity reveals that if one wants to be about 95% confident of capturing the population proportion within a certain distance of one's sample proportion, that "distance" should be about twice the standard deviation of the sampling distribution of sample proportions.

One need not use simulations to determine how sample proportions vary from sample to sample. An important theoretical result affirms what your simulations have suggested about the shape, center, and spread of this distribution:

> **Central Limit Theorem (CLT) for a Sample Proportion:**
> Suppose that a simple random sample of size *n* is taken from a large population in which the true proportion possessing the attribute of interest is θ. Then the sampling distribution of the sample proportion $\hat{p}$ is approximately normal with mean θ and standard deviation $\sqrt{(\theta(1-\theta))/n}$. This approximation becomes more and more accurate as the sample size *n* increases, and it is generally considered to be valid, provided that $n\theta \geq 10$ and $n(1-\theta) \geq 10$.

Notice that this result specifies three things about the distribution of sample proportions: shape, center, and spread.

(q) If we continue to assume that the population proportion of orange candies is θ = .45, what does the theoretical result say about the mean and standard deviation of the sampling distribution of sample proportions when the sample consists of *n* = 25 candies? Do these values come close to your simulated results in (d) above?

theoretical mean of $\hat{p}$ values:

theoretical standard deviation of $\hat{p}$ values:

(r) Repeat the previous question for a sample size of *n* = 75. Compare your theoretical answers to your simulated results in (i).

theoretical mean of $\hat{p}$ values:

theoretical standard deviation of $\hat{p}$ values:

Activity 16-4: ESP Testing

A common test for extrasensory perception (ESP) asks subjects to identify which of four shapes (star, circle, wave, or square) appears on a card unseen by the subject. Consider a test of $n = 40$ trials. If a person does not have ESP and is just guessing, he or she should specify the correct shape for 25% of the trials in the long run. In other words, the proportion of correct responses that the guessing subject would make in the long run would be $\theta = .25$. The following table and histogram reveal the results of 10,000 simulated tests in which the subject guessed randomly on each card:

# Correct	0	1	2	3	4	5	6	7	8	9	10
Proportion Correct	0	.025	.050	.075	.10	.125	.150	.175	.20	.225	.250
tally	0	1	2	39	107	279	549	887	1174	1311	1411
# Correct	11	12	13	14	15	16	17	18	19	20	21
Proportion Correct	.275	.30	.325	.35	.375	.40	.425	.450	.475	.50	.525
tally	1328	1091	746	504	306	146	70	33	14	0	2

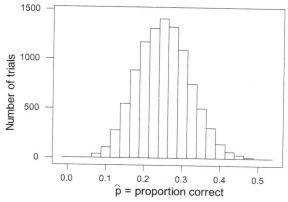

(a) Identify the observational units in this display.

(b) Is the shape of the distribution of these sample proportions similar to that with the Reese's Pieces? What differs in this distribution compared to that of the Reese's Pieces?

16

(c) In how many of these 10,000 simulated tests does the guessing subject identify *exactly* 25% of the 40 cards correctly? Does this constitute most of the 10,000 tests?

(d) How surprising would it be for a subject to get 30% (12) or more correct if he or she were just guessing? Refer to the results of the simulated tests in your answer.

(e) How surprising would it be for a subject to get 45% (18) or more correct if he or she were just guessing? Once again refer to the results of the simulated tests.

(f) Suppose that a particular subject gets 50% (20) correct in a test. How convinced would you be that she actually possesses the ability to get more than 25% correct in the long run? Explain, basing your argument on the results of the simulated tests.

(g) Repeat (f) for a subject who gets 32.5% (13) correct on the 40-card test.

This activity introduced you to the concept of ***statistical significance***. The idea is to explore the sampling distribution of a statistic, investigating how often an observed sample result would occur *just by chance*. Roughly speaking, a sample result is said to be ***statistically signifi-cant*** if it is unlikely to occur simply due to sampling variability alone. In this example, it would not be unlikely to get 32.5% correct just by guessing, but it would be highly unlikely to get 50% correct by guessing. Thus, a sample result of 50% correct is statistically significant, providing very strong evidence that the person does better than guessing. You will study this reasoning process more formally and in detail when you encounter ***tests of significance*** in Topics 21 and 22.

WRAP-UP

This topic has emphasized the fundamental distinction between a ***parameter*** and a ***statistic***. You have explored the obvious (but crucial) concept of ***sampling vari-ability*** and learned that this variability displays a very definite pattern in the long run. This pattern is known as the ***sampling distribution*** of the statistic. You have investi-gated properties of sampling distributions in the context of sample *proportions*. You have also discovered that larger sample sizes produce less variation among sample proportions, and the ***Central Limit Theorem*** provides a way of measuring that vari-ation.

In addition, you have begun to explore how sampling distributions relate to the impor-tant idea of statistical ***confidence***: that one can have a certain amount of confidence that the observed value of a sample statistic falls within a certain distance of the unknown value of a population parameter. You have also encountered the issue of ***sta-tistical significance***. You have learned that the question of statistical significance relates to how often an observed sample result would occur by sampling variability or chance alone.

In the next topic you will study sampling distributions not of sample *proportions* but of sample *means*, and you will continue to explore the connection between sampling dis-tributions and the fundamental concepts of confidence and significance.

16

HOMEWORK ACTIVITIES

Activity 16-5: Parameters vs. Statistics (*cont.*)

Identify each of the following as a parameter or a statistic, and also indicate the symbol used to denote it. In some cases you may have to form your own conclusion as to what the population of interest is.
- **(a)** the proportion of students in your class who prefer to hear "good news" before bad
- **(b)** the proportion of all students at your school who prefer to hear "good news" before bad
- **(c)** the mean number of states visited by the students in your class
- **(d)** the mean number of states visited by all students in your school
- **(e)** the standard deviation of the years of service of the 100 U.S. Senators
- **(f)** the standard deviation of the years of service of the five U.S. Senators who made up the random sample you generated using the table of random digits
- **(g)** the proportion of American voters from the 1996 election who voted for Bill Clinton
- **(h)** the proportion of people at the next party you attend who voted for Bob Dole in the 1996 election
- **(i)** the proportion of American voters from the 1996 election who voted for Ross Perot
- **(j)** the mean amount spent on Christmas presents by American adults in 1999
- **(k)** the proportion of American adults who told a Gallup pollster that they believe in witches
- **(l)** the mean number of cats in the households of the faculty members of your school
- **(m)** the mean number of cats among all American households
- **(n)** the proportion of "heads" in 100 coin flips
- **(o)** the mean weight of 20 bags of potato chips

Activity 16-6: Presidential Votes

In the 1996 U.S. Presidential election, Bill Clinton received 49% of the popular vote, compared to 41% for Bob Dole and 8% for Ross Perot. (Each of these figures has been rounded to the nearest whole percentage point.) Suppose that we take a simple random sample of 100 voters from that election and ask them for whom they voted.
- **(a)** Would it necessarily be the case that these 100 voters would include 49 Clinton supporters, 41 Bush supporters, and 8 Perot supporters?
- **(b)** Now suppose that you were to repeatedly take SRSs of 100 voters. Would you find the same sample proportion of Clinton supporters each time?
- **(c)** According to the Central Limit Theorem for a sample proportion presented above, what would be the standard deviation of the sampling distribution of the sample proportion of Clinton voters?

use
formula
from
class

(d) According to the empirical rule, about 95% of your samples would find the sample proportion of Clinton voters to be between what two values?

(e) Repeat (c) for the sample proportion of Dole supporters.

(f) Repeat (c) for the sample proportion of Perot supporters.

(g) With which candidate would you see the most variation in the sample proportions who voted for him? With which candidate would you see the least variation?

Activity 16-7: Presidential Votes (*cont.*)

Bill Clinton received 49% of the votes cast in the 1996 Presidential election. Suppose that you were to take an SRS of size n from the population of all votes cast in that election.

(a) Use the Central Limit Theorem presented above to calculate the standard deviation of the sampling distribution of these sample proportions for each of the following values of n: 50, 100, 200, 400, 500, 800, 1000, 1600, 2000.

(b) Construct a scatterplot of these standard deviations vs. the sample size n.

(c) By how many times does the sample size have to increase in order for this standard deviation to be cut in half?

Activity 16-8: Presidential Votes (*cont.*)

Let θ represent the proportion of votes received by a certain candidate in an election. Suppose that you repeatedly take SRSs of 100 voters and calculate the sample proportion who voted for the candidate.

(a) Use the Central Limit Theorem presented above to calculate the standard deviation of the sampling distribution of these sample proportions for each of the following values of θ: 0, .1, .2, .3, .4, .5, .6, .7, .8, .9, 1.

(b) Construct a scatterplot of these standard deviations vs. the population proportion θ.

(c) Which value of θ produces the most variability in sample proportions?

(d) Which values of θ produce the least variability in sample proportions? Explain in a sentence or two what happens in these cases.

Activity 16-9: Cat Households

Suppose that you want to estimate the proportion of all households in your hometown that have a pet cat. (Let us call this proportion θ.) Suppose further that you take an SRS of 200 households, while your polling competitor takes an SRS of only 50 households.

(a) Can you be certain that the sample proportion of cat households in your sample will be closer to θ than your competitor's sample proportion?

(b) Do you have a better chance than your competitor of obtaining a sample proportion of cat households that falls within ±.05 of the actual value θ? Explain.

Suppose now that the actual proportion of households that have a pet cat (among the entire population of households in your hometown) is θ = .25.

(c) Use the expression for the standard deviation of the sampling distribution of sample proportions specified by the Central Limit Theorem result to determine the standard deviation when the sample size is *n*=200. Then do the same calculation for a sample size of *n* = 50. Which sample size produces the smaller standard deviation? How many times smaller is it than the other one?

(d) Use your calculator to simulate the random selection of 500 samples from the binomial distribution (see Activity 14-4 on page 314 for a reminder of how to use the randBin command), with each sample containing 200 households. How many of these 500 samples produce a sample proportion within ±.05 of the population proportion?

(e) Use your calculator to compute the five-number summary of the distribution of these sample proportions. Record the results.

(f) Again use your calculator to simulate the random selection of 100 samples, this time with each sample containing only 50 households. How many of these 100 samples produce a sample proportion within ±.05 of the population proportion?

(g) Again use your calculator to compute the five-number summary of the distribution of these sample proportions, and record the results.

(h) Construct (by hand) boxplots of the distributions of these sample proportions. The boxplots should be on the same scale, but they need not be modified ones.

(i) Write a paragraph commenting on similarities and differences in the two boxplots.

Activity 16-10: ESP Testing (*cont.*)

Refer back to the 10,000 simulated ESP tests presented in Activity 16-4 on page 355.

(a) How many and what proportion of the 40 cards would a subject have to identify correctly so that fewer than 10% of all guessing subjects do that well?

(b) Repeat (a) with 5% instead of 10%.

(c) Repeat (a) with 1% instead of 10%.

Activity 16-11: Calling Heads or Tails

Consider the responses to the question of whether students would call "heads" or "tails" if asked to predict the result of a coin flip.

(a) What proportion said that they would call "heads"? Is this proportion a parameter or a statistic? Indicate the appropriate symbol to represent it.

Now suppose that 50% of the population would call heads.

(b) Use your calculator to simulate 1000 random samples of size n, where n is the number of students in your class who responded to the question, from the binomial distribution (see Activity 14-4 on page 314 for a reminder of how to use the rand-Bin command). Produce visual displays of the distribution of the 1000 sample proportions, and describe the shape, center, and spread of this distribution.

(c) Comment on where the sample proportion for your class falls in the distribution. Would the sample result you obtained in class be very surprising if in fact 50% of the population would call heads? Explain.

In his book *Statistics You Can't Trust*, Steve Campbell claims that people call "heads" in this situation 70% of the time.

(d) Repeat (b) and (c) under the assumption that Campbell is correct. Would your class sample be very surprising if Campbell is right? Explain.

Activity 16-12: Racquet Spinning (*cont.*)

Reconsider the situation of Activity 14-9 on page 319. A tennis racquet was spun 100 times, resulting in 46 "up" and 54 "down" results. This activity asks you to assess more formally whether these sample data provide much reason to doubt that the process would produce 50% for each result in the long run.

(a) Is .46 a parameter or a statistic? Explain.

(b) Is .50 a parameter or a statistic? Explain.

(c) Use your calculator to simulate 1000 repetitions of this experiment (spinning a tennis racquet 100 times), from the binomial distribution (see Activity 14-4 on page 314 for a reminder of how to use the randBin command) assuming that the process really is 50/50 in the long run ($\theta = .5$). Also use your calculator to compute the sample proportion of "up" outcomes in each of the 1000 repetitions.

(d) Comment on the shape of the distribution, and also calculate the mean and standard deviation of these 1000 sample proportions.

(e) Compare your answers to (d) with those predicted by the Central Limit Theorem.

(f) In how many and what proportion of your 1000 repetitions was the proportion of "up" results either 46% or smaller or 54% or larger?

(g) Does your answer to (f) suggest that the sample result would be very unlikely to occur by chance alone if in fact the process were 50/50 in the long run? Explain.

Activity 16-13: Halloween Practices

A 1999 Gallup survey of a random sample of 1005 adult Americans found that 69% planned to give out Halloween treats from the door of their home.

(a) Is this .69 a parameter or a statistic? Explain.

(b) Does this finding necessarily prove that 69% of all adult Americans planned to give out treats? Explain.

16

(c) If the population proportion planning to give out treats had really been .7, would the sample result have fallen within two standard deviations of .7 in the sampling distribution? Support your answer with the appropriate calculation. [*Hint:* To fall "within two standard deviations" means that the difference between the observed statistic $\hat{p} = .69$ and the parameter value .7 is less than two times the standard deviation of $\hat{p}$.]

(d) Repeat (c) if the population proportion had really been .6.

(e) Working only with multiples of .01 (e.g., .60, .61, .62, …), determine and list all potential values of the population proportion θ for which the observed sample proportion .69 falls within two standard deviations of θ in the sampling distribution. Show calculations to support your list. [*Hint:* Since the standard deviation of $\hat{p}$ is very similar for all of these θ values, you might want to calculate it once and then use that approximation throughout.]

(f) Summarize your findings in (e) in terms of which numbers are plausible values for the proportion of the population who planned to give out Halloween treats from the door of their home in 1999 based on this observed sample result.

Activity 16-14: Halloween Beliefs

A 1999 Gallup poll found that 22% of a sample of 493 adult Americans said that they believe in witches.

(a) Would it be possible to obtain this sample result if the proportion of all American adults who believe in witches is not .22? Explain.

(b) Would it be possible to obtain this sample result if the proportion of all American adults who believe in witches is .23? Explain.

(c) Would it be possible to obtain this sample result if the proportion of all American adults who believe in witches is .25? Explain.

(d) Would it be possible to obtain this sample result if the proportion of all American adults who believe in witches is .3? Explain.

(e) The histograms of sample proportions on the next page came from simulations of 1000 repetitions of asking 493 people whether they believe in witches. One of them was based on an assumption that 23% of the population believes in witches, one assumed that 25% do, and the third assumed that 30% do. Identify which parameter value goes with which histogram.

(f) Based on the sample proportion obtained by the Gallup pollsters, and considering these simulation results, is it very plausible to believe that 30% of all American adults believe in witches? Explain.

(g) Repeat (f) concerning the plausibility that 25% of all American adults believe in witches.

(h) Repeat (f) concerning the plausibility that 23% of all American adults believe in witches.

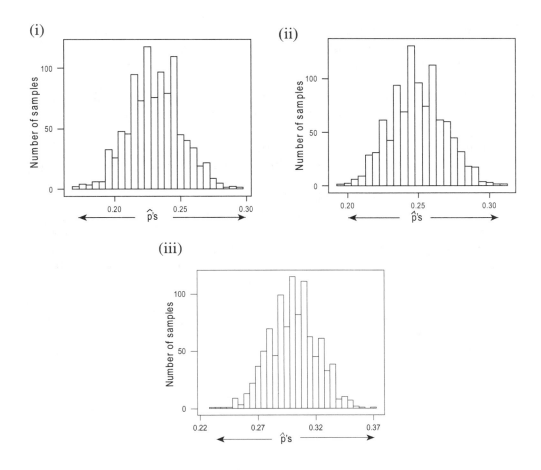

Activity 16-15: Cola Discrimination

In an experiment to determine whether people can distinguish between two brands of cola, subjects are presented with three cups. Two cups contain one brand of cola, and the third contains the other brand. Subjects are to taste from all three cups and then identify the one that differs from the other two. Suppose that an experiment consists of thirty of these trials.

(a) If a subject cannot distinguish the tastes and therefore guesses each time, what proportion would she or he get correct in the long run?

(b) Describe how you could use a fair, six-sided die to simulate this experiment over and over for a person who just guesses for each of the thirty trials.

(c) Use your calculator to simulate 1000 repetitions of this exercise for a subject who is just guessing. (*Hint:* Use the randBin command as in Activity 14-4 on page 314.)

(d) Produce a histogram of the distribution of the 1000 sample proportions of correct answers in this experiment. Comment on the shape. Is it consistent with the shape predicted by the Central Limit Theorem?

(e) Calculate the mean and standard deviation of your 1000 simulated sample proportions. Compare their values to those predicted by the CLT.

(f) In how many and what proportion of your 1000 simulated repetitions did the guessing subject get 40% or more correct?

(g) If a subject were to get 40% correct in this experiment, would you be fairly convinced that she or he does better than just guessing? Explain, relating your answer to your simulation results.

(h) If a subject were to get 60% correct in this experiment, would you be fairly convinced that she or he could do better than just guessing? Explain, relating your answer to your simulation results.

(i) Now use your calculator to simulate 1000 repetitions of this experiment for a subject who is able to distinguish correctly 2/3 of the time in the long run.

(j) Produce dotplots of both sets of simulated sample proportions on the same scale. Comment on similarities and differences in the distributions. [*Hint:* As you learned way back in Topic 3, comment on center, spread, and shape, as well as any other features of interest.] Also comment on the amount of overlap between the two distributions.

(k) In how many and what proportion of these 1000 new repetitions does the subject get at least 60% correct?

SAMPLING DISTRIBUTIONS II: MEANS

You have studied how sample *proportions* summarizing *categorical* variables vary from sample to sample. In this topic you will explore how sample *means* summarizing *quantitative* variables vary from sample to sample. The issue is a bit more complex because the shape of the underlying population comes into play, but a variety of similarities emerge. You will again find that these statistics do not vary haphazardly but according to a predictable, long-term pattern, and you will see that sample size affects the amount of variation produced. You will also notice connections between sampling distributions and the fundamental concepts of confidence and significance.

- To use simulation to investigate how sample means vary from sample to sample.
- To discover the long-term pattern that emerges from the **sampling distribution of the sample mean** when the sample size is large.
- To learn that this long-term pattern does not depend on the shape of the population when the sample size is large.
- To recognize similarities between the sampling distributions of a sample mean and of a sample proportion.
- To examine and understand the effects of **sample size** and of **population variability** on the sampling distribution of the sample mean.
- To continue to develop an understanding of the concepts of **confidence** and **significance** and their relation to sampling distributions.

PRELIMINARIES

1. How long do you think a typical U.S. penny has been in circulation?

2. Would you expect to see more variability in the sample mean ages of a penny if the sample size is 5 or if the sample size is 25?

3. How much do you think a typical American spends on Christmas gifts in one year?

4. What would you guess for the standard deviation of those expenditures?

5. In Activity 17-2 on page 371 you are asked to study samples of size $n = 20$ and $n = 40$ from a population of 1000 pennies that you will study in Activity 17-1. In order to do this simulation during your class period, download PENNY.83g and start the program PENNY now. This grouped file contains the PENNY program along with two lists named PEN and AGE. These lists are very important to the program and must not be deleted.

Choose a partner and use one of the calculators to run the PENNY program with samples of size 20 and the other to run the simulation with samples of size 40. These data will be stored in a list named MEAN.

WARNING: The PENNY program and the lists mentioned above take up a significant amount of space and memory. If you want these simulations to run without problems you *must* remove all of your programs (except PENNY and DOTPLOT) and most of your lists.

Activity 17-1: Coin Ages (*cont.*)

The following histogram displays the distribution of ages for a population of 1000 pennies in circulation and collected by one of the authors in 1999. Some summary data for this distribution of ages are:

size	mean	Std. Dev.	min	Q_1	median	Q_3	max
1000	12.264	9.613	0	4	11	19	59

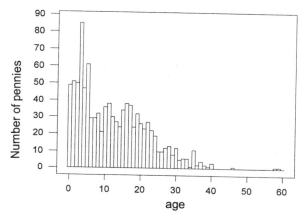

(a) Identify the observational units and variable of interest here. Is this variable quantitative or categorical?

(b) Regarding these 1000 pennies as a population from which one can take samples, are the above values parameters or statistics? What symbols would represent the mean and standard deviation?

17

(c) Does this population of coin ages roughly follow a normal distribution? If not, what shape does it have?

Rather than ask you to select actual pennies from a container with all 1000 of these pennies, you will use a table of random digits to simulate drawing random samples of pennies from this population. This requires us to assign a three-digit label to each of the 1000 pennies. The following table reports the number of pennies of each age and also assigns three-digit numbers to them.

age	count	ID#s	age	count	ID#s	age	count	ID#s
0	49	001–049	15	34	610–643	30	12	945–956
1	51	050–100	16	38	644–681	31	5	957–961
2	50	101–150	17	37	682–718	32	6	962–967
3	85	151–235	18	24	719–742	33	6	968–973
4	47	236–282	19	32	743–774	34	1	974
5	61	283–343	20	26	775–800	35	11	975–985
6	29	344–372	21	23	801–823	36	2	986–987
7	29	373–401	22	27	824–850	37	4	988–991
8	32	402–433	23	22	851–872	38	2	992–993
9	21	434–454	24	19	873–891	39	1	994
10	36	455–490	25	10	892–901	40	3	995–997
11	38	491–528	26	10	902–911	46	1	998
12	30	529–558	27	12	912–923	58	1	999
13	27	559–585	28	13	924–936	59	1	000
14	24	586–609	29	8	937–944	total	1000	

Notice that each age has a number of ID labels assigned to it equal to the number of pennies having that age in the population. Thus, for example, an age of 10 years has 36 ID labels because 36 of the 1000 pennies were 10 years old, while an age of 30 years has one-third as many ID labels because only 12 of the 1000 pennies were 30 years old.

(d) Use the table of random digits to draw a random sample of *five* penny ages from this population. (Do this by selecting three-digit numbers from the table and finding the age assigned to that number in the table above. If you happen to get the same three-digit number twice, ignore the repeat and choose another number.) Record the penny ages below, and draw a dotplot of your sample distribution on the axis provided:

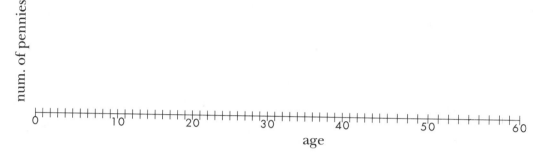

(e) Calculate the sample mean of your five penny ages.

(f) Take four more random samples of five pennies each. Calculate the sample mean each time, and record the results in the table:

Sample no.	1	2	3	4	5
Sample mean					

(g) Did you get the same value for the sample mean all five times? What phenomenon that you studied in Topic 16 does this again reveal? What is different here from the Reese's Pieces activity?

> You are again encountering the notion of **sampling variability**. Since age is a *quantitative* and not a categorical variable, you are observing sampling variability as it pertains to sample *means* and not to sample proportions. As was the case with sample proportions, sample means vary from sample to sample not in a haphazard manner but according to a predictable long-term pattern know as a **sampling distribution**.

17

(h) Use your calculator to compute the mean and standard deviation of your five sample means.

mean of $\bar{x}$ values: standard deviation of $\bar{x}$ values:

(i) Is this mean reasonably close to the population mean ($\mu = 12.264$)? Is the standard deviation greater than, less than, or about equal to the population standard deviation ($\sigma = 9.613$)?

As was the case with proportions, the sample mean is an *unbiased* estimator of the population mean. In other words, the center of its sampling distribution is the population mean. Also evident again is that variability in the sampling distribution of the statistic (sample mean, in this case) decreases with larger samples.

Now consider taking a random sample of 25 pennies. By taking five samples of five pennies each, you have essentially done so already. Consider all your observations as a random sample of size 25. (We are ignoring the possibility that a coin could be repeated in your sample of 25.) Its sample mean is exactly the mean of your five sample means recorded in (h).

(j) Pool these sample means from samples of size 25 with those of your classmates. Produce a dotplot of these sample means below:

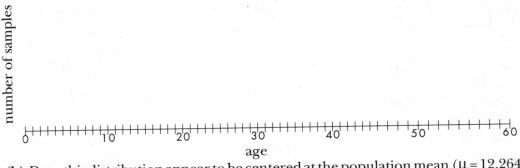

(k) Does this distribution appear to be centered at the population mean ($\mu = 12.264$)? Do the values appear to be less spread out than either the population distribution or the distribution of your five sample means of size 5?

(l) Does this distribution appear to be more normal-shaped than the distribution of ages in the original population (recall the histogram of the population distribution above question (a) on page 367)?

Activity 17-2: Coin Ages (*cont.*)

To study the sampling distribution of sample means more thoroughly requires simulating many more samples, so we turn to your calculator.

Consider again the population of 1000 penny ages described in Activity 17-1. In particular, recall that the distribution is skewed to the right, with a mean of $\mu = 12.264$ and a standard deviation of $\sigma = 9.613$. *Note:* Before running the PENNY program again (your calculator should have completed the simulation started in the "Preliminaries") copy the data stored in the list named MEAN into a list named MEAN1. Since you will be running two more simulations during this activity (one in (a) and the other in (c)), use your calculator for one of the simulations and your partner's for the other. Start both simulations at the same time.

(a) Use the program PENNY program to take 100 random samples of size $n = 1$ coin each. These data are stored in a list named MEAN (which you can view in the Stat List Editor). Look at a dotplot of the distribution of these 100 ages, and reproduce a rough sketch below. Does it resemble the distribution of ages in the population?

(b) Use your calculator to compute the mean and standard deviation of these 100 ages. Record them in the first empty row of the table below. Are they reasonably close to their population counterparts ($\mu = 12.264$, $\sigma = 9.613$)? Also describe the shape of this distribution.

17

(c) Now use your calculator to take 100 random samples of $n = 5$ coins each and to compute the sample mean age for each of the 100 samples (again these data are stored in the list named MEAN). Look at a dotplot of these 100 sample means and reproduce a rough sketch below. Comment on how this distribution differs from the population and from the distribution in (a).

(d) Use your calculator to compute the mean and standard deviation of these 100 sample means. Record them and the general shape of the distribution in the table below. Are they reasonably close to their population counterparts? If not, how do they differ?

(e) Examine your results from the "Preliminaries" section when you took 100 random samples of $n = 20$ coins each and computed the sample mean age for each of the 100 samples (MEAN1). Look at a dotplot of these 100 sample means, and sketch the shape below. Comment on how this distribution differs from the population and from the distributions in (a) and (c).

(f) Use your calculator to compute the mean and standard deviation of these 100 sample means. Record them and the general shape in the table below. Are they reasonably close to their population counterparts? If not, how do they differ?

(g) Examine your results from the "Preliminaries" section when you took 100 random samples of $n = 40$ coins each and computed the sample mean age for each of the 100 samples. Look at a dotplot of these 100 sample means, and give a rough sketch below. Comment on how this distribution differs from the population and from the distributions in (a), (c), and (e).

(h) Use your calculator to compute the mean and standard deviation of these 100 sample means. Record them and the general shape in the table below. Are they reasonably close to their population counterparts? If not, how do they differ?

	Population mean $\mu = 12.264$	Population std. dev. $\sigma = 9.613$	Population shape = skewed right
Sample size	Mean of sample means	Std. dev. of sample means	Shape of sample means
1			
5			
20			
40			

Each of these distributions approximates the sampling distribution of the sample mean from this population for that sample size. These simulations should have revealed that when the sample size is fairly large, the distribution of sample means follows approximately a normal distribution. The distribution of a sample of individual penny ages (i.e., sample size 1) resembles that of the population, but the sampling distribution becomes more and more normal as the sample size increases. The means of these distributions are all equal to the population mean, but the variability in sample means decreases as the sample size increases.

17

Below are two density curves that could represent the amount of change received in different populations:

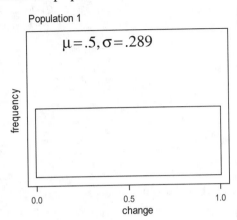
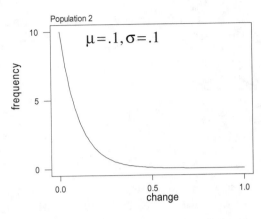

(i) Would you agree that neither of these populations has a distribution that is close to normal?

In (j) and (l) below you are asked to run simulations. To make the best use of time it is suggested that you again work with a partner and use your calculator for the one of the simulations and your partner's for the other.

(j) Population 1 is a ***uniform distribution***, with parameters 0 and 1. Download UNIFM.83p. Run this program to take random samples from this population. Specify $n = 50$ as the sample size and 100 as the number of samples. The means are stored in the list MEAN. Examine a dotplot of the 100 sample means. Comment on the shape of the distribution.

(k) Use your calculator to compute the mean and standard deviation of the 100 sample means from population 1. Record the results in the table below. Is the mean close to the population mean? Is the standard deviation much smaller than the population's?

(l) Population 2 is called an *exponential distribution*. Download EXPO.83p and use this program to repeat (j) and (k) for population 2.

	Mean	Std. dev.	Shape
Population 1	$\mu = .5$	$\sigma = .289$	uniform
Sample Means of size $n = 50$ from 1			
Population 2	$\mu = .1$	$\sigma = .1$	sharply skewed
Sample means of size $n = 50$ from 2			

Even with these very nonnormal populations, the distribution of sample means from these populations follows roughly a normal distribution. A theoretical result affirms your simulation findings:

> **Central Limit Theorem (CLT) for a Sample Mean:**
> Suppose that a simple random sample of size n is taken from a large population in which the variable of interest has mean μ and standard deviation σ. Then, provided that n is large (at least 30 as a rule of thumb), the sampling distribution of the *sample* mean $\bar{x}$ is approximately *normal* with mean μ and standard deviation $\sigma/\sqrt{n}$. The approximation holds with large sample sizes *regardless* of the shape of the population distribution. The accuracy of the approximation increases as the sample size increases. For populations that are themselves normally distributed, the result holds not approximately but exactly.

Note the similarities with the CLT for a population proportion: This result specifies the shape, center, and spread of the sampling distribution. Again the shape is normal, the mean is the population parameter of interest, and the standard deviation decreases as n increases by a factor of $1/\sqrt{n}$.

(m) For the samples of size $n = 50$ from all three populations (penny ages, 1, and 2), calculate $\sigma/\sqrt{n}$. Then confirm that its value is reasonably close to the standard deviation of your 100 simulated sample means.

Penny ages: 1: 2:

Activity 17-3: Christmas Shopping

A survey conducted by the Gallup organization during November 18–21, 1999, interviewed 922 American adults who expected to buy Christmas gifts that year. These persons were asked how much money they personally expected to spend on Christmas gifts, and the sample mean of the responses was 857 dollars.

(a) Is $857 a parameter or a statistic? Explain.

(b) For this study, identify the
   (i) population of interest

   (ii) sample selected

   (iii) parameter of interest

   (iv) statistic calculated

Let μ denote the mean amount expected to be spent on Christmas gifts among that population, and let σ represent the standard deviation of those amounts.

(c) Does μ necessarily equal $857? Is it *possible* that the sample mean could have turned out to be $857 even if μ were equal to $850? What if μ were equal to $800? What about $1000? Explain.

Assume for now that μ is $850 and that σ is known to be $250.

(d) What does the Central Limit Theorem say about how the sample mean $\bar{x}$ would vary if samples of size $n = 922$ were taken over and over? Does your answer depend on the shape of the distribution of expected expenditures in the population? Explain.

(e) Draw a sketch to represent this distribution. [Please be sure to label and scale the horizontal axis.]

(f) Would a sample mean of $857 be a very surprising result if μ were $850 and σ = $250? Explain, basing your answer on (d) and (e).

(g) Repeat (d)–(f) using $800 as the population mean of interest.

These questions should remind you of the concept of statistical *significance*, which asks whether a sample result is unlikely to have occurred by sampling variability or chance alone.

Continue to assume that $\sigma = 250$, but now return to not knowing the value of μ.

(h) What does the CLT say about the standard deviation of the sampling distribution of the sample mean?

(i) Begin to apply the empirical rule to this sampling distribution by doubling the standard deviation calculated in (h). Then subtract that amount from the sample mean to get a lower bound, and add that amount to the sample mean to get an upper bound. Report the *interval* of values that you have thereby constructed.

This question should remind you of the concept of statistical *confidence*, which creates an interval of plausible parameter values that could have generated the sample statistic actually obtained. Roughly speaking, we say that a parameter value is plausible if it is within two standard deviations of the sample statistic.

WRAP-UP

This topic has continued your study of the fundamental concept of sampling distributions. You have discovered that just as a sample proportion varies from sample to sample according to a normal distribution, so too (under the right conditions) does a sample mean. Moreover, you have learned that for large sample sizes this result is true regardless of the shape of the population from which the samples are drawn. You have again seen that the ideas of confidence and significance are closely related to the sampling distribution of a sample mean.

The next topic will ask you to consider more formally the Central Limit Theorem that you encountered in this and the previous topic.

HOMEWORK ACTIVITIES

Activity 17-4: Christmas Shopping (*cont.*)

Reconsider Activity 17-3, which examined amounts that individuals expected to spend on Christmas presents in 1999.

(a) Suppose that the population mean amount were really μ = $850. Would the sample mean actually obtained ($857) be more likely if the population standard deviation were σ = $250 or if it were σ = $1250? Explain.

(b) Suppose that the population mean amount were really μ = $800. Would the sample mean actually obtained ($857) be more likely if the population standard deviation were σ = $250 or if it were σ = $1250? Explain.

(c) Answer questions (d)–(i) of Activity 17-3 under the assumption that σ = $1250 rather than σ = $250.

(d) Comment generally on how your answers differ when σ = $1250 as opposed to σ = $250.

_____

Activity 17-5: Parents' Ages (*cont.*)

Reconsider from Activity 3-11 on page 59 the following data, which are the ages at which a sample of 35 American mothers first gave birth (MMAGE in AGECHILD.83g):

20	28	33	23	21	18	24	20	32	16	27	21	17	22	19	40	19	25
24	24	17	31	28	26	18	23	20	18	14	17	21	16	20	20	19	

(a) Create a visual display of these data, calculate summary statistics, and write a paragraph commenting on the distribution.

These sample data were actually taken from a much larger sample gathered as part of the 1998 General Social Survey. That sample involved 1199 mothers. For now, consider this large sample of mothers as the population from which the smaller sample was drawn. The ages for the population are tallied in the table below and in BIGACHILD.83g.

Age	13	14	15	16	17	18	19	20	21	22	23	24	25	26	27
Tally	1	5	18	45	76	97	134	115	124	81	85	52	80	47	51
Age	28	29	30	31	32	33	34	35	36	37	38	39	40	41	42
Tally	45	29	30	18	17	10	10	6	9	5	3	0	0	3	3

(b) Create a visual display of these data, calculate summary statistics, and write a paragraph commenting on the distribution.

(c) Is the distribution of the sample similar to that of the population? Explain.

(d) If one were to repeatedly take samples of 35 mothers from this population of 1199 mothers, what does the CLT say about how the resulting sample means would vary? [*Hint:* Be sure to refer to shape as well as to center and spread.]

(e) Draw a sketch of the (approximate) sampling distribution from (d), labeling the horizontal axes.

(f) Put a mark on the sketch indicating where the sample mean for the above sample of 35 mothers falls. Is it in a tail of the sampling distribution or near the middle?

17

Activity 17-6: Random Babies (*cont.*)

Recall from Activity 14-2 on page 309 that the (exact) probability distribution of the number of matches in the "random babies" activity is given by:

Number of matches	0	1	2	3	4
Probability	3/8	1/3	1/4	0	1/24

Recall also that the mean of this population is $\mu = 1$, and it turns out that the standard deviation is $\sigma = 1$. Below are four histograms based on simulation results similar to those you obtained in Activity 14-1 on page 305. One of them displays the number of matches for one sample of size 100. Another displays 100 sample mean number of matches for samples of size 5, and a third displays the sample mean number of matches for 100 samples of size 25. The fourth is not related to this population. Your task is to identify which histogram is which. Write a paragraph explaining your choices.

A:

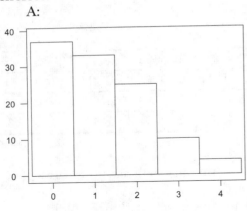

B:

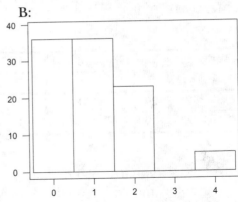

C:

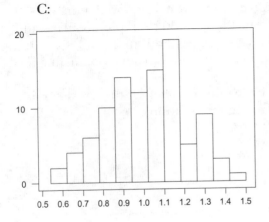

D:

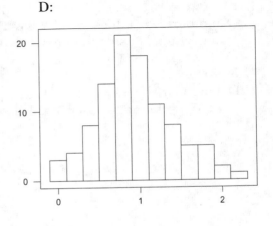

Activity 17-7: Random Babies (*cont.*)

Refer to the previous activity, which reminded you of the probability distribution, mean, and standard deviation of the number of "random baby" matches.

(a) Is the population distribution normal? Is it close to normal? Explain.

(b) Does the Central Limit Theorem say that if we repeatedly take samples of size 5 from this distribution, the resulting sample means will closely follow a normal distribution?

(c) Does the Central Limit Theorem say that if we repeatedly take samples of size 100 from this distribution, the resulting sample means will closely follow a normal distribution?

(d) What does the CLT say about the mean and standard deviation of the sampling distribution in (c)?

(e) Draw a sketch of the (approximate) sampling distribution of the sample mean number of matches when the sample size is 100, labeling the horizontal axis.

(f) If we were to repeatedly take samples of size 100 from this distribution and calculate the sample mean number of matches each time, about what percentage of these sample means would fall between 0.8 and 1.2? Explain.

(g) Continuing the scenario presented in (f), if for each of the samples we constructed an interval by subtracting .2 from the sample mean to obtain the lower bound and adding .2 to the sample mean to obtain the upper bound, about what percentage of those intervals would succeed in capturing the population mean value of $\mu = 1$? Explain.

Activity 17-8: Birth Weights (*cont.*)

Recall Activity 15-2 on page 331, where we assumed that birth weights of babies could be modeled as a normal distribution with mean $\mu = 3250$ grams and standard deviation $\sigma = 550$ grams. The following histograms display the sample mean birth weights in 1000 samples of $n = 5$ babies each and of 1000 samples of $n = 10$ babies each:

1)
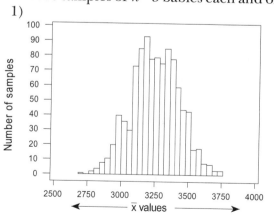

2)
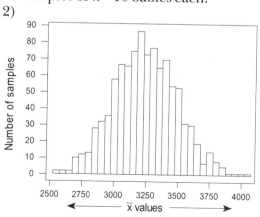

(a) Which histogram goes with which sample size? Explain.

(b) Judging from these histograms, which sample size is more likely to produce a sample mean birth weight below 2500 grams?

(c) Judging from these histograms, which sample size is more likely to produce a sample mean birth weight below 3000 grams?

(d) Judging from these histograms, which sample size is more likely to produce a sample mean birth weight above 3500 grams?

(e) Judging from these histograms, which sample size is more likely to produce a sample mean birth weight between 3000 grams and 3500 grams?

(f) Write a paragraph summarizing what this activity reveals about the effect of sample size on the sampling distribution of a sample mean.

Activity 17-9: Candy Bar Weights (*cont.*)

Recall Activity 15-9 on page 339, which asked you to assume that the actual weight of a certain candy bar, whose advertised weight is 2.13 ounces, varies according to a normal distribution with mean $\mu = 2.2$ ounces and standard deviation $\sigma = .04$ ounces.

(a) What does the CLT say about the distribution of sample mean weights if samples of size $n = 5$ are taken over and over?

(b) Draw a sketch of this sampling distribution, labeling the horizontal axis.

Suppose that you are skeptical about the manufacturer's claim that the mean is $\mu = 2.2$, so you take a random sample of $n = 5$ candy bars and weigh them. Suppose that you find a sample mean weight of 2.15 ounces.

(c) Is it possible to get a sample mean weight this small even if the manufacturer's claim that $\mu = 2.2$ is valid? Explain, referring to the graph you sketched in (b).

(d) Is it very *unlikely* to get a sample mean weight this small even if the manufacturer's claim that $\mu = 2.2$ is valid? Explain.

(e) Would finding the sample mean weight to be 2.15 provide strong evidence to doubt the manufacturer's claim that $\mu = 2.2$? Explain, referring to the sampling distribution.

(f) Would finding the sample mean weight to be 2.18 provide strong evidence to doubt the manufacturer's claim that $\mu = 2.2$? Explain, again referring to the sampling distribution.

(g) What values for the sample mean weight would provide fairly strong evidence against the manufacturer's claim that $\mu = 2.2$? Explain, once again referring to the sampling distribution. [*Hint:* Reconsider the empirical rule.]

Activity 17-10: Cars' Fuel Efficiency (*cont.*)

The highway miles per gallon rating of the 1999 Volkswagen Passat was 31 MPG. The fuel efficiency that one gets on an individual tankful of gasoline would naturally vary from tankful to tankful. Suppose that the MPG calculations per tankful have a mean of $\mu = 31$ and a standard deviation of $\sigma = 3$ MPG.

(a) Would it be surprising to obtain 30.4 MPG on one tank? Explain.

(b) Would it be surprising for a sample of 30 tankfuls to produce a sample mean of 30.4 MPG? Explain, referring to the CLT and to a sketch of the sampling distribution.

(c) Would it be surprising for a sample of 60 tankfuls to produce a sample mean of 30.4 MPG? Explain, again referring to the CLT and to a sketch of the sampling distribution.

(d) Would it be surprising for a sample of 150 tankfuls to produce a sample mean of 30.4 MPG? Explain, again referring to the CLT and to a sketch of the sampling distribution.

(e) Do any of your responses depend on knowing the shape of the population distribution? Explain.

Topic 18:

CENTRAL LIMIT THEOREM

OVERVIEW

In previous activities you have used hands-on and calculator simulations to discover that while the value of a sample proportion or a sample mean varies from sample to sample, there is a very precise long-term pattern to that variation. The **Central Limit Theorem** specifies that for large sample sizes this pattern follows a normal distribution. In this topic you will examine implications and applications of this theorem in detail, focusing on how it lays the foundation for widely used techniques of statistical inference.

OBJECTIVES

- To understand the **Central Limit Theorem (CLT)** as describing the (approximate) sampling distribution of a sample proportion and of a sample mean.
- To recognize the connection between CLT calculations and the simulations that you performed and analyzed in earlier topics.
- To become proficient with using the normal approximation to calculate probabilities of sample statistics falling in given intervals.
- To discover and appreciate the effect of the sample size on the sampling distribution and on relevant probabilities.
- To gain some insight and experience concerning the use of the CLT for drawing inferences about a population parameter based on a sample statistic.

18

385

PRELIMINARIES

1. Take a guess as to the percentage of adult Americans who smoke regularly.

2. Guess which state has the highest percentage of smokers and which has the lowest percentage.

3. Which state, California or Ohio, do you suspect has the larger proportion of residents who speak a language other than English at home?

4. Guess the proportion of California residents who speak a language other than English at home.

5. Guess the proportion of Ohio residents who speak a language other than English at home.

6. If 45% of all Reese's Pieces are orange, are you more likely to find less than 40% orange in a sample of 75 candies or in a sample of 175 candies?

7. If 45% of all Reese's Pieces are orange, are you more likely to find between 35% and 55% orange in a sample of 75 candies or in a sample of 175 candies?

IN-CLASS ACTIVITIES

In the past two topics you have discovered that sample statistics (such as proportions and means) vary from sample to sample according to a predictable long-run pattern. In many situations, this pattern follows a normal distribution centered at the population parameter. Thus, you can use your knowledge of normal distributions to calculate the probability that a sample statistic will fall within whatever values you might be interested in.

The Central Limit Theorems for a sample proportion and for a sample mean state conditions under which a sample statistic follows a normal distribution very closely. In fact, the proportion of "successes" from a binary variable can be thought of as the mean of that variable if the "successes" are coded as "1" and the "failures" as "0." Thus, these two theorems can be considered the same. Accordingly, we will refer to either version as the Central Limit Theorem, or CLT for short.

In the following activities you will apply the Central Limit Theorem to various contexts. Recall that the version for sample proportions is on page 354, and the version for sample means is on page 375.

Activity 18-1: Smoking Rates

The Center for Disease Control and Prevention reported that 22.9% of American adults smoked regularly in 1998. Treat this as the parameter value for the population of American adults.

(a) What symbol would represent this proportion of .229?

(b) Suppose that you take a random sample of 100 American adults. Will the sample proportion of smokers equal exactly .229? Explain.

(c) What does the CLT say about how this sample proportion of smokers would vary from sample to sample? [Be sure to comment on shape, center, and spread of this sampling distribution.]

(d) On the following sketch of this sampling distribution, shade the area corresponding to a sample proportion exceeding .25. Based on this area, make a guess for the
probability that the sample proportion of smokers would exceed .25.

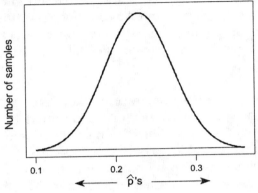

Guess:

(e) Using the CLT result, find the *z*-score corresponding to a sample proportion of
smokers equal to .25.

  mean of sampling distribution:

  standard deviation of sampling distribution:

  z-score standardizing .25:

(f) Use this *z*-score and Table II or your calculator (invNORM found in the DISTR
menu) to compute the probability that the sample proportion of smokers would
exceed .25. Is this value reasonably close to your prediction in (d)?

(g) Suppose that you take a random sample of 400 American adults. How would you
expect the probability of the sample proportion of smokers exceeding .25 to
change from when the sample size was 100? Explain your reasoning, including a
well labeled sketch of the sampling distribution when $n = 400$.

(h) Calculate the probability asked for in (g). Was your conjecture correct? Explain.

(i) Repeat (g) and (h) for a sample of size <u>1600</u>.

(j) Did the *population* size of the United States enter into your calculation?

(k) The state of Virginia was also reported to have 22.9% smokers, the same percentage as for the entire country. Suppose the random samples of only Virginia residents were selected, with the same sample sizes as above. How would the above calculations change if we wanted to find the probability of the sample proportion of Virginia smokers exceeding .25 for each sample size?

As long as the population is large relative to the size of the *sample*, the actual size of the *population* does not affect CLT calculations. However, the size of the sample does have a large impact on CLT calculations.

18

Activity 18-2: Smoking Rates (*cont.*)

The Center for Disease Control and Prevention further reported that 14.2% of Utah residents smoked regularly in 1998. Treat these as the parameter values for Utah.

(a) Suppose that you take a random sample of 100 residents from Utah and find the proportion of smokers in the sample. Sketch the sampling distribution of the sample proportion.

(b) Use the CLT to calculate the probability that this sample proportion would exceed .25.

(c) Would the probability in (b) increase, decrease, or stay the same if the sample size were larger? Explain your reasoning.

(d) If a random sample of 100 residents from an unknown state reveals 25 smokers, would you have strong reason to doubt that the state was Utah? Explain your reasoning.

This activity should reinforce the idea of statistical significance that you encountered in Topics 16 and 17. A statistically significant result is one that is unlikely to occur by random variation alone. You have used simulations to investigate this issue, and the Central Limit Theorem enables you to perform probability calculations to assess this "unlikeliness."

Activity 18-3: Candy Bar Weights (*cont.*)

Recall Activity 15-9 on page 339 and Activity 17-9 on page 382, which asked you to assume that the actual weight of a certain candy bar, whose advertised weight is 2.13 ounces, varies according to a normal distribution with mean $\mu = 2.20$ ounces and standard deviation $\sigma = .04$ ounces.

(a) What is the probability that an individual candy bar will weigh between 2.18 and 2.22 ounces?

(b) Suppose that you plan to take a sample of five candy bars and calculate the sample mean weight. What does the CLT say about how these sample means would vary from sample to sample? Draw a sketch of the sampling distribution.

(c) Shade in the region on your sketch corresponding to the probability that the sample mean weight of these five candy bars will fall between 2.18 and 2.22 ounces. Use your sketch to make a guess for the value of this probability. Is this guess greater or less than your answer to (a)?

18

(d) Use the standard normal probability table or your calculator to compute the probability that the sample mean weight of these five candy bars will fall between 2.18 and 2.22 ounces. Comment on the accuracy of your guess.

(e) How would you expect this probability to change if the sample size were 40 instead of 5? Explain your reasoning.

(f) Calculate this probability, and comment on your conjecture.

(g) Which of your above calculations ((a), (d), and (f)) would remain approximately correct even if the candy bar weights themselves had a skewed, nonnormal distribution?

Activity 18-4: Candy Bar Weights (*cont.*)

(a) Continuing the previous activity, calculate the probability that a sample of size 60 candy bars would result in a mean weight between 2.19 and 2.21 ounces, i.e., within ± .01 of the population mean $\mu = 2.20$ ounces. Accompany your calculation with a well labeled sketch.

(b) Now suppose that the population mean weight was actually $\mu = 2.30$. Calculate the probability that a sample of size 60 candy bars would result in a mean weight within ± .01 of the population mean $\mu = 2.30$ ounces, i.e., between 2.29 and 2.31 ounces. Again accompany your calculation with a sketch.

(c) How do your answers to (a) and (b) compare?

(d) Now suppose that the population mean weight was actually $\mu = 2.15$. Without doing the calculation, what do you expect to be the probability that a sample of size 60 candy bars would result in a mean weight within ± .01 of this population mean? Explain.

(e) Now return to the more realistic assumption that you do not know the value of the population mean μ. What can you say about the probability that a sample of size 60 would result in a sample mean weight within $\pm .01$ of the actual population mean μ?

This activity should remind you and make slightly more formal the notion of statistical **confidence** that you encountered in Topics 16 and 17. Even when you do not know the value of a population parameter, you can be confident that the sample statistic will fall within a certain distance of that unknown parameter value. The Central Limit Theorem allows you to determine probabilities that a sample statistic will fall within a certain distance of the population parameter.

Activity 18-5: Solitaire (*cont.*)

Recall from Activity 14-14 on page 322 that we modeled author A's games of solitaire as having a probability of success $\theta = 1/9$. Suppose he plays $n = 10$ games.

(a) Use the Central Limit Theorem to approximate the probability that he wins 10% or fewer of those games. [*Hint:* Find the mean and standard deviation of the sampling distribution for the sample proportion and standardize the observation .10.]

As we discussed in Activity 14-4 on page 314, since author A either wins or loses on each game and the games are independent of each other, we can find the exact binomial probability for each outcome of number of games won:

# of wins	0	1	2	3	4	5	6	7	8	9	10
proportion of wins	0	.1	.2	.3	.4	.5	.6	.7	.8	.9	1.0
probability	.3079	.3849	.216	.072	.015	.002	.000	.000	.000	.000	.000

(The last four probabilities listed are not exactly zero but are less than .0001.)

(b) Use these binomial probabilities to calculate his (exact) probability of winning 10% or fewer of his ten games.

(c) Are these probabilities in (a) and (b) close?

(d) Explain why the CLT provides such a poor approximation for each probability in this situation. Are the technical conditions concerning n and θ needed for the validity of the CLT met? Explain.

Always remember to check the technical conditions on which the validity of the Central Limit Theorem rests. If they are not satisfied, calculations from the CLT can be erroneous and misleading. For sample proportions, the rule of thumb is that $n\theta \geq 10$ and $n(1 - \theta) \geq 10$. For sample means, the result holds exactly when the population itself is normally distributed, and it holds approximately for large sample sizes, $n \geq 30$ as a rule of thumb.

WRAP-UP

This topic has given you practice using the result of the **Central Limit Theorem** to calculate probabilities that a sample statistic will fall in a certain interval (assuming that the population parameter is known). It has also tried to reinforce the fundamental idea that sampling distributions have less variability with larger sample sizes.

18

In practice, of course, one wishes to go in the other direction: to make inferences about a population parameter based upon the observed value of a sample statistic. Even though these two approaches seem to pull in opposite directions, the Central Limit Theorem actually provides the justification for much of **statistical inference**.

This topic has also explored ideas related to **statistical confidence** and **statistical significance**. These are the two most important concepts in statistical inference; understanding them is crucial to being able to produce and to interpret statistical reasoning. These two ideas form the focus of the next unit of the course.

HOMEWORK ACTIVITIES

Activity 18-6: Sampling Reese's Pieces (*cont.*)

As you did in Activity 16-3 on page 349, consider taking a simple random sample of size $n = 75$ from the population of Reese's Pieces. Continue to assume that 45% of this population is orange ($\theta = .45$).

(a) According to the Central Limit Theorem, how would the sample proportion of orange candies vary from sample to sample? Describe not only the shape of the distribution but also its mean and standard deviation.

(b) Sketch this sampling distribution, and shade the area under this curve corresponding to the probability that a sample proportion of orange candies will be less than .4. Judging from this area, guess the value of the probability.

(c) Use Table II or your calculator to compute the probability that a sample proportion of orange candies will be less than .4.

(d) Calculate the probability that a sample proportion of orange candies will fall between .35 and .55 (i.e., within ±.10 of .45).

(e) Compare this probability with question (k) of Activity 16-3, in which you found the percentage of 500 simulated sample proportions that fell within ±.10 of .45. Are they reasonably close?

Activity 18-7: Sampling Reese's Pieces (*cont.*)

Now suppose that you take simple random samples of size $n = 175$ from the population of Reese's Pieces, still assuming that 45% of this population is orange ($\theta = .45$).

(a) According to the Central Limit Theorem, how would the sample proportion of orange candies vary from sample to sample? Describe not only the shape of the distribution but also its mean and standard deviation. What is different from the situation when the sample size was 75?

(b) Sketch this sampling distribution, and shade the area corresponding to the probability that a sample proportion of orange candies will be less than .4 (with a sample size of 175). Judging from this area, guess the value of the probability.

(c) Calculate the probability that (with a sample size of 175) a sample proportion of orange candies will be less than .4.

(d) How does this probability differ from that when the sample size is 75? Explain why this makes sense.

(e) Calculate the probability that a sample proportion of orange candies (with a sample size of 175) will fall between .35 and .55 (i.e., within ±.10 of .45). How does this probability differ from that when the sample size is 75?

Activity 18-8: ESP Testing (*cont.*)

Consider the ESP test described in Activity 16-4 on page 355. Continue to assume that the subject is presented with 40 cards and that she blindly guesses on each card.

(a) If a subject is just guessing, what does the Central Limit Theorem say about the sampling distribution of the sample proportion of correct responses? Report the mean and standard deviation of this distribution and describe its shape as well.

(b) Draw a sketch of this sampling distribution, and shade the area corresponding to the probability of a guessing subject getting 40% or more correct responses. Based on this area, guess the value of this probability.

(c) Calculate the *z*-score corresponding to .40, and then use the standard normal probability table or your calculator to compute the probability that a guessing subject will give 40% or more correct responses.

(d) Look back at the 10,000 simulated ESP tests presented in Activity 16-4 on page 355. Compare the probability from (c) with the proportion of the 10,000 simulated tests in which a guessing subject got 40% or more correct responses. Are they reasonably close?

(e) How surprising would it be for a subject to get 40% or more correct responses if he or she were just guessing? Would such an event happen less than 10% of the time in the long run? Would it happen less than 1% of the time in the long run?

Activity 18-9: ESP Testing (*cont.*)

Reconsider the ESP test described in Activity 16-4 on page 355. Suppose now that the subject is presented with $n = 100$ cards.

(a) If the subject is just guessing, will he or she *always* get *exactly* 25 correct? Explain.

(b) According to the Central Limit Theorem, how would the proportion of correct responses vary from subject to subject if all of the subjects were just guessing? Specify the shape of the distribution as well as its mean and standard deviation in presenting your answer. Also sketch the distribution (with labels).

(c) Use the CLT and Table II or your calculator to find the probability that a guessing subject would get 27% or more correct out of 100 cards. Also fill in the appropriate area to represent this probability on a sketch of the sampling distribution.

(d) Based on the magnitude of this probability, would you say that it would be very surprising, somewhat surprising, or not terribly surprising for someone to get 27% or more correct even if he or she is blindly guessing on each card?

(e) Repeat (c) and (d) with regard to getting 31% or more correct by guessing.

(f) Repeat (c) and (d) with regard to getting 35% or more correct by guessing.

(g) Try to calculate how many cards a subject would have to identify correctly in order for the probability of having done that well or better by sheer guessing to be only .025.

Activity 18-10: Smoking Rates (*cont.*)

The proportion of smokers among adult residents of Kentucky in 1998 was .308. Treat this as the population proportion, and suppose that you take a random sample of 400 Kentucky residents.

(a) Draw a sketch of the sampling distribution of the sample proportion of smokers.

(b) Determine the probability of obtaining a sample proportion of Kentucky smokers more than .025 away from the population proportion $\theta = .308$ (i.e., either greater than .333 or less than .283). [*Hint:* Either calculate both pieces separately and add them, or calculate the middle piece and subtract from 1, or calculate one piece and explain why it is valid to double it.]

(c) Repeat (b) with a distance of .05 instead of .025.

(d) If you are presented with a random sample of 400 residents from an unknown state, and you find that 25% of the sample are smokers, would you have reason to believe that the state is not Kentucky? Explain.

(e) Repeat (d) if you find that 30% of the sample are smokers.

Activity 18-11: Christmas Shopping (*cont.*)

Consider the population of American households that purchase Christmas presents, and consider the variable "amount expected to be spent on Christmas presents as reported in late November." Suppose that this population has mean $\mu = \$850$ and standard deviation $\sigma = \$250$.

(a) If a random sample of 5 households is selected, is it valid to use the Central Limit Theorem to describe the sampling distribution of the sample mean? Explain.

(b) If a random sample of 500 households is selected, would the Central Limit Theorem tell you the sampling distribution of the sample mean? Draw a sketch of the sampling distribution.

(c) With a random sample of 500 households, determine the probability that the sample mean would fall within ±$18.39 of the population mean $850 (i.e., between $831.61 and $868.39)?

(d) Repeat (c) for the probability that the sample mean would fall within ±$21.91 of $850.

(e) Repeat (c) for the probability that the sample mean would fall within ±$28.80 of $850.

(f) Try to find a value k such that there would be a .8 probability of the sample mean falling within ±k of $850. Fill in the appropriate area on a sketch of the sampling distribution.

(g) If the population mean were actually $\mu = \$1000$, what would be the probability that a random sample of 500 households would produce a sample mean within ±$18.39 of $1000? Does this probability look familiar? Explain.

Activity 18-12: Presidential Votes (*cont.*)

Recall that 49% of the voters in the 1996 Presidential election voted for Bill Clinton. Suppose that you take a simple random sample of 500 voters from this population.

(a) Is .49 a parameter or a statistic?

(b) Determine the probability that the sample proportion of Clinton voters turns out to be less than .45.

(c) Determine the probability that the sample proportion of Clinton voters exceeds .50.

(d) Determine the probability that the sample proportion of Clinton voters falls between .46 and .52.

(e) Determine the probability that the sample proportion of Clinton voters falls between .43 and .55.

(f) Determine the probability that the sample proportion of Clinton voters falls between .49 and .89.

(g) Without doing the actual calculations, indicate how your answers to (b)–(f) would change (get smaller, get larger, or stay the same) if the sample size were 1500 instead of 500.

Activity 18-13: Hospital Births (*cont.*)

Reconsider Activity 14-4 on page 314, which asked about births in two local hospitals: hospital A, which has 100 deliveries per day, and hospital B, which has 20 deliveries per day. Continue to assume that each birth is equally likely to be a boy or a girl.

(a) Use the Central Limit Theorem to sketch the sampling distributions of daily sample proportions of boy births in each hospital. Sketch them on the same scale.

(b) Calculate the probability of hospital A having 60% or more boy births in a day.

(c) Calculate the probability of hospital B having 60% or more boy births in a day.

(d) Again answer which hospital is more likely to have a day for which 60% or more of its births are boys. Compare your answer now to your answer in Activity 14-5 on page 316.

Activity 18-14: Non-English Speakers

In the state of California in 1990, 31.5% of the residents spoke a language other than English at home. Suppose that you take a simple random sample of 100 California residents.

(a) Draw a sketch of the California sampling distribution.
(b) Determine the probability that more than half of the residents sampled would speak a language other than English at home.
(c) Determine the probability that fewer than one-quarter of the residents sampled would speak a language other than English at home.
(d) Determine the probability that between one-fifth and one-half of the residents sampled would speak a language other than English at home.
(e) In the state of Ohio, 5.4% of the 1990 residents spoke a language other than English at home. Suppose that an SRS of 100 Ohio residents is selected. Draw a sketch of the sampling distribution of Ohio on the same scale as that for California from (a).
(f) Without doing the calculations, indicate how you would expect the answers to (b), (c), and (d) to change for Ohio as opposed to California.

Activity 18-15: Solitaire (*cont.*)

Refer to Activity 18-5 on page 394, where you found that the Central Limit Theorem did not provide an accurate approximation for the probabilities of author A's proportion of wins in $n = 10$ games.

(a) According to the rule of thumb presented after Activity 18-5, about how many games would author A have to play before the CLT would make reasonably accurate approximations? Explain.
(b) Recall from Activity 14-15 on page 323 that author B has a probability of about 1/6 of winning a game of solitaire. About how many games would she have to play in order for the CLT to make reasonably accurate approximations? Explain.
(c) For a basketball player who makes 80% of her foul shots, about how many shots would she have to take in order for the CLT to make reasonably accurate approximations? Explain.

Unit V

Inference from Data: Principles

Topic 19:

CONFIDENCE INTERVALS I: PROPORTIONS

OVERVIEW

In the last unit you explored how sample statistics vary from sample to sample. You studied this phenomenon empirically through simulations and theoretically with the Central Limit Theorem. You learned that this variation has a predictable long-term pattern. This pattern enables you to make probability statements about sample statistics, provided that you know the value of the population parameter. The much more common problem is to estimate or to make a decision about an unknown population parameter based on an observed sample statistic. These are the goals of *statistical inference*.

There are two major techniques of classical statistical inference: *confidence intervals* and *tests of significance*. Confidence intervals seek to estimate a population parameter with an interval of values calculated from an observed sample statistic. Tests of significance assess the extent to which sample data support a particular hypothesis concerning the population parameter. This topic extends your study of the concept of statistical confidence by introducing you to the construction of confidence intervals for estimating a population proportion.

OBJECTIVES

- To understand the purpose of *confidence intervals* for estimating population parameters.
- To learn how to construct confidence intervals for estimating a population proportion.
- To appreciate the distinctions between correct and incorrect interpretations of confidence intervals.

- To explore some properties of confidence intervals related to confidence level and sample size.
- To calculate confidence intervals for a variety of genuine applications and interpret the results.

PRELIMINARIES

1. If a new penny is spun on its side (rather than tossed in the air), about how often would you expect it to land "heads" in the long run?

2. Spin a new penny on its side five times. Make sure that the penny spins freely and falls naturally, without hitting anything or falling off the table. How many heads did you get in the five spins?

3. Pool the results of the penny spinning experiment for the entire class; record the total number of spins and the total number of heads.

4. Which magazine would you expect to have advertisements on a higher proportion of its pages, *Sports Illustrated* or *Soap Opera Digest?*

5. Select one of these two magazines. Record which it is, and make a guess concerning the proportion of its pages that contain an advertisement.

6. Mark on the number line below an interval, as small as possible, that you believe with 80% confidence to contain the actual proportion of the magazine's pages that contain an advertisement.

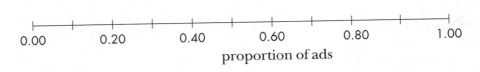

proportion of ads

7. Mark on the number line below an interval, as small as possible, that you believe with 99% confidence to contain the actual proportion of the magazine's pages that contain an advertisement.

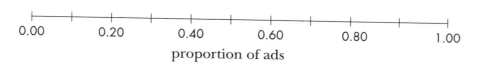

proportion of ads

8. Which of these two intervals is wider, the 80% interval or the 99% interval?

9. Take a guess concerning the proportion of American households that made a financial contribution to a charity in 1995.

10. In Activity 19-5 on page 413 you are asked to analyze the results of a calculator simulation of 200 samples of 75 Reese's Pieces candies. In order for your calculator to conduct this simulation, you need to download SIMINT.83p into your calculator.

Choose a partner and use one of the calculators to run SIMINT.83p and have the other available for calculations. Enter 75 for the sample size, 200 for the number of samples, 0.95 for the confidence level (which you will learn about during the In-Class Activities), and .45 for θ, the assumed population proportion. The information given on your home screen while the program is running tells you the number of samples that have been completed, the upper and lower bounds of the confidence interval (again, which you will learn about), and whether the population proportion falls within the interval.

IN-CLASS ACTIVITIES

Activity 19-1: Penny Spinning

As you spun pennies, you kept track of whether each spin resulted in a Head or a Tail.

(a) What are the observational units in this study?

(b) Is the variable a quantitative or categorical one?

Because the variable of interest is categorical and binary, the relevant parameter of interest is a *proportion*.

(c) Determine the proportion of the class penny spins that landed heads.

(d) Is this proportion a parameter or a statistic? What symbol would you use to represent it?

The parameter in this example, denoted by θ, is the true proportion of spins that would land "heads" among the (hypothetically infinite) population of all possible spins of a penny.

(e) Does the class exercise allow us to determine θ exactly?

(f) Is θ more likely to be close to the observed proportion of heads than to be far from it?

These questions deal with the issue of **statistical confidence** that you explored in Topic 16. While a sample statistic from a simple random sample provides a reasonable estimate of a population parameter, one certainly does not expect the sample statistic to equal the (unknown) population parameter exactly. It is likely, however, that the unknown parameter value is "in the ballpark" of the sample statistic. The purpose of confidence intervals is to use the sample statistic to construct an interval of values that one can be reasonably confident contains the true (unknown) parameter.

You discovered earlier through simulations that going two standard deviations on either side of the sample proportion $\hat{p}$ produced an interval that succeeded in capturing the value of the population proportion θ for about 95% of samples. The Central Limit Theorem revealed this standard deviation of $\hat{p}$ to be $\sqrt{\theta(1-\theta)/n}$.

One complication is that one cannot use this formula in practice because θ is unknown. (Indeed, the whole point of the interval is to estimate θ.)

(g) What seems like a reasonable replacement to use as an estimate of θ in this expresion?

> The estimated standard deviation $\sqrt{\hat{p}(1-\hat{p})/n}$ of the sample statistic $\hat{p}$ is called the **standard error** of $\hat{p}$.

(h) Calculate the standard error of $\hat{p}$ for the penny spinning data collected in class.

(i) Now, to go two standard errors on either side of $\hat{p}$, double this standard error, and then add the result to $\hat{p}$ and subtract it from $\hat{p}$ to form a reasonable interval estimate of θ.

(j) Do you know for sure whether θ is in this interval?

All we can say is that if we were to construct many intervals this way, about 95% of these intervals would contain θ. This has led statisticians to coin a new term, **confidence**, to describe this uncertainty. Thus, our conclusion is that we are "95% confident" that the interval we calculate contains θ.

The other complication is that we are so far limiting ourselves to the value 2 for the number of standard errors and the resulting approximate 95% confidence. Returning to the empirical rule, we know that going just one standard error on either side would correspond to 68%, while going three standard errors would correspond to 99.7%. By using a different multiple of the standard error to vary the distance that we go from $\hat{p}$, we can vary the **confidence level**, which is the measure of how confident one is that the interval does in fact contain the true parameter value. We find this multiplier, called the **critical value**, using the normal distribution, and we will denote it by z^*. One specifies the confidence level by deciding the level of confidence that is necessary for the given situation; common values are 90%, 95%, and 99%. The confidence level, in turn, determines the critical value.

> **Confidence interval for a population proportion θ:**
>
> $$\hat{p} \pm z^* \sqrt{\frac{\hat{p}(1 - \hat{p})}{n}},$$
>
> where $\hat{p}$ denotes the sample proportion, n is the sample size, and z^* represents the critical value from the standard normal distribution for the confidence level desired.

The "±" means that we will add and subtract that quantity from $\hat{p}$. One subtracts the term following it from the term preceding it to get the lower endpoint of the interval and then adds the term following it to the term preceding it to get the upper endpoint of the interval:

$$\left(\hat{p} - z^* \sqrt{\frac{\hat{p}(1 - \hat{p})}{n}}, \ \hat{p} + z^* \sqrt{\frac{\hat{p}(1 - \hat{p})}{n}} \right).$$

> The **technical conditions** necessary for this procedure to be valid are that:
>
> - the sample is a simple random sample from the population of interest;
> - the sample size is large relative to the value of the sample proportion.

A rule of thumb for this second condition is that $n\hat{p} \geq 10$ and $n(1 - \hat{p}) \geq 10$, which says that there must be at least ten "successes" and ten "failures" in the sample. We saw earlier in Topic 12 that we need the simple random sample so that the estimate is unbiased. We also saw that with a large sample size, the sampling distribution of $\hat{p}$ is approximately normal.

Activity 19-2: Critical Values

Before calculating confidence intervals, we need to determine how to calculate critical values z^*. Find the value (call it z^*) such that the area under the standard normal curve between $-z^*$ and z^* is equal to .98 by following the steps below:

(a) Draw a sketch of the standard normal curve and shade the area corresponding to the middle 98% of the distribution. Also indicate where $-z^*$ and z^* roughly fall in your sketch.

(b) Based on your sketch, what is the *total* area under the standard normal curve to the *left* of the value z^*?

(c) Use the table of standard normal probabilities (Table II) or calculator (invNorm) to find the value (z^*) that has this area (your answer to (b)) to its left under the standard normal curve.

This value z^*, which you should have found to be approximately 2.33, is called the ***upper .010 critical value*** of the standard normal distribution because the area to its right under the standard normal curve is .010. This critical value is used for 98% confidence intervals.

(d) Repeat (a)–(c) to find the critical value corresponding to a 95% confidence interval.

(Note: 1.96 is the more exact value that we previously approximated by 2.) Critical values for any confidence level can be found in this manner. Some commonly used critical values are listed here:

confidence level	80%	90%	95%	99%	99.9%
area to left	.90	.95	.975	.995	.9995
area to right	.10	.05	.025	.005	.0005
critical value z^*	1.282	1.645	1.960	2.576	3.291

Activity 19-3: Penny Spinning (*cont.*)

(a) For the penny spinning data collected in class, construct a 95% confidence interval for θ, the true proportion of all (hypothetical) penny spins that would land heads.

(b) Can you be *certain* that this interval contains the actual value of θ?

(c) What is the *width* of this interval? (The width is the difference between the upper and lower endpoints of the interval.)

(d) What is the *half-width* of this interval? (The half-width is the width divided by 2.)

(e) Explain how you could have determined the half-width of the interval without first having calculated the width.

(f) What technical conditions underlie the validity of the procedure? Do they seem to be satisfied here?

Activity 19-4: Halloween Practices and Beliefs (*cont.*)

As mentioned in Activity 16-13 and Activity 16-14, the Gallup organization conducted a poll about Halloween practices and beliefs during October 21–24, 1999. A sample of 1005 adult Americans were asked whether someone in their family would give out Halloween treats from the door of their home, and 69% answered in the affirmative.

(a) Is .69 a parameter or a statistic?

(b) Does the question asked by Gallup constitute a quantitative or a categorical variable?

(c) Construct a 95% confidence interval for θ, the proportion of all adult Americans who planned to give out Halloween treats from their home in 1999.

(d) Indicate on the axis below the values that fall within this interval.

(e) Report the half-width and the width of this interval.

half-width: width:

(f) Use the Gallup survey results to create a 95% confidence interval for the proportion of all adult Americans who did *not* plan to give out Halloween treats from their home in 1999. Also report the half-width and width of this interval.
   interval:

half-width: width:

(g) Are these two half-widths the same? Are the two intervals the same? Are they related? Explain.

Remember that an interval covers a range of values. It is not just its width and not just a pair of numbers but an *interval*. The half-width of a 95% confidence interval is often called the survey's **margin of error**.

(h) How would you expect the interval in (c) to differ if the confidence level were 99% instead of 95%? Explain.

(i) Calculate a 99% confidence interval for the proportion of American adults who planned to give out Halloween treats from their home in 1999. Also report the width and half-width of the interval.

  interval: half-width:

(j) How does the 99% interval compare with the 95% interval? Compare their midpoints as well as their widths.

> As one increases the confidence level desired, the width of the confidence interval increases.

(k) In the same Gallup poll, a sample of 493 adults were asked whether they believe in witches, and 22% said yes. Use your calculator to find a 95% confidence interval for the proportion of the population that believes in witches:
  • Go to the STAT TESTS menu (which you can find by pressing the STAT button) and scroll down until you find option 1-PropZInt.
  • Input the number of successes (people who said yes), sample size, and confidence level as follows:

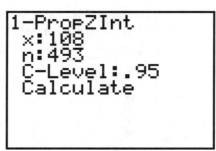

```
1-PropZInt
 x:108
 n:493
 C-Level:.95
 Calculate
```

Note that you could enter `int(.22*493)` on the x line where int(is found in the MATH NUM menu. If the result is an integer you can type .22*493. Highlight "Calculate" and press ENTER. Report the interval below, including the midpoint and half-width.

interval: midpoint: half-width:

(l) For the poll question about witches, suppose that the sample size had been 1005 instead of 493, and that 22% had answered yes. Would you expect a 95% confidence interval based on this larger sample to be wider or narrower than the one you found in (k)? Would you expect the midpoint of the interval to be the same or different? Explain.

wider/narrower: midpoint:

(m) Use your calculator to compute the 95% confidence interval asked for in (l). [Remember to convert to the *number of* yeses first.] Also report the midpoint and the half-width. How do they compare to the case where the sample size was 493?

interval: midpoint: half-width:

comparison?

A larger sample size produces a narrower confidence interval whenever other factors remain the same.

Activity 19-5: Reese's Pieces (*cont.*)

Assume for the moment that 45% of all Reese's Pieces are orange, i.e., that the population proportion of orange candies is $\theta = .45$. To examine more closely what the interpretation of "confidence" means in this statistical context, we asked you in the preliminaries section to use your calculator to simulate 200 samples of 75 candies each, assuming that $\theta = .45$. The calculator constructed a 95% confidence interval for each of the sample proportions of orange candies obtained. The results of this simulation are automatically stored within the Stat List Editor. The list names are LOWER, UPPER, and PHAT.

(a) How many of the 200 confidence intervals actually contain the known value (.45) of θ? What percentage of the 200 samples produces an interval that contains .45?

(b) For each interval that does *not* contain the actual value (.45) of θ, record its sample proportion $\hat{p}$ of oranges obtained (this information is stored in a list named PROP).

(c) If you had taken a *single* sample of 75 candies, would you have any *definitive* way of knowing whether your 95% confidence interval contained the actual value of θ?

(d) Based on what you have seen in this simulation, explain what it means to be 95% confident that your interval contains the actual value of θ.

This simulation illustrates the following about the proper use and interpretation of confidence intervals:

- One interprets a 95% (for instance) confidence interval by saying that one is 95% *confident* that the interval contains the true value of the population proportion.
- More specifically, this interpretation means that if one repeatedly takes simple random samples from the population and constructs 95% confidence intervals for each sample, then in the long run 95% of these confidence intervals will capture the true population proportion.
- It is incorrect to say that θ has probability .95 of falling within the 95% confidence interval. It is also technically incorrect to say that the probability is .95 that a particular 95% confidence interval contains the actual value of θ. The technicality here is that θ is not random; it is some fixed (but unknown) value. What is random, changing from sample to sample, is the sample proportion and thus the interval based on it. Thus the probability statement applies to what values an interval will take prior to the sample being collected, not whether a particular interval contains the fixed parameter value.
- Confidence intervals estimate the value of a parameter; they do not estimate the value of a statistic or of an individual observation.

(e) If we conducted this simulation again, using 95% confidence intervals based on samples of size <u>250</u> instead of samples of size 75, how many and what proportion of the 200 intervals would you expect to succeed in capturing the actual value of θ? What would be the difference between these intervals and the intervals with samples of size 75?

(f) Repeat (e), this time using 80% confidence intervals with a sample size of 75.

Activity 19-6: *Literary Digest* Poll (*cont.*)

Recall from Activity 12-2 on page 249 that in 1936 the *Literary Digest* received 2.4 million responses to their survey, with 57% of the sample indicating support for Alf Landon over Franklin Roosevelt in the upcoming election.

(a) Use this sample result to form a 99.9% confidence interval for θ, the actual proportion of all adult Americans who preferred Landon over Roosevelt.

(b) In the actual election, 37% of the population voted for Landon. Explain why the above confidence interval did such a poor job of predicting the election result.

> This activity serves as a reminder of the critical importance of collecting data by sound strategies. Inference procedures such as confidence intervals are invalid, and can produce very misleading results, when applied to data obtained through a biased sampling plan.

WRAP-UP

This topic has provided your first exposure to statistical inference by introducing you to **confidence intervals**, a very widely used inference technique. In addition to discovering how to construct confidence intervals for a population proportion, you have seen proper and improper interpretations of them and explored the effect of the confidence level and the sample size on the interval.

In the next topic you will continue to study confidence intervals, turning your attention to quantitative variables and therefore to a population mean as the parameter of interest. You will see that while the details of implementing the procedure change, the underlying structure and interpretation of confidence intervals remain the same.

HOMEWORK ACTIVITIES

Activity 19-7: Magazine Advertisements

The September 13, 1999 issue of *Sports Illustrated* had 116 pages, and 54 of them contained an advertisement. The September 14, 1999 issue of *Soap Opera Digest* consisted of 130 pages, including 28 with advertisements.

(a) What are the observational units in this study?

(b) Consider whichever of these magazines you selected in question 5 of the "Preliminaries" section. Treat this issue's pages as a sample from the population of all of the magazine's pages, and construct a 95% confidence interval for the population proportion of pages that contain ads.

(c) Write a sentence or two interpreting this interval. [Be sure to relate your interpretation to the context.]

(d) Clearly explain what is meant by the phrase "95% confidence" in this context.

(e) Does the interval contain the sample proportion of pages with ads?

(f) Explain why the previous question is silly and did not require you to look at the interval at all.

(g) Find a recent issue of another magazine and repeat this analysis on it to produce a confidence interval for the proportion of its pages that contain ads.

Activity 19-8: Phone Book Gender (*cont.*)

Suppose that you want to estimate the proportion of women among the residents of San Luis Obispo County, California. Recall from Activity 12-17 on page 266 that a random sample of columns in the phone book revealed 36 listings with both male and female names, 77 listings with a male name only, 14 listings with a female name only, 34 listings with only initials, and 5 listings with a pair of initials.

(a) How many first names are supplied in these listings altogether? [*Hint:* Ignore the listings with initials, and remember to count both male and female names for the couples.]
(b) How many and what proportion of those names are female?
(c) Use these sample data to form a 90% confidence interval for the proportion of women in San Luis Obispo County.
(d) Do you have any concerns about the sampling method that might render this interval invalid? Explain.
(e) Suggest a more reasonable but still practical sampling method for estimating the proportion of women in the county.

Activity 19-9: Closeness of Baseball Games

The following table tallies the margin of victory in the 190 Major League Baseball games played between July 26 and August 8, 1999 (MARGN in BASEBALL.83g):

Margin	1	2	3	4	5	6	7	8	9	10	11
tally	38	37	27	23	12	16	13	14	1	5	4

(a) Identify the observational units and variable of interest in this study.
(b) Is this "margin of victory" variable quantitative or categorical? Does the confidence interval procedure of this topic apply?
(c) Consider a new variable: whether or not a game had a victory margin of exactly one run. Is this variable quantitative or categorical? Does the procedure of this topic apply?
(d) What proportion of these games were decided by one run? Is this number a parameter or a statistic?
(e) Let θ denote the true proportion of *all* Major League Baseball games that are decided by one run. Is θ a parameter or a statistic?
(f) Use the sample result to estimate θ with a 90% confidence interval, with a 95% confidence interval, and with a 99% confidence interval.
(g) Write a sentence or two describing what these intervals reveal about the proportion of baseball games decided by a single run.

(h) Is this sample a simple random one from the population of all Major League Base-ball games? If not, is the sample likely to be biased in a certain direction with respect to margin of victory? Explain.

(i) If these games were again used as the sample but game-time temperature were the variable of interest, would the sample likely be biased in a certain direction? Explain.

Activity 19-10: Home Field Advantage

For the sample of 190 baseball games described in Activity 19-9, another variable recorded for each game was whether the home or visiting team won. A 95% confidence interval for the population proportion of games won by the home team turns out to be (.450, .592).

(a) Use this information to determine the sample proportion of games won by the home team.

(b) How many of the 190 games in the sample must have been won by the home team?

(c) Find a 99% confidence interval for the population proportion of games won by the home team.

(d) What is suggested by the fact that the interval includes the value .5? Explain.

Activity 19-11: Random Babies (*cont.*)

Reconsider the simulation data collected by the class in Activity 14-1 on page 305.

(a) What proportion of the simulated repetitions resulted in no mothers getting the correct baby?

(b) Use this sample information to form a 95% confidence interval for the long-term proportion of times that no mother would get the right baby.

(c) Write a sentence or two interpreting what this interval says. [Be sure to relate your interpretation to the context.]

This is a rare instance in which you can actually calculate the population parameter and check whether the confidence interval succeeds in capturing it. The parameter here is the theoretical probability of zero matches, which you calculated in Activity 14-2 on page 309.

(d) Report the value of this population parameter.

(e) Does the 95% confidence interval succeed in capturing the population parameter?

(f) If 1000 different statistics classes carried out this simulation and calculated a 95% confidence interval as you did in (b), about how many of their intervals would you expect to succeed in capturing the parameter value? Explain.

(g) Use the simulated sample data to form an 80% confidence interval for the parameter. Does this interval succeed in capturing its value?

Activity 19-12: Television Magic

In June of 1994, the *Nick-at-Nite* television network conducted a phone-in survey of its viewers with regard to the question, "Which classic sitcom character, Jeannie from *I Dream of Jeannie* or Samantha from *Bewitched*, possessed more magic powers?" Samantha received 810,000 votes to Jeannie's 614,000.
(a) Identify the population and the parameter of interest in this case.
(b) Is this a simple random sample from the population of interest?
(c) Find a 97% confidence interval for the population parameter of interest.
(d) Explain why this confidence interval is so narrow.

Activity 19-13: Cat Households (*cont.*)

A survey of 80,000 American households in 1996 found that 27.3% of the households in the sample owned a pet cat.
(a) Use this sample information to form a 99% confidence interval to estimate the proportion of all American households that owned a pet cat in 1995.
(b) Write a sentence or two interpreting what this interval means.
(c) Explain what the phrase "99% confidence" means in this context.
(d) What additional information about this survey would you need in order to comment on whether the technical conditions underlying the confidence interval procedure are satisfied in this case?

Activity 19-14: Charitable Contributions

In a survey of American households, 68.5% of the households claimed to have made a financial contribution to charity in 1995.
(a) Describe the parameter value of interest in this situation.
(b) If the survey had involved 250 households, what would a 99% confidence interval for this parameter be?
(c) Repeat (b), supposing that the survey had involved 500 households.
(d) Repeat (b), supposing that the survey had involved 1000 households.
(e) Repeat (b), supposing that the survey had involved 2000 households.
(f) Compare the half-widths of these four intervals. Describe how they are related. [Be as specific as possible.]
(g) Does doubling the sample size cut the half-width in half? If not, by what factor does the sample size have to be increased in order to cut the half-width in half?
(h) The actual survey involved 2719 households. Find and interpret a 99% confidence interval for the parameter.

Activity 19-15: Marriage Ages (*cont.*)

Reconsider Activity 3-13 on page 60, in which you analyzed the ages of couples who applied for marriage licenses. Consider now the issue of estimating the proportion of all marriages in this county for which the bride is younger than the groom. These 24 couples are actually a subsample from a larger sample of 100 couples who applied for marriage licenses in Cumberland County, Pennsylvania, in 1993. Marriage ages for this larger sample appear below:

husb	wife	husb	wife	husb	wife	husb	wife	husb	wife
22	21	40	46	23	22	31	33	24	25
38	42	26	25	51	47	23	21	25	24
31	35	29	27	38	33	25	25	46	37
42	24	32	39	30	27	27	25	24	23
23	21	36	35	36	27	24	24	18	20
55	53	68	52	50	55	62	60	26	27
24	23	19	16	24	21	35	22	25	22
41	40	52	39	27	34	26	27	29	24
26	24	24	22	22	20	24	23	34	39
24	23	22	23	29	28	37	36	26	18
19	19	29	30	36	34	22	20	51	50
42	38	54	44	22	26	24	27	21	20
34	32	35	36	32	32	27	21	23	23
31	36	22	21	51	39	23	22	26	24
45	38	44	44	28	24	31	30	20	22
33	27	33	37	66	53	32	37	25	32
54	47	21	20	20	21	23	21	32	31
20	18	31	23	29	26	41	34	48	43
43	39	21	22	25	20	71	73	54	47
24	23	35	42	54	51	26	33	60	45

(a) For how many of these 100 marriages can you determine which partner was younger? (In other words, eliminate the cases in which both bride and groom listed the same age on the marriage license, since you cannot tell which partner is younger in those cases.)

(b) In how many of *these* marriages is the bride younger than the groom?

(c) Find a 90% confidence interval for the proportion of all marriages in this county for which the bride is younger than the groom.

(d) Repeat (c) with 95% confidence.

(e) Repeat (c) with 99% confidence.

(f) Do any of these intervals include the value .5?

(g) Comment on whether the sample data suggest that the bride is younger than the groom in more than half of the marriages in this county.

Activity 19-16: Newspaper Reading

The 1998 General Social Survey asked a sample of adult Americans how often they read a newspaper. Of the 1870 who responded, 805 said that they read a newspaper daily.

(a) Produce a 90% confidence interval for the proportion of all American adults who read a newspaper daily.

(b) Does this interval include the value .5?

(c) Comment on whether this sample provides any reason to doubt the proposition that half of all American adults read a newspaper daily.

Activity 19-17: Emotional Support (*cont.*)

Recall from Activity 12-11 on page 264 that Shere Hite received mail-in questionnaires from 4500 women, 96% of whom claimed that they give more emotional support than they receive from their husbands or boyfriends. Also recall that an ABC News/ *Washington Post* poll of a random sample of 767 women found that 44% claimed to give more emotional support than they receive.

(a) Determine the margin of error for each of these surveys. Also report each survey's 95% confidence interval for the proportion of all American women who feel that they give more emotional support than they receive.

(b) Are these two confidence intervals similar? Do they overlap at all?

(c) Which survey has the smaller margin of error, i.e., the narrower confidence interval?

(d) Which of these two confidence intervals do you have more confidence in? Explain.

Activity 19-18: Veterans' Marital Problems

Researchers found that in a sample of 2101 Vietnam veterans, 777 had been divorced at least once.

(a) Use this sample information to form a confidence interval for the proportion of divorced men among *all* Vietnam veterans.

(b) Would it be legitimate to use this interval as an estimate of the divorce rate among all middle-aged American men? Explain.

Activity 19-19: E-Mail Usage (*cont.*)

Reconsider again the data collected in Topic 12 on students' e-mail usage.
(a) What proportion of the students sampled had read or sent at least one e-mail message in the preceding twenty-four hours?
(b) Use these sample data to construct a 90% confidence interval to estimate the proportion of all students at your school who had read or sent an e-mail message in those 24 hours.
(c) Are the technical conditions that underlie the validity of this procedure satisfied? Explain.

Activity 19-20: Critical Values

Determine the critical value z^* corresponding to:
(a) 85% confidence
(b) 97.5% confidence
(c) 51.6% confidence

Activity 19-21: Incorrect Conclusions (*cont.*)

Suppose that Andrew and Becky both study a random sample that has a sample proportion of $\hat{p} = .4$. Each uses the sample data to produce a confidence interval for the population proportion, with Andrew obtaining the interval (.346, .474) and Becky obtaining the interval (.286, .514).
(a) One of these intervals has to be wrong. Identify which, and explain why.
Suppose that Andrew and Becky decide to gather new random samples, with one using a sample size of 100 and the other using a sample size of 200. Andrew's confidence interval turns out to be (.558, .682), and Becky's is (.611, .779).
(b) Report the sample proportion $\hat{p}$ obtained by each researcher. (Assume that both of these intervals were calculated correctly.)
(c) Report the half-width of each interval.
(d) Explain why the information provided does not enable you to tell which sample size was used by which researcher.
(e) Suppose that Andrew had the sample size of 100 and Becky 200. Determine the confidence level used by each. [*Hint:* First determine the value of the z^* critical value used by each.]
Suppose that Andrew and Becky decide to study another issue with new samples, one with a sample size of 250 and one with a sample size of 1000. Both decide to form 90%

confidence intervals from their samples. Andrew obtains the interval (.533, .635), and Becky gets (.550, .602).

(f) Report the sample proportion and the half-width for each interval.

(g) Which sample size goes with which researcher? Explain.

(h) Which interval is more likely to succeed in capturing the population proportion? Explain.

## 20

Topic 20:

CONFIDENCE INTERVALS II: MEANS

You have explored confidence intervals for a population proportion. We now turn from binary categorical variables to quantitative variables, so that the population mean is the parameter of interest. In this topic you will investigate and apply confidence interval procedures for estimating a population mean. Some of the details of the procedure will change, and you will work with a new probability model called a ***t-distribution***, but you will find that the reasoning and interpretation of confidence intervals remain unchanged.

OBJECTIVES

- To learn to apply a confidence interval procedure for estimating a population mean.
- To recognize similarities of structure and interpretation between confidence interval procedures for a population mean and for a population proportion.
- To become familiar with the ***t-distribution*** for finding critical values.
- To discover the roles played by sample size and sample standard deviation as well as by sample mean concerning confidence intervals.
- To see how to check whether the *t*-procedure can be validly applied to a given sample of data.
- To continue to recognize the utility of graphical analyses and the limitation of summarizing a distribution of data exclusively by its mean.

PRELIMINARIES

1. Take a guess as to the least amount of sleep that a student in your class got last night. Also make a guess for the most sleep enjoyed by a student in your class last night.

2. Mark on the scale below an interval that you believe with 90% confidence to include the mean amount of sleep (in hours) that students at your school got last night.

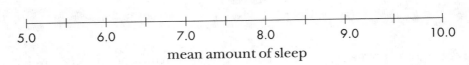

mean amount of sleep

3. Mark on the scale below an interval that you believe with 99% confidence to include the mean amount of sleep (in hours) that students at your school got last night.

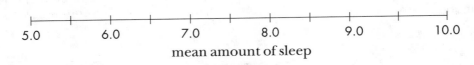

mean amount of sleep

4. Which of these two intervals is wider?

5. Record the bedtimes and wake times of the students in your class last night. Also calculate and record the sleeping times, in hours (e.g. 2.2 hours = 2 hours and 12 minutes).

student	bed time	wake time	sleep time
1			
2			
3			
...			

6. Take a guess as to the average number of hours per week that third or fourth graders spend watching television.

IN-CLASS ACTIVITIES

Activity 20-1: Christmas Shopping (*cont.*)

Recall from Activity 17-3 on page 376 that the Gallup organization conducted a survey about how much money people expected to spend on Christmas presents in 1999. The sample mean of the 922 responses was 857 dollars.

(a) Is the variable "amount expected to be spent on Christmas presents in 1999" categorical or quantitative?

(b) Is $857 a parameter or a statistic? Explain. What symbol would represent it?

(c) Identify the parameter of interest in this study. What symbol would represent it?

(d) Do you know the value of the parameter in this study? Is it more likely to be close to $857 than far from it?

As you have seen, we can form an interval estimate of a parameter by starting with the sample statistic and going two standard deviations of that statistic on either side of it. The Central Limit Theorem for a sample mean tells us that the standard deviation of a sample mean $\bar{x}$ is $\sigma/\sqrt{n}$, where σ represents the population standard deviation and n the sample size.

(e) Suppose that for these expected shopping expenditures the population standard deviation is σ = 250. Calculate the standard deviation of the sample mean $\bar{x}$.

(f) Go two standard deviations on either side of the sample mean to form a reasonable interval estimate of the population mean μ.

One problem with this strategy is that we had to supply a guess for the population standard deviation σ, which is almost never known. (After all, if the population standard deviation were known, isn't it likely that the population mean would also be known?)

(g) What is a reasonable substitute for σ that can be calculated from sample data?

> The estimated standard deviation of a sample mean $\bar{x}$, also known as the **standard error** of the sample mean, is $s/\sqrt{n}$, where s denotes the sample standard deviation. Since this estimate introduces more uncertainty into the process, the normal distribution is no longer the appropriate distribution from which to determine critical values. We will use what is called a **t-distribution** for this purpose.

This t-distribution is actually an entire family of distribution curves, not unlike the normal distributions. Whereas a normal distribution is identified by its mean and standard deviation, a t-distribution is characterized by an integer number called its **degrees of freedom** (abbreviated d.f.). These t-distributions are mound-shaped and centered at zero, but they are more spread out (i.e., they have wider, fatter tails) than a normal distribution. As the number of degrees of freedom increases, the tails get narrower and the t-distribution gets closer and closer to a normal distribution.

Activity 20-2: Parents' Ages (*cont.*)

Recall from Activity 3-11 on page 59 and Activity 17-5 on page 379 the data on ages at which a sample of 35 mothers had their first child. The sample mean turns out to be 22.31 years, and the sample standard deviation is 5.60 years. Before we can construct a confidence interval for the population mean, we need to find the appropriate critical value from the *t*-distribution.

Previously, you have used a normal probability table to find critical values; now you will see how to use a *t*-table. A *t*-table can be found in Table III on page 593. Notice that each line of the table corresponds to a different value of the degrees of freedom. Always start by going to the line that is relevant for the degrees of freedom with which you are working. Next note that the table gives various values of "area to the right" across the top of the table. Finally, observe that the body of the table gives values such that the probability of lying to the right of that value (equivalent to the area to the right of that value under the *t*-distribution) is given at the top of the column.

In conducting inferences about a population mean, the degrees of freedom can be found by subtracting 1 from the sample size, d.f.=$n-1$.

(a) Draw a rough sketch of the *t*-distribution with $35 - 1 = 34$ degrees of freedom. (It should look very much like a standard normal curve.)

(b) The critical value t^* for a 95% confidence interval is the value such that 95% of the area under the curve is between $-t^*$ and t^*. Shade this area on your sketch above.

(c) What is the area to the *right* of t^* under the curve?

(d) Look at the *t*-table to find the value t^* that has that area to its right under a *t*-distribution with 34 degrees of freedom.

(e) Is this critical value less than or greater than the critical value z^* from the standard normal distribution for a 95% confidence interval?

Critical values t^* from the t-distribution are always greater than their counterparts from the z- (normal) distribution.

Confidence interval for a population mean μ:

$$\bar{x} \pm (t^*_{n-1}) \frac{s}{\sqrt{n}},$$

where t^*_{n-1} is the appropriate critical value from the t-distribution with $n-1$ **degrees of freedom** (rounding down if the exact value does not appear in the table).

Two **technical conditions** must be satisfied in order for this procedure to be valid:
- the sample is a simple random sample from the population of interest;
- *either* the sample size is large ($n \geq 30$ as a rule of thumb) *or* the population is normally distributed.

The second technical condition stems from what you saw in Topic 17: that the sampling distribution of the sample mean will be normal if the population itself is normal or approximately normal for any population shape as long as the sample size is large.

(f) Calculate a 95% confidence interval for the population mean age at birth of first child for this population based on the sample results for these 35 mothers.

(g) Now use your calculator to find a 95% confidence interval for the population mean age at birth of first child for this population based on the sample results for these 35 mothers. To do this use the TInterval feature (found in the STAT TESTS menu of the calculator). Select Stats as the Input option, and use the down arrow to enter all of the relevant information. Highlight Calculuate and press ENTER. How does this compare with your answer in (f)?

(h) Do you believe that the distribution of ages at which the population of mothers had their first child follows a normal distribution? Is normality of the population required for the use of this procedure to be valid?

The *t*-procedures are fairly **robust** in that they tend to give reasonable results even for small sample sizes as long as the population is not severely skewed or has extreme outliers.

(i) Review the plot of the sample data you constructed in Activity 3-11 on page 59. Does this sample behave reasonably well enough for you to believe the above procedure to have been valid even with a smaller sample size?

(j) Write one or two sentences interpreting the interval constructed in (f).

Finally, be aware that the reasoning behind and the interpretation of these confidence intervals are the same as always: If one were to repeat the procedure over and over for many samples, in the long run 95% of the intervals so generated would contain the population mean. This allows us to say that we are 95% confident that the interval we constructed contains the true value of the population mean.

Another way to think of confidence is to suppose that we were to take all possible samples from this population and to construct a confidence interval from each. Now suppose that we were to write each interval on a slip of paper and put the slips in a bag. Then if we were to randomly draw one slip from the bag, there would be a .95 probability that the interval we draw would contain the true parameter value.

Activity 20-3: Parents' Ages (*cont.*)

Recall that this sample of 35 mothers was actually taken from a much larger sample of 1199 mothers, for whom the mean age of having the first child was 22.52 years, with a standard deviation of 4.89 years.

(a) If you were to use this larger sample to form a 95% confidence interval for μ, would you expect the interval to be narrower or wider than the one based on a sample of 35 mothers? Explain.

(b) Use the larger sample to construct a 95% confidence interval for μ. How does this interval compare to the confidence interval from Activity 20-2?

(c) How would you expect a 90% confidence interval based on the larger sample to compare with the interval from (b)? Explain.

(d) Calculate this 90% confidence interval, and comment on whether your prediction in (c) is confirmed.

These questions should convince you that confidence intervals for a population mean behave like those for a population proportion. Increasing the sample size makes the interval narrower, and increasing the confidence level makes the interval wider.

Activity 20-4: Sleeping Times

Suppose that you want to estimate the mean sleeping time of *all* students at your school last night. Consider the four different (hypothetical) samples of sleeping times presented in the dotplots and boxplots below:

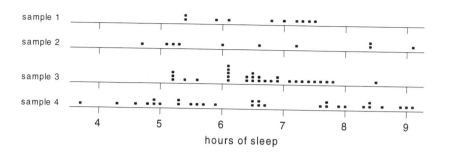

hours of sleep

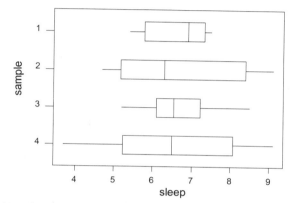

sleep

(a) The following descriptive statistics were calculated from these sample data. Fill in the "sample number" column by figuring out which statistics go with which plots.

Sample number	Sample size	Sample mean	Sample std. dev.
	30	6.6	.825
	10	6.6	.825
	10	6.6	1.597
	30	6.6	1.597

(b) What do all of these samples have in common?

(c) What strikes you as the most important difference between the distribution of sleeping times in sample 1 and in sample 2?

(d) What strikes you as the most important difference between the distribution of sleeping times in sample 1 and in sample 3?

Below, the 95% confidence interval from each sample is given.

Sample number	Sample size	Sample mean	Sample std. dev.	95% conf. int.
3	30	6.6	.825	(6.292, 6.908)
1	10	6.6	.825	(6.010, 7.190)
2	10	6.6	1.597	(5.457, 7.743)
4	30	6.6	1.597	(6.004, 7.196)

(e) Which of these two samples (1 or 2) produces a more accurate estimate of μ, i.e., a narrower confidence interval for μ? Explain why this makes sense. [*Hint:* Refer to your answer in (c).]

(f) Which of these two samples (1 or 3) produces a more accurate estimate of μ, i.e., a narrower confidence interval for μ? Explain why this makes sense. [*Hint:* Refer to your answer in (d).]

> In addition to sample size and confidence level, the sample standard deviation plays a role in determining the width of a confidence interval. Samples with more variability produce wider confidence intervals.

Activity 20-5: Sleeping Times (*cont.*)

Now consider the data on sleeping times collected from the students in your class.

(a) Enter the sleeping times (in hours) into your calculator. Then produce visual displays of the distribution. Write a few sentences commenting on key features of this distribution.

(b) Use your calculator to compute the sample size, sample mean, and sample standard deviation. Record these values below, and indicate the symbol used to represent each.

(c) Let μ denote the mean sleeping time for *all* students at your school on that particular night. Use your calculator to find a 90% confidence interval for μ:

- Use the TInterval feature (found in the STAT TESTS menu of the calculator) and select Data.
- On the List line, select the list you created. Change the value for C-Level to .90. Highlight "Calculate" and press ENTER.

Report the resulting interval and write a sentence interpreting the interval.

(d) Count how many of the sample values fall within the interval. What percentage of the sample is this?

(e) Is this percentage in (d) close to 90%? Should it be? Explain.

Confidence intervals of this type estimate the value of a population *mean*. They do not estimate the values of *individual* observations in the population or in the sample.

WRAP-UP

This topic has continued your study of confidence intervals by introducing you to confidence intervals for a population mean. You have found that the structure, reasoning, and interpretation of these intervals are the same as for a population proportion. You have also encountered the ***t-distribution*** and examined properties of these confidence intervals. Effects of sample size and confidence level are as before, and you have learned to distinguish between an interval estimate for a population mean and for an individual observation.

In the next two topics you will turn your attention to the second principal technique of statistical inference: ***tests of significance***. You will again find that these procedures build on the concept of a sampling distribution that you studied earlier. In the next topic you will investigate tests of significance concerning a population proportion, and the topic after that will extend the procedure to considering a population mean.

HOMEWORK ACTIVITIES

Activity 20-6: Sleeping Times (*cont.*)

Consider again the sleeping times collected in class and the confidence interval that you produced from them. Describe how the confidence interval would have been different if the only change had been:
(a) a larger standard deviation among the sample sleeping times.
(b) a smaller sample size.
(c) a larger sample mean by 0.5 hours.
(d) each person's sleeping time had been 15 minutes longer than reported.

Activity 20-7: Critical Values (*cont.*)

(a) Use the *t*-table to find the critical values t^* corresponding to the following confidence levels and degrees of freedom, filling in a table as the one below with those critical values:

d.f. \ conf. level	80%	90%	95%	99%
4				
11				
23				

80				
infinity				

(b) Does the critical value t^* get larger or smaller as the confidence level gets larger (if the number of degrees of freedom remains the same)?

(c) Does the critical value t^* get larger or smaller as the number of degrees of freedom gets larger (if the confidence level remains the same)?

(d) Do the critical values from the t-distribution corresponding to infinitely many degrees of freedom look familiar? Explain. (Refer back to Topic 19 if they do not.)

Activity 20-8: Sentence Lengths

The following data are the lengths (measured as numbers of words) in a sample of 28 sentences from chapter 3 of John Grisham's novel *The Testament* (GRISH.831).

17	21	8	32	13	16	17	37	27	20	30	15	64	34
18	26	23	17	5	10	29	9	22	18	7	16	13	10

(a) Look at a dotplot, histogram, and boxplot of these data. Write a few sentences commenting on key features of the distribution.

(b) Use these sample data to produce a 95% confidence interval for the mean length among all sentences in this book.

(c) Explain in your own words what is meant by the phrase "95% confidence" in this context.

(d) Comment on whether the technical conditions necessary for the validity of this procedure seem to be satisfied.

Activity 20-9: Coins' Ages (*cont.*)

Recall from Activity 17-1 on page 367 the population of 1000 coins from which you drew random samples.

(a) As you did in Activity 17-1, use a table of random digits to select a random sample of ten pennies from this population. Record their ages.

(b) Use this sample to construct a 90% confidence interval for the mean age of the pennies in this population.

(c) Use your calculator to determine the mean age in that population of 1000 pennies.

(d) Do you think the technical conditions for this procedure are met? Explain.

(e) Does the interval in (b) succeed in capturing this population mean?

(f) If you had constructed a 95% interval, would it have been more likely to capture the population mean than your interval in (b)?

(g) If you had taken a random sample of 40 pennies and calculated a 90% interval, would it have been more likely to capture the population mean than your interval in (b)? Explain.

_____

Activity 20-10: Children's Television Viewing (*cont.*)

Recall from Activity 13-11 on page 294 the study that investigated a relationship between watching television and obesity in third and fourth grade children. Prior to assigning children to treatment groups, researchers gathered baseline data on their television viewing habits. Children were asked to report how many hours of television they watch in a typical week. The 198 responses had a mean of 15.41 hours and a standard deviation of 14.16 hours.

(a) Are these values parameters or statistics? Explain.

(b) Do you think the technical conditions for the confidence interval for μ were met? Explain.

(c) Use this sample information to determine 90%, 95%, and 99% confidence intervals for the mean hours of television watched per week among all third and fourth graders.

(d) Do any of these intervals include your guess from the "Preliminaries" section?

(e) In this situation would it make much difference if z-intervals were used in place of t-intervals? Explain.

_____

Activity 20-11: Closeness of Baseball Games (*cont.*)

Reconsider Activity 19-9 on page 417, which reported the following table of counts concerning margins of victory in the 190 Major League Baseball games played between July 26 and August 8, 1999:

Margin	1	2	3	4	5	6	7	8	9	10	11
count	38	37	27	23	12	16	13	14	1	5	4

(a) The sample mean is $\bar{x} = 3.93$ runs per game, and the sample standard deviation is $s = 2.66$ runs. Find a 99% confidence interval for μ, the mean margin of victory among all Major League Baseball games.

(b) How many and what percentage of these 190 games have a margin of victory that falls within this interval? Is this close to 99%?

(c) Should the percentage asked for in (b) be close to 99%? Explain.

(d) Produce a visual display of the distribution of margin of victories. Does the skewness of the distribution of margins of victory cast doubt on the validity of the procedure? Explain.

(e) Which of the following is the best interpretation of this confidence level:
- 99% of all games in this sample had a margin of victory within this interval.
- If we took another sample of 190 games, there is a 99% chance that its sample mean would fall within this interval.
- If we repeatedly took random samples of 190 games, this interval would contain 99% of their sample means in the long run.
- If we repeatedly took random samples of 190 games and constructed intervals in this same manner, 99% of the intervals in the long run would include the population mean margin of victory.
- This interval captures the margin of victory for 99% of the games in the population.

20

Activity 20-12: Parents' Ages (*cont.*)

Recall from Activity 6-11 on page 130 the following data on the ages at which a sample of 35 *fathers* had their first child (FAGE in AGECHILD.83g):

23	32	35	19	20	13	23	23	26	25	25	25	24	29	24	26	39	31
22	24	22	21	25	29	25	28	19	30	23	30	30	26	26	32	16	

(a) Define the population parameter of interest in this context.

(b) Do you think that the technical conditions for a confidence of interest for the population mean have been met? Explain.

(c) Use these sample data to form a 90% confidence interval for the mean age of having a first child among all fathers.

(d) Repeat (c) with a 95% confidence interval.

(e) Repeat (c) with a 99% confidence interval.

(f) Calculate and compare the half-widths and the midpoints of these intervals.

(g) Compare your 95% interval in (d) with the corresponding one for mothers that you calculated in Activity 20-3 on page 432 above. Are they similar? Do they overlap?

As was the case with the mothers, this sample of 35 fathers is a subsample from the larger sample of 800 fathers that were part of the 1998 General Social Survey. This larger sample had a mean of 25.214 years and a standard deviation of 5.419 years. Regard for now this larger sample as the population from which the smaller sample was taken.

(h) Do your intervals in (c)–(e) succeed in including this mean value?

Activity 20-13 : Word Lengths (*cont.*)

Reconsider the data that you collected in Topic 1 concerning the numbers of letters in the words of a sentence that you wrote.

(a) Treating these words as a sample from the population of all words that you have written in your scholastic career, use your calculator to produce an 80% confidence interval for the mean of this population.

(b) Write a sentence explaining what this interval says and what the phrase "80% confidence" means.

(c) Is there any reason to doubt the validity of the technical conditions underlying this procedure? Explain.

(d) How many and what percentage of your sample's words fall within the interval from (a)? Should this percentage be close to 80%? Explain.

(e) How would the interval in (a) have differed if your sample size had been larger (and everything else remained the same)?

(f) How would the interval in (a) have differed if your sample standard deviation had been larger (and everything else remained the same)?

(g) How would the interval in (a) have differed if your sample mean had been smaller (and everything else remained the same)?

(h) How would the interval in (a) have differed if every word in your sample had contained two more letters than it actually did? [*Hint*: Be specific in your answer.]

Activity 20-14: Sleeping Times (*cont.*)

Consider your analysis of students' sleeping times from Activity 20-4 on page 432. Identify each of the following statements as legitimate or illegitimate interpretations of your results.

(a) One can be 95% confident that the interval contains the true value of μ.

(b) If one repeatedly took random samples of college students and generated 95% confidence intervals in this manner, then in the long run 95% of the intervals so generated would contain the true value of μ.

(c) The probability is .95 that μ lies in the interval.

(d) One can be 95% confident that the sleeping time for any particular student falls within the interval.

(e) 95% of the students in the population would have sleeping times that fall within the interval.

Activity 20-15: E-Mail Messages (*cont.*)

Reconsider the data that you collected in Topic 12 concerning the number of e-mail messages that had been received and read in the preceding 24 hours by you and your classmates.

(a) Use the sample data to form a 95% confidence interval for the mean number of messages received and read in a 24-hour period among all students at your school.
(b) Does your class constitute a simple random sample from the population of students at your school? If not, do you suspect that your sample is biased in one direction or the other with respect to this variable? Explain.
(c) Does the requirement of a large sample size or a normally distributed population seem to be satisfied here? Explain.

Activity 20-16: Student-Generated Data (*cont.*)

Choose one *quantitative* variable on which student data have been collected (for example, ideal temperatures or age guesses or Scrabble points or ...). Select a confidence level, and calculate a confidence interval for the population mean of that variable. Write a few sentences describing your findings. Also address the question of the population to which your sample generalizes.

Activity 20-17: Parents' Ages (*cont.*)

Recall from Activity 3-11 on page 59 that the ages at which a sample of 35 mothers had their first child are as follows (MMAGE in AGECHILD.83g):

20	28	33	23	21	18	24	20	32	16	27	21	17	22
19	40	19	24	24	24	17	31	28	26	18	23	20	18
14	16	21	16	20	20	19							

In Activity 20-2 on page 429 you found a confidence interval for the mean age in this population. Now suppose that you want to find a confidence interval for the proportion of the population who have their first child after age 25.
(a) Calculate the proportion of mothers in the sample who had their first child after age 25.
(b) Find a 95% confidence interval for θ, the proportion of the population who have their first child after age 25.

Activity 20-18: Coins' Ages (*cont.*)

Recall from Activity 17-1 on page 367 the population of 1000 penny ages with mean age $\mu = 12.26$ years.

(a) Use a table of random digits or your calculator to draw a random sample of $n = 40$ pennies from this population. Report the sample mean age and the sample standard deviation of the ages.

(b) Use your calculator to construct a 90% confidence interval for μ based on these sample ages. Report the lower and upper endpoints of the interval. Also indicate whether the interval succeeds in capturing the known population mean value ($\mu = 12.26$).

(c) If you were to repeatedly draw random samples from this population and construct 90% confidence intervals for μ from each, what proportion of those intervals would you expect to contain μ in the long run? Explain.

(d) Use your calculator (PENNY.83g) to repeat (a) and (b) a total of 100 times. Download CONFD.83p and use this program to find the proportion of the 100 confidence intervals that succeed in containing the value 12.26. Is this close to what you expected?

(e) For the samples whose interval fails to capture the population mean, report their sample mean and sample standard deviation. What do you notice about these samples? Explain why this makes sense.

Activity 20-19: Confidence Interval of Personal Interest (*cont.*)

Think of a real situation in which you would be interested in producing a confidence interval to estimate a population *mean*. Describe precisely the observational units, population, and parameter involved. Also describe how you might select a sample from the population. [*Hint:* Be sure to think of a *quantitative* variable, so that a mean is a sensible parameter to deal with.]

Topic 21:

TESTS OF SIGNIFICANCE I: PROPORTIONS

OVERVIEW

The last two topics have introduced you to **statistical inference**, where the goal is to make a statement about a population parameter based on a sample statistic. The **confidence intervals** that you have studied allow one to estimate a population parameter at a certain confidence level with an interval of values.

In this topic you will discover the other major type of statistical inference: **tests of significance**. This procedure assesses the degree to which sample data support a particular conjecture about the value of the population parameter of interest. By exploring the concept of **statistical significance** in Topics 16 and 17, you have already studied the reasoning procedure behind these tests. This topic will introduce you to the formal structure of tests of significance about a population proportion.

OBJECTIVES

- To develop an intuitive understanding of the reasoning process used in **tests of significance**.
- To become familiar with the formal structure of tests of significance and to learn to translate appropriate questions into that structure.
- To discover how to perform calculations relevant to a test of significance concerning a population proportion.
- To acquire the abilities to interpret and explain the results of tests of significance.
- To explore how the statistical significance of a sample result is related to the sample size of the study.

PRELIMINARIES

1. Recall the cola discrimination test in which a subject has to identify which one of three cups contains a different brand of cola than the other two. In a test consisting of 30 trials, how convinced would you be that subjects do better than just guessing if they get 9 correct?

2. What if the subjects get 12 out of 30 correct?

3. What if the subjects get 15 out of 30 correct?

4. If a subject identifies 40% of a sample of trials correctly, would you be more impressed if it were a sample of 200 trials or a sample of 20 trials?

5. A campus legend tells the story of two friends who lied to their professor by blaming a flat tire for their having missed an exam. On the make-up exam the professor sent them to separate rooms and asked "which tire was it?" If you were faced with this situation, which of the four tires would you say it was?

6. Record the counts of responses for you and your classmates:

    Left front: Right front:

    Left rear: Right rear:

IN-CLASS ACTIVITIES

Activity 21-1: Cola Discrimination (*cont.*)

Consider again from Activity 16-15 on page 363 the description of an experiment to assess whether people can distinguish between the tastes of two brands of cola. Subjects are presented with three cups. Two cups contain one brand of cola, and the third cup contains the other brand. Subjects are to taste from all three cups and then identify the one that differs from the other two. Suppose that an experiment consists of thirty of these trials.

(a) If subjects cannot distinguish among the colas and therefore guess on each trial, for what proportion of trials will they correctly distinguish in the long run? In other words, what is the probability that a *guessing* subject will correctly identify the one cup that differs from the other two?

(b) Is this value a parameter or a statistic? Explain. What symbol represents it?

(c) According to the Central Limit Theorem, as discussed in Topic 16, if θ actually has this value, what pattern will the sampling distribution of the sample proportion follow? (Describe its shape, center, and spread, and also draw a sketch of this distribution.)

(d) Are the conditions met for the Central Limit Theorem to be valid? Explain.

(e) Shade the area under this curve corresponding to the probability of getting at least 20 correct identifications ($\hat{p}=.667$) by guessing in the thirty trials. Calculate the probability, as in Topic 18, that a subject who is just guessing would guess correctly on 20 or more of the 30 trials.

(f) Based on this probability, would you consider such a sample result surprising if someone is just guessing? Would you consider it so surprising that you believe the subject is not just guessing but really does have some ability to discriminate among the sodas?

If you consider this result surprising enough to convince you that the subject is doing better than guessing, we say that the result is **statistically significant**. We now formalize this process of determining whether a sample result provides statistically significant evidence against a conjecture about the population parameter. The resulting procedure is called a **test of significance**.

> The **null hypothesis** is denoted by H_o. It states that the parameter of interest is equal to a specific, hypothesized value. In the context of a population proportion, H_o has the form
>
> $$H_o: \theta = \theta_o,$$
>
> where θ is the population proportion of interest and θ_o is replaced by the conjectured value of interest. The null hypothesis is typically a statement of "no effect" or "no difference." The significance test is designed to assess the strength of evidence *against* the null hypothesis.

(g) Clearly define θ in the cola discrimination study:

Let $\theta =$

I am
sorry!

(h) The null hypothesis here is that the subject is just guessing. Translate this into a null hypothesis statement by referring back to your answer to (a):

$$H_o: \theta =$$

(i) Now suppose that subjects have some (but not perfect) ability to distinguish and therefore do better than just guessing. In this case, in the long run, the proportion that they identify correctly will exceed what value?

> The **alternative hypothesis** is denoted by H_a. It states what the researchers suspect or hope to be true about the parameter of interest. It depends on the purpose of the study and must be specified *before* the data are examined. The alternative hypothesis can take one of three forms:
>
> (a) $H_a: \theta < \theta_o$ or (b) $H_a: \theta > \theta_o$ or (c) $H_a: \theta \neq \theta_o$.
>
> The first two forms are called **one-sided** alternatives, while the last is a **two-sided** alternative.

(j) Translate what you said in (i) into the form of an alternative hypothesis by completing the following statement about θ:

$$H_a: \theta$$

(k) Would a guessing subject always identify the odd cup on exactly 1/3 of an experiment's trials? What term describes the phenomenon involved here?

Consider the following results of three of these experiments. Suppose that Alicia identifies the differing cup correctly in 10 of the 30 trials. Suppose Brenda identifies the different cup correctly for 12 of the 30 trials, whereas Celia gets 15 correct in 30 trials.

(l) Calculate the sample proportion of correct identifications for Alicia, Brenda, and Celia.

A: B: C:

(m) Are these proportions parameters or statistics? Explain.

(n) Is it possible that *guessing* subjects could get as many as 12 out of 30 or even 15 out of 30 correct just by chance?

(o) If a subject is just guessing, what will the standard deviation of the sample proportion of $\hat{p}$ be?

(p) Draw a sketch of this sampling distribution of $\hat{p}$ for a guessing subject, labeling the horizontal axis. Mark where the sample proportions for Alicia, Brenda, and Celia fall on the sketch.

(q) Calculate the *z*-score corresponding to Brenda's sample proportion of .4.

> The **test statistic** is a value computed by *standardizing* the observed sample statistic on the basis of the hypothesized parameter value. It is used to assess the evidence against the null hypothesis. In this context of a population proportion, it is denoted by z and calculated as follows:
>
> $$z = \frac{\hat{p} - \theta_o}{\sqrt{\dfrac{\theta_o(1 - \theta_o)}{n}}}.$$

Note: During the calculations of the significance test we assume that the null hypothesis is true, so we use that value for θ in the denominator of this test statistic.

(r) Use the normal probability table (Table II) to determine the probability that Brenda would obtain a sample proportion of correct identifications equal to .4 or higher if she were guessing. Is this value consistent with the area under the sampling distribution curve lying to the right of .4?

> The **p-value** is the probability, assuming the null hypothesis to be true, of obtaining a test statistic at least as extreme as the one actually observed. "Extreme" means "in the direction of the alternative hypothesis," so the p-value takes one of three forms (corresponding to the appropriate form of H_a):
>
> (a) $\Pr(Z \le z)$ (area below z-score) or
> (b) $\Pr(Z \ge z)$ (area above z-score) or
> (c) $2\Pr(Z \ge |z|)$ (area more extreme than the z-score in both directions)

(s) Does this probability suggest that it is very *unlikely* for Brenda to get 12 or more correct identifications out of 30 if she is guessing? For instance, would such an event happen less than 10% of the time in the long run? Less than 5%? Less than 1%?

One judges the strength of the evidence that the data provide against the null hypothesis by examining the p-value. The *smaller* the p-value, the stronger the evidence *against* H_o (and thus the stronger the evidence *in favor of* H_a). For instance, typical evaluations are:

p-value >.1: *little* or *no* evidence against H_o

.05 < p-value ≤ .10: *some* evidence against H_o

.01 < p-value ≤ .05: *moderate* evidence against H_o

.001 < p-value ≤ .01: *strong* evidence against H_o

p-value ≤ .001: *very strong* evidence against H_o

The **significance level**, denoted by α, is an optional "cut-off" level for the p-value that one decides to regard as decisive. The experimenter specifies the significance level in advance; common values are $\alpha = .10, \alpha = .05,$ and $\alpha = .01$. The smaller the significance level, the more evidence you require in order to be convinced that H_o is not true.

If the p-value of the test is less than or equal to the significance level α, the **test decision** is to *reject* H_o; otherwise, the decision is to *fail to reject* H_o. Notice that failing to reject H_o is not the same as affirming its truth; it is simply to declare that the evidence was not convincing enough to reject it. Another very common expression is to say that the data are **statistically significant** at the α level if the p-value is less than or equal to α. Thus, a result is statistically significant if it is unlikely to have occurred by chance or sampling variability alone.

[handwritten: not likely to happen by chance]

(t) Calculate the test statistic and p-value for Celia, who obtained 15 of 30 ($\hat{p} = .5$) correct identifications in her sample.

(u) Based on this p-value, would you reject the null hypothesis that Celia is just guessing at the .05 significance level? Explain.

(v) Based on this p-value, is the result 15/30 statistically significantly greater than 1/3 at the .01 significance level? Explain.

(w) Recall that Alicia had ten correct identifications in 30 trials. Do we need to conduct a test of significance to decide whether this sample provides strong evidence that she does better than guessing in the long run? Explain.

(x) Write a few sentences describing and explaining how your conclusions differ for Alicia, Brenda, and Celia.

The **technical conditions** needed to establish the validity of this significance testing procedure are similar to those for the confidence interval procedure:

- the data are a simple random sample from the population of interest, and
- the sample size is large relative to the proportions involved. The rule of thumb for this second condition is that $n\theta_o \geq 10$ and $n(1 - \theta_o) \geq 10$, which says that the conjectured numbers of "successes" and "failures" in the sample should each be at least ten.

(y) Verify that this condition about the sample size is satisfied for these tests.

Activity 21-2: Baseball "Big Bang"

A reader wrote in to the "Ask Marilyn" column in *Parade Magazine* to say that his grand-father told him that in 3/4 of all baseball games, the winning team scores more runs in one inning than the losing team scores in the entire game. (This phenomenon is known as a "big bang.") Marilyn responded that this proportion seemed to be too high to be believable. Let θ be the proportion of all Major League Baseball games in which a "big bang" occurs.

(a) Restate the grandfather's assertion as the null hypothesis, in symbols and in words.

(b) Given Marilyn's conjecture, state the alternative hypothesis, in symbols and in words.

For the sample of 190 Major League Baseball games played during July 26–August 8, 1999, 98 of them contained a "big bang."

(c) Sketch the sampling distribution for the sample proportion, as predicted by the Central Limit Theorem, assuming that the null hypothesis is true.

(d) Calculate the sample proportion of games in which a big bang occurred.

(e) Is this sample proportion less than 3/4 and therefore consistent with Marilyn's (alternative) hypothesis? Shade the area under your sampling distribution curve corresponding to this sample result in the direction conjectured by Marilyn.

(f) Compute the test statistic and use Table II to find its p-value.

$z =$ p-value =

(g) Confirm your previous sketch and calculations using your calculator to determine the test statistic and p-value:
   • From the STAT TESTS menu, select 1-PropZTest.
   • Enter the hypothesized value of θ (note that the TI calls this p_0 instead of θ_0), the number of successes, the sample size, and the form of the alternative. Your setup should look like the following:

```
1-PropZTest
 P0:.75
 x:98
 n:190
 Prop≠P0  <P0  >P0
 Calculate Draw
```

Highlight "Draw" and press ENTER .

Based on this p-value, would you say that the sample data provide strong evidence to support Marilyn's contention that the proportion cited by the grandfather is too high to be the actual value? Explain.

Marilyn went on in her response to claim that the actual proportion of "big bang" games is one-half.

21

(h) Using a two-sided alternative, state the null and alternative hypotheses (in symbols and in words) for testing Marilyn's claim.

(i) Use your calculator to determine the test statistic and p-value for this test.

  $z =$ p-value =

(j) Based on this p-value, would you say that the sample data provide strong evidence to reject Marilyn's claim? Explain.

Activity 21-3: Flat Tires

It has been suggested that when asked to say which tire on a car is likely to have gone flat, people choose the right front tire more often than they would if the four tires were equally likely.

(a) Specify the variable of interest in this activity. Is this variable categorical or quantitative?

(b) Define the parameter of interest, including a statement of the population you hope to generalize to with this parameter.

(c) If the four tires were, in fact, equally likely, how often would the front right tire be selected in the long run?

(d) Use your answer to (c) to specify the null hypothesis of the appropriate significance test.

(e) Use the conjecture suggested above to state the alternative hypothesis.

(f) Carry out the test for your sample data collected in class. Include a sketch of the sampling distribution for the sample proportion based on the conjecture stated in the null hypothesis. Shade the portion of this curve beyond the observed sample proportion in the direction specified by the alternative hypothesis. Report the test statistic and p-value. Write a sentence interpreting the p-value in context. Then write a few sentences summarizing and explaining your conclusion about whether your sample data provide strong evidence that people tend to pick the "right front" tire more than 1/4 of the time in the long run.

(g) Do you think the technical conditions of this procedure are met? If not, which condition(s) do you think are violated? Would specifying a different population of interest help?

Activity 21-4: Flat Tires (*cont.*)

Reconsider the "name a tire" situation. Now suppose that in a random sample of people, 30% select the front right tire.

(a) Does this sample result constitute strong evidence against the null hypothesis that the front right tire would be chosen 1/4 of the time in the long run? Do you need more information to answer this question? Explain.

(b) Suppose that this sample result (30% answering "right front") had come from a sample of $n = 50$ people. Use your calculator to conduct the appropriate test of significance. [*TI Hint:* Remember that you can enter .30*50 in the x line.] Record in the first row of the table below the value of the test statistic and the p-value of the test. Also indicate (yes or no) whether the test is significant at each of the significance levels listed.

(c) Repeat (b) for the other sample sizes listed in the table.

sample size	"right front"	$\hat{p}$	z statistic	p-value	$\alpha = .10$?	$\alpha = .05$?	$\alpha = .01$?	$\alpha = .001$?
50	15	.30						
100	30	.30						
150	45	.30						
250	75	.30						
500	150	.30						
1000	300	.30						

(d) Write a few sentences summarizing what your analysis reveals about whether a sample result of 30% is significantly greater than a hypothesized value of 25%.

This activity reveals that sample size plays a key role in tests of significance. The statistical significance of a sample result depends largely on the sample size involved. With large sample sizes, even small differences can be statistically significant because they are unlikely to occur by chance.

WRAP-UP

This topic has introduced you to the very important technique of **tests of signifi-cance** by asking you to understand the reasoning process underlying the procedure. You have also encountered the formal structure of tests of significance and explored further the concept of **statistical significance**.

The basic structure of a test of significance involves the following steps:
1. Define the parameter of interest.
2. State the null and alternative hypotheses based on the study question.
3. Calculate the test statistic, which measures the distance between the observed sample statistic and the hypothesized value.
4. Calculate the p-value, which reports the probability of obtaining a test statis-tic value at least this extreme when the null hypothesis is true.
5. Make a conclusion about the study question based on the magnitude of this p-value. Small p-values constitute evidence against the null hypothesis.

You have concentrated exclusively on tests involving a population *proportion* in this topic. The next topic will continue your study of significance testing by asking you to examine tests involving a population *mean*.

HOMEWORK ACTIVITIES

Activity 21-5: Calling Heads or Tails (*cont.*)

Refer to the data collected in Topic 16 and analyzed in Activity 16-11 on page 360 about whether you would call "heads" or "tails" if asked to predict the result of a coin flip.

(a) What proportion of the responses were "heads"?

(b) Is this proportion a parameter or a statistic? Explain.

(c) Write a sentence identifying the parameter of interest in this situation.

(d) Specify the null and alternative hypotheses, in words and in symbols, for testing whether the sample result differs significantly from .5.

(e) Sketch the sampling distribution for the sample proportion specified by the null hypothesis. Shade the region corresponding to the p-value.

(f) Calculate the test statistic and p-value for this test. Also provide a check of the tech-nical conditions.

(g) Is the sample result statistically significantly different from .5 at the .10 level? At the .05 level? At the .01 level?

(h) Write a few sentences summarizing and explaining your conclusion.

(i) Describe specifically what would have changed in this analysis if you had worked with the proportion of "tails" responses rather than "heads."

Activity 21-6: Calling Heads or Tails (*cont.*)

Refer to Activity 16-11 on page 360, where you examined data on the proportion of students who would respond "heads" if asked to predict a coin toss. In his book *Statistics You Can't Trust*, Steve Campbell claims that people call "heads" 70% of the time when asked to predict the result of a coin flip. Conduct a test of whether your sample data provide evidence against Campbell's hypothesis. Report the hypotheses, sampling distribution sketch specified by the null hypothesis, check of the technical conditions, test statistic, and p-value. Write a few sentences describing your conclusion.

Activity 21-7: Flat Tires (*cont.*)

Reconsider the hypothetical results presented in Activity 21-4 on page 455 for the "tire" question. Determine the smallest sample size n for which a sample result of 30% answering "right front" would be significant at the .10 level. [*Hint:* You may either use trial and error or work analytically with the normal table and the formula for the test statistic.]

Activity 21-8: Baseball "Big Bang" (*cont.*)

Consider again the "big bang" phenomenon described in Activity 21-2 on page 452. Statistician Hal Stern examined all 968 baseball games played in the National League in 1986 and found that 419 of them contained a "big bang."

(a) Perform the appropriate test to see whether this sample proportion differs significantly from .5 at the $\alpha = .02$ level. Report your hypotheses in symbols and in words, your sketch of the sampling distribution under the null hypothesis, your check of the technical conditions, the test statistic, and the p-value, in addition to stating and explaining your conclusion.

(b) If one redefines "big bang" to mean that the winning team scores *at least* as many (instead of more) runs in one inning as the losing team scores in the entire game, then 651 of those 968 games contained a big bang. Does this sample proportion differ significantly from the "Ask Marilyn" reader's grandfather's assertion of .75 at the $\alpha = .08$ level? Again report the details of your analysis.

Activity 21-9: Racquet Spinning (*cont.*)

Recall from Activity 14-9 on page 319 that a spun tennis racquet yielded 46 "up" results in 100 spins.

(a) Does this sample result provide strong evidence that the racquet would not land up 50% of the time in the long run? Conduct the appropriate test, checking the technical conditions, and show the details of your work. Explain your conclusion in your own words.

(b) Would you reject the null hypothesis at the .05 level?

(c) Does the test result indicate that the racquet would definitely land up 50% of the time in the long run? Explain.

(d) What is the smallest significance level at which you would reject the null hypothesis? [*Hint:* Do not confine your consideration to common α levels.]

(e) Explain precisely how your analysis would change depending on whether you work with the proportion landing "up" or the proportion landing "down."

(f) Use the sample data to find a 95% confidence interval for the long-run proportion of times that the racquet would land "up."

(g) Does this interval include the value .5?

(h) Explain the consistency between your answers to (b) and (g).

Activity 21-10: Therapeutic Touch (*cont.*)

In the "therapeutic touch" experiment described in Activity 13-23 on page 298, subjects were to identify which of their hands the experimenter had placed her hand over.

(a) Identify the null and alternative hypotheses, in symbols and in words, for testing whether the practitioners could distinguish more often than not over which hand the experimenter's hand was held. Also clearly identify the parameter of interest in words.

(b) Combining the results of the 21 subjects, there was a total of 123 correct identifications in 280 repetitions of the experiment. Use these sample data to conduct the test of the hypotheses specified in (a). Report the test statistic and p-value.

(c) Explain why it makes sense that the p-value is greater than .5 in this situation.

(d) Is it fair to conclude that you should "accept the null hypothesis" in this situation? Explain.

(e) What conclusion do you draw from this study about the effectiveness of therapeutic touch?

Activity 21-11: Magazine Advertisements (*cont.*)

Recall from Activity 19-7 on page 416 that the September 13, 1999 issue of *Sports Illustrated* had 116 pages, and 54 of them contained an advertisement. Prior to collecting these data, a subscriber Frank Chance conjectured that 30% of the magazine's pages contain ads.

(a) Write a sentence identifying the parameter of interest here. Also indicate a symbol used to represent it.

(b) Express the subscriber's conjecture in symbols. Is this the null or the alternative hypothesis?

(c) Treat the pages of this issue as a sample of the population of all *Sports Illustrated* pages. Conduct a significance test of whether the sample data provide strong evidence against the subscriber's conjecture.

(d) Write a few sentences describing and explaining your conclusion.

(e) Do you think that the technical conditions for the above procedure to be valid have been met? Explain.

Activity 21-12: Volunteer Work

The 1998 *Statistical Abstract of the United States* reports that 48.8% of a sample of 2719 people claimed to have done some volunteer work during 1996.

(a) Does this sample result provide evidence that the proportion of the population who claimed to have done volunteer work in 1996 differs from 50%? Conduct the appropriate test and report your conclusion. Be sure to discuss the technical conditions.

(b) Determine a sample size for which a sample proportion of $\hat{p} = .488$ does differ significantly from .5 at the $\alpha = .05$ significance level. Report the details of the test results for this sample size.

Activity 21-13: Hiring Discrimination

In the case of *Hazelwood School District vs. United States* (1977), the U.S. government sued the city of Hazelwood, a suburb of St. Louis, on the grounds that it discriminated against blacks in its hiring of school teachers. The statistical evidence introduced noted that of the 405 teachers hired in 1972 and 1973 (the years following the passage of the Civil Rights Act), only 15 had been black. The proportion of black teachers living in the county of St. Louis at the time was 15.4% if one includes the city of St. Louis and 5.7% if one does not include the city.

(a) Identify the parameter of interest here in words.

(b) Conduct a significance test to assess whether the proportion of black teachers hired by the school district is statistically significantly less than 15.4% (the percentage of black teachers in the county). Use the .01 significance level. Along with your conclusion, report the null and alternative hypotheses, a sketch of the sampling distribution specified by the null hypothesis, a check of the technical conditions, the test statistic, and the p-value. (You may use your calculator.)

(c) Conduct a significance test to assess whether the proportion of black teachers hired by the school district is statistically significantly less than 5.7% (the percentage of black teachers in the county if one excludes the city of St. Louis). Again use the .01 significance level and report the null and alternative hypotheses, test statistic, and p-value along with your conclusion.

(d) Write a few sentences comparing and contrasting the conclusions of these tests with regard to the issue of whether the Hazelwood School District was practicing discrimination.

Activity 21-14: Television Magic (*cont.*)

Recall from Activity 19-12 on page 419 that a sample of *Nick-at-Nite* television viewers produced 810,000 votes for Samantha and 614,000 votes for Jeannie as having more magic powers. Conduct the appropriate test of significance to assess the extent to which these sample data provide evidence that more than half of the population of all *Nick-at-Nite* viewers favor Samantha over Jeannie. Report the null and alternative hypotheses, sketch of the sampling distribution specified by the null hypothesis, a check of the technical conditions, the test statistic, and the p-value of the test. Also indicate whether the sample result is statistically significant at the .001 significance level.

Activity 21-15: Marriage Ages (*cont.*)   by   hand

Reconsider Activity 19-18 on page 419, in which you analyzed sample data and found the sample proportion of marriages in which the bride was younger than the groom. Conduct a test of significance to address whether the sample data support the theory that the bride is younger than the groom in more than half of all the marriages in the county. Report the details of the test and write a short paragraph describing and explaining your findings. $n = 94$ $6 > $ the bride is younger

Let $\theta =$

Activity 21-16: Veterans' Marital Problems (*cont.*)   calc

Refer to the study mentioned in Activity 19-19 on page 422. U.S. Census figures indicate that among all American men aged 30–44 in 1985, 27% had been divorced at least once. Conduct a test of significance to assess whether the sample data from the study provide strong evidence that the divorce rate among Vietnam veterans is higher than 27%. Report the null and alternative hypotheses (identifying whatever symbols you introduce) and the test statistic and p-value. Also write a one-sentence conclusion and explain why the conclusion follows from the test results.

$$\hat{p} = \frac{777 = x}{2101 = n}$$

$$p_0 = .27$$

Topic 22:

TESTS OF SIGNIFICANCE II: MEANS

In the last topic you studied tests of significance, which along with confidence intervals constitute one of the two most widely used groups of techniques in statistical inference. The procedure that you examined applies only to a population *proportion*, so in this topic you will investigate tests about the *mean* of a population. You will encounter and explore one of the most famous of statistical techniques: the ***t-test***. You will also see how to analyze data resulting from a ***matched pairs*** experimental design.

OBJECTIVES

- To learn to apply the ***t-test*** concerning a population mean by hand and with your calculator.
- To recognize the extensive similarities of structure and interpretation between different test procedures.
- To become familiar with the *t*-distribution for finding p-values.
- To discover the roles played by sample size and sample standard deviation as well as by sample mean concerning tests of significance about a population mean.
- To understand the purpose of a ***matched pairs*** experimental design as well as the application of a *t*-test to the resulting data.
- To see how to check whether the *t*-test can be validly applied to a given sample of data.
- To continue to recognize the utility of graphical analyses and the limitation of summarizing a distribution of data exclusively by its mean.

1. How many points do you think are scored (by the two teams combined) in a typical professional basketball game?

2. If you are looking to find evidence that rules changes have increased basketball scoring from the previous season, do you think that a sample of one day's games or a sample of one week's games would be more convincing?

3. What would you guess for the difference in ages of a typical marriage couple?

IN-CLASS ACTIVITIES

Activity 22-1: Basketball Scoring

Prior to the 1999–2000 season in the National Basketball Association, the league made several rule changes designed to increase scoring. The average number of points scored per game in the previous season had been 183.2. Let μ denote the mean number of points per game in the 1999–2000 NBA season.

(a) If the rule changes had *no effect* on scoring, what value would μ have? Is this a null or an alternative hypothesis?

(b) If the rule changes had the desired effect on scoring, what would be true about the value of μ? Is this a null or an alternative hypothesis?

The following data are the number of points scored in the 25 NBA games played during December 10–12, 1999, and are stored in the NBAPT.831 list.

196	198	205	163	184	224	206	190	140	204	200	190	195
180	200	180	198	243	235	200	188	197	191	194	196	

(c) Use your calculator to examine visual displays of this distribution. Write a few sentences commenting on these points per game, particularly addressing the issue of whether scoring seems to have increased over the previous season's mean of 183.2 points per game.

(d) Use your calculator to calculate the mean and standard deviation of this sample of points per game.

sample mean, $\bar{x}$:   sample standard deviation, s:

(e) Is the sample mean in the direction specified in the alternative hypothesis? In other words, is the mean of the points per game in this sample higher than the 1998–99 season's mean?

(f) Is it *possible* to have gotten such a large sample mean even if the new rules had no effect on scoring?

The structure, reasoning, and interpretation of significance tests about a population mean are the same as for a population proportion. The null hypothesis asserts that the population mean equals some value of interest, while the alternative hypothesis reflects the researcher's conjecture. The test statistic is calculated by taking the difference between the observed sample mean and the hypothesized population mean and dividing by the standard error of the sample mean. When the null hypothesis is true and the population has a normal distribution, the test statistic follows a *t*-distribution with $n-1$ degrees of freedom, so the p-value is calculated from the *t*-distribution. This p-value again reports the probability of getting such an extreme sample if the null hypothesis were true. Thus, the smaller the p-value, the stronger the evidence against the null hypothesis supplied by the sample data.

Test of significance for a population mean μ:

$H_o: \mu = \mu_o$

$H_a: \mu < \mu_o$ or $H_a: \mu > \mu_o$ or $H_a: \mu \neq \mu_o$

test statistic: $t = \dfrac{\bar{x} - \mu_o}{s/\sqrt{n}}$

p-value: $\Pr(T_{n-1} \leq t)$ or $\Pr(T_{n-1} \geq t)$ or $2\Pr(T_{n-1} \geq |t|)$

where μ_o represents the hypothesized value of the population mean and T_{n-1} represents a *t*-distribution with $n-1$ degrees of freedom.

The **technical conditions** for this procedure to be valid are that:
- the data are a simple random sample from the population of interest, and
- either the population values follow a normal distribution or the sample size is large ($n \geq 30$ as a rule of thumb).

Activity 22-2: Basketball Scoring (*cont.*)

(a) Return to the analysis that you began in Activity 22-1 on page 464. Translate the statements you made in (a) and (b) into null and alternative hypotheses about μ by completing the following:

         Let $\mu =$

         $H_o: \mu =$
         $H_a: \mu$

(b) Use the sample statistics you found in (d) of Activity 22-1 to compute the value of the *t*-test statistic from this sample.

(c) Draw a rough sketch of the *t*-distribution, with $n-1$ degrees of freedom, representing the sampling distribution of the *t*-test statistic when the null hypothesis is true (remembering to label the horizontal axis). Shade the area under this curve lying to the right of the test statistic value you calculated in (b).

As you used the *t*-table (Table III) to find critical values for confidence intervals, you will also use it to find p-values for significance tests.

(d) Use the *t*-table to find the area to the right of your test statistic value with $n-1$ degrees of freedom. [Note: Use the appropriate row of the *t*-table to find two values on either side of your test statistic. Read off the probabilities from the top of the chart that correspond to these two values. You can report that the p-value is between these two probabilities. If the test statistic value is off the chart, determine the bound on the p-value from the last probability listed. Your sketches should be very helpful here.]

(e) Interpret the p-value in the context of these data and hypotheses.

(f) Would you reject the null hypothesis at the .10 level? At the .05 level? At the .01 level? At the .005 level?

(g) Use your calculator to verify your sketch and calculations for this test:
- From the STAT TESTS menu, select T-Test.
- You can select either Data or Stats. With Data, enter NBAPT as the List. With Stats, enter the sample mean, standard deviation, and sample size that you determined in part (d) of Activity 22-1 on page 464.
- Enter the hypothesized value of the population mean (183.2) and $> \mu_0$ as the form of the alternative. Highlight "Draw" and press ENTER.

(h) Are you able to verify that the technical conditions for the validity of the significance test of the population mean have been met? Explain. [*Hint:* You may want to refer to the graphical summary you constructed in Activity 22-1(c).]

(i) Write a sentence or two summarizing your conclusion about whether the sample data provide evidence that the mean points per game in the 1999–2000 season is higher than in the previous season. Include an explanation of how your conclusion follows from the test result.

Activity 22-3: Sleeping Times (*cont.*)

Reconsider the data that you collected on the sleeping times of you and your classmates and analyzed in Activity 20-5 on page 434.

(a) Conduct a significance test of whether your class data provide strong evidence that the mean sleeping time among all students at your school that night was less than seven hours. Clearly define the parameter of interest and report the hypotheses, test statistic, and p-value. [*Hint:* Use the symmetry of the *t*-distribution in finding the p-value.] Would you reject the null hypothesis at the .05 level?

(b) Do you think that the technical conditions for the validity of this procedure have been met? Explain.

Now consider the four samples of hypothetical sleeping times presented in Activity 20-4 on page 432. Dotplots and summary statistics are reproduced here:

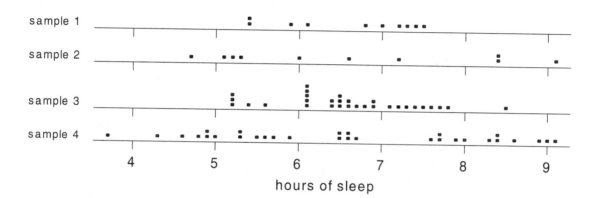

Sample number	Sample size	Sample mean	Sample std. dev.	p-value
1	10	6.6	.825	(e)
2	10	6.6	1.597	(e)
3	30	6.6	.825	(e)
4	30	6.6	1.597	(e)

(c) Between samples 1 and 2, which do you think supplies stronger evidence that $\mu < 7$; i.e., which sample would produce a smaller p-value of the appropriate test of significance? Explain.

(d) Between samples 1 and 3, which do you think supplies stronger evidence that $\mu < 7$; i.e., which sample would produce a smaller p-value of the appropriate test of significance? Explain.

(e) The data are stored in HYPOSLEEP.83g. Use the T-Test feature of your calculator (located in the STAT TESTS menu) to calculate the p-values for testing that μ is less than 7 with each of these four samples. Record the p-values in the table above.

(f) With which of the samples do you have enough evidence to reject the null hypothesis at the .05 level and conclude that the mean sleeping time is in fact less than seven hours?

(g) Comment on whether your conjectures in (c) and (d) are confirmed by the test results.

> This activity should reinforce what you have discovered earlier about effects of sample size and variation. An observed difference between a sample mean and a hypothesized mean is more statistically significant (unlikely to occur by chance) with a larger sample than with a smaller one. Also, that difference is more significant if the data are less spread out as opposed to more spread out.

Activity 22-4: Marriage Ages (*cont.*)

Reconsider the data presented in Activity 3-13 on page 60 concerning the ages at marriage for a sample of 24 couples who obtained their marriage licenses in Cumberland County, Pennsylvania, in 1993. The data are recorded below and in MARRIAGE.83g:

couple	husband	wife	difference (husb–wife)	couple	husband	wife	difference (husb–wife)
1	25	22		13	25	24	
2	25	32		14	23	22	
3	51	50		15	19	16	
4	25	25		16	71	73	

5	38	33		17	26	27	
6	30	27		18	31	36	
7	60	45		19	26	24	
8	54	47		20	62	60	
9	31	30		21	29	26	
10	54	44		22	31	23	
11	23	23		23	29	28	
12	34	39		24	35	36	

(a) For each couple in the sample, subtract the wife's age from the husband's age. Record the results in the table above.

(b) Create a dotplot of the differences in ages for these 24 couples. Describe the distribution.

(c) Use your calculator to calculate the relevant summary statistics regarding the sample of *age differences*; record them below along with the symbols used to represent them.

(d) Use your calculator to conduct a significance test of whether the *mean* of the *population* of *age differences* exceeds zero, suggesting that the husband tends to be older than the wife. Clearly define the parameter of interest, state the null and alternative hypotheses, sketch the sampling distribution for the *t*-test statistic specified by the null hypothesis, check the technical conditions, and record the test statistic and p-value. Also indicate whether the sample data are statistically significant at the .05 level.

(e) Use your calculator to find a 90% confidence interval for the mean of the *population* of *age differences*. Comment on whether the interval includes zero.

22

(f) Write a one- or two-sentence conclusion about the difference in ages between husbands and wives based on your analysis of these data.

This example illustrates a **matched pairs** experimental design. One can *control* some of the variation in marriage ages by considering *couples* rather than *individuals*. A less sensible design for this study would have been to obtain the ages of one sample of 24 brides and of a separate sample of 24 grooms.

One analyzes matched pairs data by analyzing *differences* using the inference procedures that you have studied concerning a population mean. In this case, the mean of the population of age differences is what one makes inferences about.

WRAP-UP

This topic has introduced you to tests of significance concerning a population mean. You have found that the structure, reasoning, and interpretation of these tests are the same with a mean as they were for a proportion. Two differences as compared to tests about a proportion are that more factors are involved in the procedure, the sample standard deviation as well as the sample mean and sample size, and that the p-value is calculated from the *t*-distribution and not from the normal distribution. You have also encountered and learned how to analyze data from an important experimental design known as matched pairs.

Now that you have studied both confidence intervals and tests of significance for both a population proportion and a population mean, the next topic will lead you to discover a variety of important considerations related to these inference procedures.

HOMEWORK ACTIVITIES

Activity 22-5: Exploring the *t*-Distribution (*cont.*)

(a) Use the *t*-table to find the p-values (as accurately as possible) corresponding to the following test statistic values and degrees of freedom, filling in a table as below with those p-values:

d.f.	$\Pr(T \geq 1.415)$	$\Pr(T \geq 1.960)$	$\Pr(T \geq 2.517)$	$\Pr(T \geq 3.168)$
4				
11				
23				
80				
infinity				

(b) Does the p-value get larger or smaller as the value of the test statistic gets larger (if the number of degrees of freedom remains the same)?

(c) Does the p-value get larger or smaller as the number of degrees of freedom gets larger (if the value of the test statistic remains the same)?

Activity 22-6: Exploring the *t*-Distribution (*cont.*)

(a) Use the *t*-table and/or your answers to Activity 22-5 to find the following p-values:

d.f.	$\Pr(T \leq -1.415)$	$\Pr(T \leq -1.960)$	$2\Pr(T \geq |-2.517|)$	$2\Pr(T \geq |-3.168|)$
4				
11				
23				
80				
infinity				

(b) Describe how the p-values of the first two columns of the table compare with those of the first two columns from Activity 22-5. Explain why this makes sense.

(c) Describe how the p-values of the last two columns of the table compare with those of the last two columns from Activity 22-5. Explain why this makes sense.

22

(d) You can also compute probabilities from the *t*-distribution using your calculator. From the DISTR menu select tcdf(. As with the normalcdf command, enter a lower limit (which may be -1EE99) and an upper limit. Then enter the degrees of freedom: `tcdf(-1EE99,-1.415, 4)`. You can use 100,000 to represent infinite degrees of freedom.

Activity 22-7: Sleeping Times (*cont.*)

Reconsider the hypothetical samples of sleeping times presented in Activity 20-4 on page 432 and analyzed in Activity 22-3 on page 468 (HYPOSLEEP.83g). Suppose now that you were interested in testing whether the sample data provide evidence that the population mean sleeping time *differs from* seven hours. Use your calculator to conduct the test of significance for each of the four samples with a *two-sided* alternative hypothesis. Record the p-value for each sample and comment on how these p-values compare with the one-sided ones found in Activity 22-3.

Activity 22-8: UFO Sighters' Personalities

In a 1993 study, researchers took a sample of people who claimed to have had an intense experience with an unidentified flying object (UFO) and a sample of people who did not claim to have had such an experience. They then compared the two groups on a wide variety of variables, including IQ. The sample mean IQ of the 25 people in the study who claimed to have had an intense experience with a UFO was 101.6; the standard deviation of these IQs was 8.9. As 100 is a common mean value for IQ scores, suppose that you want to test $H_o: \mu = 100$ vs. $H_a: \mu \neq 100$.

(a) Identify clearly what the symbol μ represents in this context.

(b) Is this a one-sided or a two-sided test? Explain how you can tell.

(c) Sketch the sampling distribution of the *t*-test statistic specified by the null hypothesis.

(d) Calculate the test statistic and shade the area corresponding to obtaining a test statistic as extreme or more extreme as this one observed for the sample of 25 UFO observers.

(e) Use the *t*-table to determine (as accurately as possible) the p-value.

(f) Are you able to completely check the technical conditions? What would need to be true for this procedure to be valid?

(g) Write a sentence interpreting the p-value in the context of this sample and these hypotheses. Summarize the conclusion of your test in context.

Activity 22-9: Basketball Scoring (*cont.*)

Recall from Activity 22-1 on page 464 that the NBA made rule changes prior to the 1999–2000 season that were intended to boost scoring from the average of 183.2

points per game in the previous season. The November 29, 1999 issue of *Sports Illustrated* reported that for the first 149 games of the 1999–2000 season, the mean number of points per game was 196.2.

(a) State in words and in symbols the hypotheses for testing whether the sample data provide strong evidence that the mean for the entire 1999–2000 season exceeds 183.2.

(b) Do you have enough information to calculate the test statistic? Explain.

(c) How large would the test statistic have to be in order to reject the null hypothesis at the .01 level?

(d) If the sample standard deviation for these 149 games were close to that for the 25 games that you analyzed in Activity 22-1 on page 464, would the test statistic exceed the critical value in (c)? By a lot? Explain.

(e) Even though the magazine did not provide all of the information necessary to conduct a significance test, can you reasonably predict whether the test result would be significant at the .01 level? Explain, based on your answers to (c) and (d).

(f) Does the validity of the test procedure depend on the scores being normally distributed? Explain.

Activity 22-10: Exam Score Improvements

The following data are scores on the first and second exams for a sample of students in an introductory statistics course. (The * denotes a missing value; that student did not take the second exam.) The data are stored in EXAMS.83g.

student	exam 1	exam 2	improvement	student	exam 1	exam 2	improvement
1	98	80		13	91	92	
2	76	71		14	83	80	
3	90	82		15	83	84	
4	95	68		16	93	96	
5	97	96		17	96	90	
6	89	93		18	98	97	
7	77	50		19	84	95	
8	94	64		20	76	67	
9	88	*	*	21	97	77	
10	95	84		22	72	56	
11	87	76		23	80	78	
12	84	69		24	74	64	

Disregard for now the student with the missing value; consider this a sample of 23 students.

(a) Construct a dotplot of the *improvements* in scores from exam 1 to exam 2; those who scored lower on exam 2 than on exam 1 have negative improvements.

(b) What was the largest improvement from exam 1 to exam 2?

(c) What was the biggest decline from exam 1 to exam 2?

(d) What proportion of these students scored higher on exam 1 than on exam 2?

(e) Use your calculator to calculate the sample mean and sample standard deviation of the *improvements*.

(f) Perform the test of whether the mean improvement *differs* significantly from zero. State the null and alternative hypotheses and record the test statistic and p-value.

(g) Based on the significance test conducted in (f), would you reject the null hypothesis at the $\alpha = .10$ significance level? Are the sample data statistically significant at the .05 level? How about at the .01 level?

(h) Find a 95% confidence interval for the mean of the population of *improvements*.

(i) Write a few sentences commenting on the question of whether there seems to have been a significant difference in scores between the two exams.

(j) Now treat the missing value as a score of 0 on the second exam. Recalculate the p-value of the test from (f) and the confidence interval in (h).

(k) Comment on whether disregarding the missing value or treating it as a 0 makes much difference in this analysis.

(l) Explain how you would carry out a significance test to investigate whether fewer than half of the test takers improved. [Define the new population parameter, null and alternative hypotheses, and appropriate test procedure.]

(m) Carry out the test described in (l). In addition to stating your conclusion, comment on the validity of the technical conditions.

Activity 22-11: Marriage Ages (*cont.*)

Recall that the sample of 24 marriage ages that you analyzed in Activity 22-4 on page 470 is actually a subsample from a larger sample of 100 marriages for which data appear in Activity 19-18 on page 419 (MARRIAGE100.83g). Use your calculator to reproduce your analysis of Activity 22-4 using all 100 couples in the larger sample. Write a paragraph reporting on your findings and how they differ from those found by analyzing only 24 couples' ages.

Activity 22-12: Comparison Shopping (*cont.*)

Consider again the data presented in Activity 8-19 on page 182 concerning prices of grocery items at two different stores, Lucky's and Von's (SHOPPING.83g).

(a) Explain why these data are from a matched pairs design.

(b) Calculate the differences in the prices of these items between the two stores.

(c) Examine visual displays and numerical summaries of these price differences. Write a few sentences commenting on the question of whether one store seems to have lower prices than the other.

(d) Conduct the appropriate test of significance to assess whether the sample data provide significant evidence that the prices tend to differ between these two stores. Report your hypotheses, sampling distribution specified by the null hypothesis, test statistic, and p-value. Also look into and comment on whether the technical conditions seem to be satisfied. Write a few sentences summarizing your conclusion.

(e) Determine a 95% confidence interval for the mean price difference between the two stores. What does this interval tell you about how much you expect to save?

(f) If you had subtracted in the opposite order to calculate the differences in (b), what would have changed about the test and interval? Describe specifically how they would have changed.

(g) Remove the outlier and repeat this analysis. Comment on the degree to which your conclusions change depending on whether one includes the suspicious value.

Activity 22-13: Comparison Shopping (*cont.*)

Consider again the shopping data of Activity 8-19 on page 182 that you analyzed in Activity 22-12 (SHOPPING.83g).

(a) Suppose that for each of the grocery items you were to indicate whether it cost more at Lucky's, at Von's, or cost the same at both stores. Would this be a categorical or a quantitative variable? Explain.

(b) Determine how many of these items cost more at Lucky's, how many cost more at Von's, and how many cost the same at both stores.

(c) Ignoring the items that cost the same at both stores, how many items remain? What proportion of them cost more at Lucky's?

(d) Consider testing the null hypothesis that half of all items cost more at Lucky's vs. the alternative that the proportion of all items that cost more at Lucky's is not one-half. Clearly identify the population and parameter of interest in this test.

(e) Carry out the test described in (d). Write a paragraph describing your conclusion, including details of your calculations and a check of the technical conditions.

(f) Determine a 96% confidence interval for the proportion of all items that cost more at Lucky's.

Activity 22-14: Age Guesses (*cont.*)

Consider the data collected in Topic 4 and analyzed in Activity 4-6 on page 81 concerning guesses of your instructor's age. Conduct a full analysis of the data with regard to the question of whether the guesses tend to average out to the actual age. (If your instructor prefers not to reveal the actual age, address whether the guesses tend to average out to the age that you personally guessed.) Include an appropriate test of significance, and be sure to identify the population and parameter of interest very clearly. Write a paragraph or two describing and explaining your analysis and findings.

Activity 22-15: Random Babies (*cont.*)

Recall the simulated data that you collected by shuffling and dealing cards to represent assigning babies to mothers at random in Activity 14-1 on page 305. Consider the variable "number of matches."

(a) Report (by either recalling or recalculating) the sample size (number of repetitions), the sample mean number of matches per repetition, and the sample standard deviation of those numbers of matches.

(b) Recall that your theoretical analysis in question (g) of Activity 14-2 on page 309 revealed the long-run average number of matches to be exactly 1.

(c) Use the sample statistics from (a) to test whether the simulated data provide strong reason to doubt that the population mean equals 1. Report the hypotheses (in words and in symbols), sampling distribution specified by the null hypothesis, test statistic, and p-value. Also comment on the technical conditions and write a conclusion.

Activity 22-16: Children's Television Viewing (*cont.*)

Suppose that you want to test whether children watch more than fourteen hours of television per week (or two hours per day) on average.

(a) State (in words and in symbols) the null and alternative hypotheses in terms of hours per week.

Recall from Activity 20-10 on page 438 that researchers gathered baseline data on the television viewing habits of third and fourth graders. The 198 children reported watching a mean of 15.41 hours and a standard deviation of 14.16 hours of television per week.

(b) Sketch the sampling distribution specified by the null hypothesis and use the above sample information to calculate the test statistic.

(c) Report the p-value of the test.

(d) Do the data provide evidence at the .05 level for concluding that third and fourth graders watch more than two hours of television per day on average? Explain.

(e) Do you think the technical conditions for the validity of this procedure have been met? Explain.

Topic 23:

MORE INFERENCE CONSIDERATIONS

In the last few topics you have explored and applied the two principal techniques of statistical inference: confidence intervals and tests of significance. Along the way you have been cautioned about some inappropriate uses and interpretations of these procedures. This topic asks you to consider the relationship between intervals and tests and to discover some more ways in which these techniques can be misapplied.

23

- To continue to develop an understanding of the reasoning, structure, and interpretation of confidence intervals and tests of significance.
- To investigate and understand the **duality** between confidence intervals and two-sided significance tests.
- To understand the inappropriateness of treating commonly used significance levels as "sacred."
- To recognize the distinction between **practical significance** and statistical significance.
- To understand the distinction between **type I** and **type II** errors in significance testing situations.
- To explore the concept of **power** and recognize the importance of considering power when interpreting test results.
- To learn how to determine the sample size necessary to achieve interval estimates with desired levels of accuracy and confidence.

- To continue to appreciate the importance of random sampling as an underpinning for statistical inference procedures.

PRELIMINARIES

1. If 24,643 of a random sample of 50,000 people own a digital or cellular telephone, would you be fairly convinced that fewer than half of all people own such a phone?

2. If 24,643 of a random sample of 50,000 people own a digital or cellular phone, would you be fairly convinced that *much* fewer than half of all people own such a phone?

3. Suppose that a baseball player who has always been a .250 career hitter suddenly improves over one winter to the point where he now has a 1/3 probability of getting a hit during an at-bat. Do you think he would be likely to convince the manager of his improvement in a trial consisting of 30 at-bats?

4. Do you think this improved baseball player would be more likely to convince the manager of his improvement in a trial of 100 at-bats?

5. Do you have your own credit card? Do you have your own digital or cellular phone? Record this information for you and your classmates. [Be sure to record the raw data and not just summaries so that association can be explored later.]

6. In Activity 23-4 you will be asked to study the scenario mentioned in 3 above by running a simulation. The name of the program needed to do this simulation is SIM-SAMP.83p. Download this program into your calculator (if you have not done so already).

You and a partner should both enter 30 for sample size and 999 for the number of samples. One of you should enter .25 for the true population proportion and the other should enter .3333. This program places all of the simulated sample proportions into a list named PROP.

WARNING: This program requires a significant amount of memory. You should delete any lists from your calculator that are no longer necessary.

IN-CLASS ACTIVITIES

Activity 23-1: Racquet Spinning (*cont.*)

Recall from Activity 14-9 on page 319 that 100 spins of a tennis racquet produced 46 "up" outcomes.

(a) Use your calculator to produce a 95% confidence interval for θ, the long-run proportion of racquet spins that would land "up."

(b) Does the value .5 fall within this interval? Record your answer in the third row of the table below.

(c) Based on your answer to (b), would you expect a significance test of whether θ differs from .5 to be significant at the .05 level? Explain. [*Hint*: Remember that a confidence interval contains plausible values of the parameter.]

(d) Use your calculator to perform a two-sided test of whether the sample provides strong evidence that θ differs from one-half. Report the test statistic and p-value; also indicate whether the sample proportion differs significantly from one-half at the $\alpha=.05$ significance level. Record the results in the table below. Was your expectation in (c) realized?

23

(e) Use your calculator to perform a two-sided test of whether the sample provides strong evidence that θ differs from .35. Again record the results in the table below, and also indicate whether the 95% confidence interval for θ includes the value .35.

(f) Repeat (e) where the value of interest is .4 rather than .35.

(g) Repeat (e) for the value .55.

(h) Repeat (e) for the value .6.

Hypothesized value	Contained in 95% CI?	Test statistic	p-value	Significant at .05?
.35	(e)	(e)	(e)	(e)
.40	(f)	(f)	(f)	(f)
.50	(b)	(d)	(d)	(d)
.55	(g)	(g)	(g)	(g)
.60	(h)	(h)	(h)	(h)

(i) Do you notice any connection between whether a 95% confidence interval for θ includes a particular value and whether the sample proportion differs significantly from that particular value at the α = .05 level? Explain.

This activity reveals a **duality** between confidence intervals for estimating a parameter and a two-sided test of significance about the value of the parameter. Roughly speaking, if a 95% confidence interval for a parameter does not include a particular value, then a two-sided test of whether the parameter equals that particular value will be statistically significant at the α = .05 level.

Confidence intervals and tests of significance are complementary procedures. While tests of significance can establish strong evidence that an effect exists, confidence intervals serve to estimate the magnitude of that effect.

Activity 23-2: Racquet Spinning (*cont.*)

Consider again the racquet spinning exercise, and continue to suppose that a goal is to determine whether the sample data provide evidence that the proportion of "up" results would differ from .5 in the long run.

(a) Suppose that you were to spin the racquet 200 times and obtain 113 "up" results. Use your calculator to determine the appropriate test statistic and p-value. Also report the value of the sample proportion of "up" landings and whether or not that sample proportion differs significantly from .5 at the $\alpha = .05$ level.

sample proportion: test statistic: p-value: significant at .05?

(b) Repeat (a) supposing that you obtained 115 "up" results in 200 spins.

sample proportion: test statistic: p-value: significant at .05?

(c) Repeat (a) supposing that you obtained 130 "up" results in 200 spins.

sample proportion: test statistic: p-value: significant at .05?

(d) In which pair of cases (a and b, a and c, b and c) are the sample results most similar?

(e) In which pair of cases (a and b, a and c, b and c) are the decisions about significance at the .05 level the same?

The moral here is that it is unwise to treat standard significance levels as "sacred." It is much more informative to consider the p-value of the test and to base one's decision on it. There is no sharp border between "significant" and "insignificant," only increasingly strong evidence as the p-value decreases. Reports of significance tests should include the sample information and p-value, not just a statement of significance or a decision about rejecting a hypothesis.

Activity 23-3: Cat Households (*cont.*)

Reconsider Activity 19-13 on page 419, which noted that 27.3% of the households in a 1995 survey of 80,000 owned a pet cat.

(a) Use your calculator to construct a 99.9% confidence interval for the proportion of all American households that own a pet cat and to conduct a test of whether the sample data provide very strong evidence that more than one-quarter of all American households own a pet cat. List the hypotheses, and report the test statistic and p-value. Is the sample result significant at the .001 level?

99.9% confidence interval:

hypotheses:

test statistic: p-value:

(b) Is the test result consistent with the confidence interval with respect to the plausibility that 25% of households own a pet cat? Explain.

(c) Would you say that the sample data provide strong evidence that the proportion of all American households that own a pet cat is very much more than 25%? Explain.

(d) Which procedure, the confidence interval or the test of significance, addresses the question in (c)?

This example illustrates that *statistical* significance is not the same thing as *practical* significance. A statistically significant result is simply one that is unlikely to have occurred by chance alone; that does not necessarily mean that the result is substantial or important in a practical sense. While there is strong reason to believe that the true proportion of households with a pet cat does indeed exceed 25%, the true proportion is actually *quite close to* 25%. When one works with very large samples, an unimportant result can be statistically significant nonetheless. Confidence intervals are useful for estimating the size of the effect involved and should be used in conjunction with significance tests.

Activity 23-4: Hypothetical Baseball Improvements

Suppose that a baseball player who has always been a .250 career hitter works very hard during the winter off-season and genuinely improves to the point where he now has a .333 probability of getting a hit during an at-bat. Suppose that his manager offers him a trial of 30 at-bats with which to demonstrate his improvement.

(a) One of the SIMSAMP programs that you ran in the Preliminaries simulated 999 repetitions of 30 at-bats using .250 as the success probability. Look at visual displays, including a histogram, of the distribution of simulated proportions of success stored in the PROP list, and describe it below.

(b) Based on the histogram of the simulated data, about how many hits would the player have to get in 30 at-bats so that the probability of a .250 hitter doing that well by chance alone is less than .05? Explain.

(c) The other SIMSAMP program that you ran earlier simulated 999 repetitions of 30 at-bats using .333 as the success probability. Look at visual displays of this distribution and the one for the .250 hitter on the same scale. Comment on the amount of overlap between the two distributions.

(d) In roughly what percentage of the 999 repetitions did the .333 hitter exceed the number of hits that you identified in (b)? (This would be the percentage of the 1000 repetitions in which the .333 hitter would do well enough in 30 at-bats to convince the manager that a .250 hitter would have been very unlikely to do that well just by chance.)

(e) Is it very likely that a .333 hitter will be able to establish that he is better than a .250 hitter in a sample of 30 at-bats? Explain. [*Hint:* Refer to your answer to (d) and to the overlap between these two distributions that you noted in (c).]

Two kinds of errors can be made with a hypothesis test: The null hypothesis can be rejected when it is actually true (called **type I error**), and the null hypothesis can fail to be rejected when it is actually false (**type II error**). The significance level α of a test puts an upper bound on the probability of a type I error.

The **power** of a statistical test is the probability that the null hypothesis will be rejected when it is actually false (and therefore should be rejected). Especially with small sample sizes, a test may have low power, so it is important to recognize that failing to reject the null hypothesis does not mean accepting it to be true.

(f) Use your simulation results above to report the approximate power of this significance test involving the baseball player. [*Hint:* Focus on your answer to (d).]

(g) If you were to repeat (a)–(f) above assuming that the player has a sample of 100 at-bats in which to establish his improvement, what differences (if any) do you think might occur? Write a paragraph summarizing your thoughts and discuss the power of the test with this larger sample size.

Increasing the sample size is one way to obtain a more powerful test, i.e., one that is more likely to detect a difference from the hypothesized value when a difference is actually there. With very large sample sizes, even minor differences can be detected, which reinforces the distinction between statistical and practical significance.

(h) If the player had improved to the point of being a .400 hitter, would you expect the test to be more or less powerful than when his improvement was to the .333 level? Explain.

(i) If your analysis had used the $\alpha = .10$ significance level rather than $\alpha = .05$, would you expect the test to be more or less powerful? Explain.

(j) In addition to sample size, list two other factors that are directly related to the power of a test.

Activity 23-5: Halloween Practices (*cont.*)

Recall from Activity 16-13 on page 361 and Activity 19-4 on page 411 that a Gallup survey of 1005 adult Americans revealed that 69% intended to give out Halloween treats from their home.

(a) Report the margin of error of this result, as you found with the half-width of a 95% confidence interval in (e) of Activity 19-4.

(b) If you wanted to estimate the population proportion to within ±0.01 with 95% confidence, would you expect to need a larger or a smaller sample? Explain.

(c) Use the formula for a confidence interval for a population proportion to determine how many people would need to be sampled in order for the 95% confidence interval to have a half-width of .01. [*Hints*: Work only with the half-width part of the formula. Use the sample proportion .69 from this study as your estimate of $\hat{p}$. Treat *n* as the unknown, and use algebra to solve for *n*. Be careful not to let rounding errors enter into your intermediate calculations. Always round your final answer up to express the sample size as an integer.]

(d) Would you expect to need more or fewer people in order to estimate the population proportion to within ± .01 with 99% confidence? Explain.

(e) Determine the sample size required in (d).

This activity reveals that one can plan ahead to determine the sample size needed to achieve a desired level of accuracy and confidence. One does need to supply an estimate for the value that the sample statistic will assume.

(f) How (if at all) did the population size enter into these calculations? How would your answers to (c) and (e) differ if the population of interest were all California adults rather than all American adults?

This last question again demonstrates that as long as one is dealing with a population that is much larger than the sample involved, the margin of error of a survey does not depend on the *population* size.

(g) How many people would have to be interviewed to determine the value of the population proportion exactly, with 100% confidence?

Activity 23-6: Hypothetical ATM Withdrawals (*cont.*)

Refer to the hypothetical ATM withdrawals described in Activity 5-24 on page 110. These pertain to a sample of withdrawal amounts from three different automatic teller machines (HYPOATM.83g).

(a) Use your calculator to compute the sample size, sample mean, and sample standard deviation of the withdrawal amounts for each machine. Also use your calculator to determine a 95% confidence interval for the mean withdrawal amount among all withdrawals for each machine. Record the results in the table below.

	Sample size	Sample mean	Sample std. dev.	95% confidence interval for μ
machine 1				
machine 2				
machine 3				

(b) Use your calculator to look at visual displays of the three distributions of withdrawal amounts. Do the distributions look the same or even similar? Write a paragraph comparing and contrasting the three distributions of ATM withdrawals.

This activity should remind you that a mean summarizes just one aspect of a distribution. While the mean is often very important and the focus of most inference procedures, it does not completely describe a distribution.

Activity 23-7: Female Senators (*cont.*)

Suppose that an alien lands on Earth, notices that there are two different sexes of the human species, and sets out to estimate the proportion of humans who are female. If the alien happened upon the members of the 1999 U.S. Senate as its sample of human beings, it would find 9 women and 91 men.

(a) Use this sample information to form a 95% confidence interval for the actual proportion of all humans who are female.

(b) Is this confidence interval a reasonable estimate of the actual proportion of all humans who are female?

(c) Explain why the confidence interval procedure fails to produce an accurate estimate of the population parameter in this situation.

(d) It clearly does not make sense to use the confidence interval in (a) to estimate the proportion of women on Earth, but does the interval make sense for estimating the proportion of women in the 1999 U.S. Senate? Explain your answer.

This example illustrates some important limitations of inference procedures. First, they do not compensate for the problems of a biased sampling procedure. If the sample is collected from the population in a biased manner, the ensuing confidence interval will be a biased estimate of the population parameter of interest.

A second important point to remember is that confidence intervals and significance tests use *sample* statistics to estimate *population* parameters. If the data at hand constitute the entire population of interest, then constructing a confidence interval from these data is meaningless. In this case, you know precisely that the proportion of women in the population of the 1999 U.S. Senators is .09, so it is senseless to construct a confidence interval from these data.

WRAP-UP

This topic has aimed to deepen your understanding of confidence intervals and tests of significance so that you may better understand some common misinterpretations of them. You have discovered the **duality** between intervals and tests and the distinction between practical and statistical significance. You have also learned that significance levels should not be regarded as fixed and have investigated the **power** of statistical tests. Finally, you have been reminded that exploratory examinations of data are crucial and that statistical inference applies only when a sample has been drawn from a population.

In the next unit you will learn to apply inferential techniques to situations that call for comparing two groups and detecting relationships between variables. These methods are particularly important for analyzing data from controlled experiments. You will again find that while the details of implementing the procedures necessarily change, the basic concepts do not.

HOMEWORK ACTIVITIES

Activity 23-8: Charitable Contributions (*cont.*)

Recall from Activity 19-14 on page 419 that in a 1996 survey of 2719 American households, 68.5% claimed to have made a financial contribution to charity in 1995.

(a) Find a 90% confidence interval for the proportion of all American households that made a financial contribution to charity in 1995.

(b) Repeat (a) with a 99% confidence interval.

(c) Based on these confidence intervals, without carrying out a test of significance, indicate whether this sample proportion differs significantly from 70% at the $\alpha = .01$ level. Explain your reasoning.

(d) Based on these confidence intervals, without doing a test of significance, indicate whether this sample proportion differs significantly from 65% at the $\alpha = .10$ level. Explain your reasoning.

Activity 23-9: Charitable Contributions (*cont.*)

Reconsider again the survey from Activity 19-14 on page 419, in which 68.5% of a sample of 2719 households said that they made a financial contribution to charity.
(a) Test whether the sample data provide evidence that the population proportion differs from 2/3. Report the test statistic and p-value of the test. Also indicate whether the result is statistically significant at the $\alpha = .05$ level.
(b) Do the sample data provide evidence that the population proportion differs *substantially* from 2/3? Explain.
(c) If the p-value had been smaller, would that constitute evidence that the population proportion differs substantially from 2/3? Explain.

Activity 23-10: Random Babies (*cont.*)

Recall again from Activity 14-1 on page 305 and Activity 14-2 on page 309 the class simulation of the "random babies" activity. Let θ be the long-term proportion of repetitions that result in no matches, and let μ be the long-term mean number of matches per repetition.
(a) Using the simulated sample data, conduct a significance test of whether θ differs from .4. Report the hypotheses, test statistic, and p-value. Is the result significant at the $\alpha = .05$ level?
(b) If the test result in (a) is not significant, does it follow that you accept that θ equals .4 exactly?
(c) Recall from your theoretical analysis in Activity 14-2 what the exact value of θ is. Explain how this relates to question (b).
(d) Using the simulated sample data, conduct a significance test of whether μ differs from 1.1. Report the hypotheses, test statistic, and p-value. Is the result significant at the $\alpha = .05$ level?
(e) If the test result in (a) is not significant, does it follow that you accept that μ equals 1.1 exactly?
(f) Recall from your theoretical analysis in Activity 14-2 what the exact value of μ is. Explain how this relates to question (e).

Activity 23-11: Planetary Measurements (*cont.*)

Reconsider the data presented in Activity 4-5 on page 80 that listed (among other things) the distance from the sun for each of the nine planets in our solar system. The mean of the distances turns out to be 1102 million miles, and the standard deviation is 1341 million miles.
(a) Use these numbers to construct a 95% confidence interval.
(b) Does this interval make any sense at all? If so, what population parameter does it estimate? Do you know the exact value of that parameter? Explain.

Activity 23-12: Women Senators (*cont.*)

Recall from Activity 7-5 on page 148 that the 1999 U.S. Senate consisted of 9 women and 91 men.
(a) Treat these numbers as sample data and calculate the test statistic for the significance test of whether the population proportion of women is less than .50.
(b) Use the test statistic and the table of standard normal probabilities to calculate the p-value of the test.
(c) If the goal is to decide whether women make up less than half of the entire U.S. Senate of 1999, does this test of significance have any meaning? Explain.

Activity 23-13: Cola Discrimination (*cont.*)

Recall the cola discrimination experiment described in Activity 16-15 on page 363 and the corresponding test of significance detailed in Activity 21-1 on page 445.
(a) Report (in symbols and in words) the null hypothesis of the test, corresponding to the conjecture that a subject is just guessing at identifying the one cup of soda that differs from the other two.
(b) Suppose that one particular subject (Randy) is actually able to identify the one differing cup 50% of the time in the long run. Would this subject's sample data necessarily lead to rejecting the null hypothesis? Explain.
(c) Suppose that an experiment for Randy consists of $n = 50$ trials. Verify that the null hypothesis would be rejected at the .05 significance level if the subject obtained a sample proportion of correct identifications of .46 or higher.
(d) Is there better than a 50/50 chance that Randy with his actual 50% success rate will get a sample proportion this high or higher? Explain, based on a CLT calculation.
(e) Would this probability increase, decrease, or remain the same if the sample size were 100 instead of 50? Explain.
(f) Would this probability increase, decrease, or remain the same if Randy's probability of a correct identification were 2/3 instead of 1/2? Explain.

Activity 23-14: Home Field Advantage (*cont.*)

Recall from Activity 19-10 on page 418 that the home team won 99 of 190 Major League Baseball games played between July 26 and August 8, 1999.
(a) Conduct a significance test of whether the sample data support the hypothesis of a home field advantage. Report the hypotheses (in words and in symbols), the test statistic, and the p-value, as well as your conclusion.
(b) Find a 90% confidence interval for the proportion of all games won by the home team.
(c) Does this sample information and your inferential analysis convince you that there is a home field advantage in Major League Baseball? Does it convince you

that there is *not* a home field advantage? Explain both your answer and the difference between these two statements.

Activity 23-15: Home Field Advantage (*cont.*)

Recall that 99 of the 190 baseball games in a sample were won by the home team.

(a) Report the p-value from Activity 23-14 for testing whether this constitutes evidence of a home field advantage. Would you reject the null hypothesis of no home field advantage at the .05 level?

(b) Repeat (a), supposing that it had turned out that 106 of the 190 games had been won by the home team.

(c) Repeat (a), supposing that it had turned out that 107 of the 190 games had been won by the home team.

(d) Now suppose that your assistant performed these tests and reported to you not the p-value but only whether the null hypothesis was rejected at the .05 level. With which two scenarios (a and b, a and c, or b and c) would the assistant report the same results from carrying out the test of significance?

(e) With which two scenarios are the sample results most similar?

(f) Are your answers to (d) and (e) the same?

(g) Explain what this activity reveals about the relative information provided by reporting only the decision to reject or fail to reject the null hypothesis as opposed to reporting a p-value.

Activity 23-16: Voter Turnout (*cont.*)

Recall from Activity 12-20 on page 267 that a random sample of 2613 adult Americans in 1998 revealed that 1783 claimed to have voted in the 1996 Presidential election.

(a) Use these sample data to construct a 99.9% confidence interval for θ, the proportion of all adult Americans who voted in that election.

(b) Even though this truly was a random sample, do you really have 99.9% confidence that this interval captures the actual proportion who voted in 1996? Explain.

(c) The Federal Election Commission reported that 49.0% of those eligible to vote in the 1996 election had actually voted. Is this value included within your interval?

(d) Do you think this interval succeeds in capturing the proportion of all adult Americans who would claim to have voted in 1996? Explain how this parameter differs from that in (c).

(e) Explain why the confidence interval renders it unnecessary to conduct a significance test of whether θ differs from .49 at the .001 level.

Activity 23-17: Basketball Scoring (*cont.*)

Reconsider the data that you analyzed in Activity 22-9 on page 474 concerning whether the mean points per game in the NBA was higher in the 1999–2000 season than it had been in the previous season. Suppose that your sample had consisted of only the ten games played on December 10, 1999, whose total points scored were as follows:

196	198	205	163	184	224	206	190	140	204

(a) Examine visual displays and summary statistics for these data. Write a few sentences describing the distribution with respect to the question of whether games average more than 183.2 points, which had been the previous season's average.

(b) Use these sample data to test whether there is reason to believe that the mean points per game during the 1999–2000 NBA season exceeds 183.2. Report your hypotheses, test statistic, and p-value.

(c) Would you reject the null hypothesis at the $\alpha = .05$ level? Write a sentence indicating what this signifies in this context.

(d) Would you *accept* the null hypothesis and conclude that the mean points per game in the 1999–2000 season is in fact 183.2? Explain.

(e) Remove the outlier from the sample data and repeat this analysis. Report your results, and also comment on what effect the removal of the outlier has on your findings.

Activity 23-18: Phone Book Gender (*cont.*)

Recall from Activity 12-17 on page 266 the sample data collected from a random page of the San Luis Obispo County telephone book: 36 listings had both male and female names, 77 had male names, 14 had female names, 34 had initials only, and 5 had pairs of initials. Before collecting the data, the authors conjectured that fewer than half of the names in the phone book would be female.

(a) A total of how many *first names* were studied (ignore listings with only initials)? How many of them were female names? What proportion of the names were female?

(b) Identify the observational units in this study.

(c) Identify the population and the parameter of interest.

(d) Do the sample data support the authors' conjecture that fewer than half of the names in the phone book would be female at the $\alpha = .05$ level? Report the details of the test and write a short paragraph describing your findings and explaining how your conclusions follow from the test results. Also discuss the technical conditions.

(e) Accompany your test with a 95% confidence interval. Interpret the interval, and comment on how it relates to the test result.

(f) Do the sample data provide evidence that fewer than half of the residents of San Luis Obispo County are female? Explain your answer, and be sure to discuss how this question differs from the one you addressed in (d).

Activity 23-19: Cat Households (*cont.*)

Recall from Activity 19-13 on page 419 that in a survey of 80,000 households conducted by the American Veterinary Medical Association in 1996, 27.3% of households reported that they owned a pet cat. Of those 21,840 households that did own a pet cat, the mean number of cats per household was 2.2.

(a) State (in words and in symbols) the hypotheses for testing whether the sample data provide strong evidence that the mean number of cats per cat-owning household exceeds 2.

(b) What additional sample information do you need in order to conduct this test?

(c) Supply a reasonable estimate for the missing sample statistic, and calculate the test statistic and p-value. Does the sample provide strong evidence that the mean number of cats exceeds two? Explain.

(d) As a check on the sensitivity of this test, double your estimate for the missing sample statistic and repeat (c). Does your conclusion change substantially? Explain.

(e) Using your estimate from (c), find a 99% confidence interval for the population mean number of cats per cat-owning household. Would you say that this mean value greatly exceeds two? Explain.

Activity 23-20: Parent Ages (*cont.*)

Recall from Activity 3-11 on page 59 the data on the ages at which a sample of 35 mothers gave birth, which have a mean of $\bar{x} = 22.31$ years and a standard deviation of $s = 5.60$ years. Also recall that this sample was taken from a much larger sample of 1199 mothers described in Activity 17-5 on page 379, for which the mean age is 22.52 years.

(a) Use the data from the small sample to test whether the population mean age at which a mother has her first child differs from 22.52 years. State the hypotheses, and report the test statistic and p-value. Would you reject the null hypothesis at the $\alpha = .05$ significance level?

(b) Repeat (a) but test whether the population mean differs from 23.0 years.

(c) Repeat (a) but test whether the population mean differs from 23.5 years.

(d) Repeat (a) but test whether the population mean differs from 21.0 years.

(e) Discuss what the results of the tests in (a)–(d) reveal about the power of this test.

(f) Use the small sample to find a 95% confidence interval for the population mean age of first having a child.

(g) Discuss how this confidence interval is consistent with the test results in (a)–(d).

(h) State a hypothesized value for the population mean that would be rejected at the $\alpha = .05$ level. Explain your choice.

Activity 23-21: Penny Spinning (*cont.*)

Suppose that you spin a penny as described in Activity 19-1 on page 405 with the intention of determining whether a spun penny is equally likely to land heads or tails, i.e., whether it would land heads 50% of the time in the long run.

(a) Identify in a symbol and in words the *parameter* of interest in this experiment.

(b) Considering the stated goal of the study, is the alternative hypothesis one-sided or two-sided in this case? Explain.

(c) Specify the null and alternative hypotheses for this study, both in symbols and in words.

(d) Suppose that you spin the penny 150 times, obtaining 65 heads and 85 tails. Use your calculator to compute the test statistic and p-value of the test.

(e) Use your calculator to determine how the test statistic and p-value would have been different if the 150 spins had produced 85 heads and 65 tails.

(f) In either of these cases (85 of one outcome and 65 of the other), do the sample data provide strong evidence that a spun penny would *not* land heads 50% of the time in the long run? Would you reject the null hypothesis at the .05 significance level?

(g) Now suppose that the goal of the experiment is to investigate whether a spun penny tends to land heads less than 50% of the time. Restate the null and alternative hypotheses (in symbols).

(h) Use your calculator to determine the test statistic and p-value of the one-sided test, assuming that 65 of the 150 sample spins landed heads. Report the test statistic and p-value of the test, and comment on how they compare to the ones found with the two-sided test.

(i) Still supposing that the goal of the experiment is to investigate whether a spun penny tends to land heads less than 50% of the time, use your calculator to determine the test statistic and p-value of the one-sided test assuming that 85 of the 150 sample spins landed heads. Again report the test statistic and p-value of the test, and comment on how they compare to the ones found with the two-sided test.

(j) For the sample results of question (i), explain why the formal test of significance is unnecessary.

Activity 23-22: Penny Spinning (*cont.*)

Reconsider the penny spinning experiment conducted in Activity 19-1 on page 405. Use the sample results collected in class to test whether the data provide evidence that θ, the long-term proportion of heads that would result from *all* hypothetical penny spins, *differs from* one-half. Be sure to report the null and alternative hypotheses, the test statistic, and the p-value. Also write a brief conclusion.

Activity 23-23: Penny Spinning (*cont.*)

(a) Suppose that you want to estimate the proportion of penny spins that would land heads to within ±0.05 with 95% confidence. Determine how many spins would be required. [*Hint*: Supply your own estimate for the sample proportion of heads, and clearly identify it as such.]

(b) Repeat (a) for estimating that proportion to within ±0.05 with 90% confidence.

(c) Repeat (a) for estimating that proportion to within ±0.02 with 90% confidence.

(d) Repeat (a) for estimating that proportion to within ±0.02 with 99% confidence.

(e) Repeat (a) using the following estimates of the sample proportion of heads: .3, .5, and .6. Which of these estimates produces the largest required sample size?

Activity 23-24: Christmas Shopping (*cont.*)

Suppose that you want to estimate the mean amount that American adults spend on Christmas presents in one year.

(a) Use the formula for a confidence interval for a population mean to determine how many people would need to be sampled in order to estimate this mean to within ±$100 with 95% confidence. [*Hints*: Use $s = \$250$ as an estimate of the standard deviation, and use the z^* critical value to approximate the t^* critical value.]

(b) Repeat (a) for estimating the mean amount to within ±$10 with 95% confidence.

(c) Repeat (a) for estimating the mean amount to within ±$10 with 80% confidence.

(d) Would you need to sample more or fewer people than in (a) in order to estimate the mean amount to within ±$10 with 95% confidence. Explain why this makes sense.

(e) Would you need to sample more or fewer people than in (c) in order to estimate the mean amount to within ±$10 with 90% confidence. Explain why this makes sense.

Activity 23-25: Hypothetical Baseball Improvements (*cont.*)

Reconsider the baseball player from Activity 23-4 on page 485. Suppose that he had actually become a .400 hitter (i.e., had a .4 probability of getting a hit during an at-bat). Use the SIMSAMP program to investigate whether the power for a significance test with a sample size of 30 at-bats is larger or smaller than it was when he became a .333 hitter by repeating your simulation analysis of Activity 23-4 for a .400 hitter. Write a paragraph reporting on your findings concerning the change in the test's power.

Activity 23-26: Hypothetical Baseball Improvements (*cont.*)

Consider once again the baseball player from Activity 23-4 on page 485 who had become a .333 hitter. Investigate whether the test is more or less powerful at higher significance levels by repeating your simulation analysis of Activity 23-4 with the SIM-SAMP program using $\alpha = .10$ rather than $\alpha = .05$ as the significance level. Write a paragraph reporting on your findings concerning the change in the test's power resulting from this change in significance level.

Activity 23-27: Credit Cards and Cellular Phones

Consider the data that you collected in class about whether students have their own credit card and/or cellular phone.

(a) To what population do you think it is valid to generalize from your sample results, all students at our school, all students in the country, all statistics students, all people? In other words, do you think that your sample is representative of any of these populations with respect to this issue? Justify your choice.

(b) Write out (in words and in symbols) the hypotheses for testing whether the population proportion of credit card owners differs from one-half.

(c) Is your sample size large enough for the technical conditions of this test to be satisfied? If not, try to sample more students in an unbiased manner until the sample size is large enough.

(d) Carry out the test described in (b). Show the details of the calculations, and write a few sentences summarizing your findings.

(e) Produce a 90% confidence interval to estimate the population proportion who own a credit card. Interpret the interval, and comment on how the interval relates to the test result.

(f) Repeat (b)–(e) with regard to having a cellular phone.

Activity 23-28: Credit Cards and Cellular Phones (*cont.*)

Suppose that a simple random sample of 50,000 people revealed that 24,643 own a digital or cellular phone.

(a) Does this sample constitute strong evidence that fewer than half of all people in the population own such a phone? Apply an appropriate inferential technique to support your answer.

(b) Does this sample constitute evidence that the proportion of the population who own such a phone is substantially less than one-half? Again apply an appropriate inferential technique to support your answer.

23

Unit VI:

Inference from Data: Comparisons and Relationships

Topic 24:

COMPARING TWO PROPORTIONS

OVERVIEW

In the previous unit you discovered and explored techniques of statistical inference for drawing conclusions about population parameters on the basis of sample statistics. All of those procedures applied to a single parameter (proportion or mean) from a single population. In the next two topics you will investigate and apply inference procedures for *comparing* parameters between two groups. These techniques are especially important due to the crucial role of comparison in experimental design. The randomization involved in experiments enables us to make inferences about the effects of the explanatory variable on the response variable. In this topic you will investigate inference procedures for *comparing* proportions between two experimental groups. You will see that the basic reasoning, structure, and interpretation of these tests and intervals are the same as in the case of inference about a single population parameter.

OBJECTIVES

- To understand the reasoning behind tests of significance for comparing sample proportions between two experimental groups.
- To explore properties of tests of significance for comparing two proportions.
- To learn a confidence interval procedure for estimating the magnitude of the difference between two population proportions.
- To continue to recognize the distinction between controlled experiments and observational studies, and to appreciate the different types of conclusions that can be drawn from each.

- To gain experience with applying these significance tests and confidence intervals to actual data and with interpreting the results.

PRELIMINARIES

1. Would you expect to perform better on a cognitive task if you were being observed by someone with a vested interest in your performance or by a neutral observer?

2. Suppose that 10 patients are randomly assigned to receive a new medical treatment, while another 10 are randomly assigned to receive the old standard treatment. If 7 patients in the "new" group and 5 patients in the "old" group recover, would you be fairly convinced that the new treatment is superior to the old?

3. Suppose that 10 patients are randomly assigned to receive a new medical treatment, while another 10 are randomly assigned to receive the old standard treatment. If 9 patients in the "new" group and 3 patients in the "old" group recover, would you be fairly convinced that the new treatment is superior to the old?

4. Take a guess concerning the proportion of Americans who say that they are satisfied with their physical appearance.

5. Would you expect the proportion who say that they are satisfied with their physical appearance to be higher among men or among women?

6. Take a guess concerning the proportion of Americans who would respond in the affirmative to the question, "Do you think the United States should forbid public speeches in favor of communism?"

7. Suppose that one group of subjects is asked the question in 6, while another group is asked, "Do you think the United States should allow public speeches in favor of communism?" Would you expect these groups to differ in terms of their opposition to communist speeches? If so, which group do you think would more likely oppose communist speeches?

IN-CLASS ACTIVITIES

Activity 24-1: Friendly Observers (*cont.*)

Recall from Activity 13-10 on page 293 that in a study published in the *Journal of Personality and Social Psychology* (Butler and Baumeister, 1998), researchers investigated a conjecture that having an observer with a vested interest would decrease subjects' performance on a skill-based task. Subjects were given time to practice playing a video game that required them to navigate an obstacle course as quickly as possible. They were then told to play the game one final time with an observer present. Subjects were randomly assigned to one of two groups. One group (call it A) was told that the participant and observer would each win $3 if the participant beat a certain threshold time, and the other group (B) was told only that the participant would win the prize if the threshold were beaten. It turned out that 3 of the 12 subjects in group A beat the threshold, while 8 of the 11 subjects in group B did so.

	A: observer shares prize	B: no sharing of prize	Total
Beat threshold	3	8	11
Did not beat threshold	9	3	12
Total	12	11	23

(a) Calculate the sample proportions of success for each group. (Denote these by $\hat{p}_A$ and $\hat{p}_B$.)

(b) Do these sample proportions differ in the direction conjectured by the researchers?

(c) Even if there were absolutely *no effect* of the observer's interest, is it possible to have gotten such a big difference between the two groups just due to chance variation?

In keeping with the reasoning of significance tests, we will ask how likely the sample results would have been if in fact the observer's incentive had no effect on the subject's performance. One way to analyze this question is to assume that those 11 successes would have been successes regardless of which group the subject had been assigned to. We can then *simulate* the process of assigning subjects at random to the two groups, observing how often we obtain a sample result at least as extreme (3 or fewer successes assigned to A) in the actual sample. Repeating this a large number of times will give us a sense for how unusual it would be for this sample result to occur by chance alone.

(**d**) Mark 11 cards as "success" and 12 as "failure", shuffle them well, and randomly deal out 12 to represent the cases assigned to group A. How many of these 12 are successes? Is this result at least as extreme as in the actual sample?

(**e**) Repeat this simulation a total of five times, recording your results in the table:

Repetition #	1	2	3	4	5
"successes" assigned to group A					
as extreme as actual study?					

(**f**) Combine your results with the rest of the class, forming a dotplot of the number of successes randomly assigned to group A:

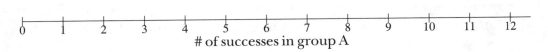

(**g**) What value or values are the most common? Explain why this makes sense.

(h) How would the above dotplot change if we instead graphed the proportion of successes in group A minus the proportion of successes in group B? What value do you expect to be in the center of this distribution given that in the simulation the "successes" are equally likely to be assigned to either group?

(i) How many repetitions were performed by the class as a whole? How many of them gave a result at least as extreme as the actual sample (3 or fewer successes in group A)? What proportion of the repetitions is this?

(j) Remember that your random shuffling and dealing assumed that the observer's incentive had no effect on the participant's performance. Based on these simulated results, does it appear that it is very unlikely for random assignment to produce a result as extreme as the actual sample when the observer has no effect?

(k) In light of your answer to the previous question, considering that the actual sample is what the researchers found, would you say that the data provide reasonably strong evidence in support of the researchers' conjecture? Explain.

24

> This activity should reinforce the important idea that statistical significance assesses the likelihood of a sample result by asking how often such an extreme result would occur by chance alone. When the sample result is unlikely to occur by chance, it is said to be statistically significant. The probability of obtaining a result at least as extreme as the sample by chance alone is known as the *p-value* of the test. Our simulation has approximated this p-value.

Rather than rely on simulations, we can use a formal test of significance to assess how likely a sample result would have occurred by chance alone. The reasoning and interpretation of these tests are the same as when you applied them to a single population proportion. In this situation we are interested in comparing population proportions between two groups; denote these population proportions by θ_1 and θ_2.

The null hypothesis asserts that θ_1 and θ_2 are equal. The alternative hypothesis can again take one of three forms depending on the researcher's conjecture about θ_1 and θ_2 prior to conducting the study. The test statistic is calculated by comparing the two sample proportions and standardizing appropriately. The p-value is again found from the standard normal table. One interprets this p-value as before: It represents the probability of having gotten a sample result as extreme as the actual result if the null hypothesis were true. Thus, the smaller the p-value, the stronger the evidence against the null hypothesis.

Significance test of equality of θ_1 and θ_2:

$H_o: \theta_1 = \theta_2$

$H_a: \theta_1 < \theta_2$ or $H_a: \theta_1 > \theta_2$ or $H_a: \theta_1 \neq \theta_2$

test statistic:
$$z = \frac{\hat{p}_1 - \hat{p}_2}{\sqrt{\hat{p}_c(1 - \hat{p}_c)\left(\dfrac{1}{n_1} + \dfrac{1}{n_2}\right)}}$$

p-value $= \Pr(Z \leq z)$ or $\Pr(Z \geq z)$ or $2\Pr(Z \geq |z|)$,

where n_1 and n_2 are the respective sample sizes from the two groups, $\hat{p}$ and $\hat{p}$ are the respective *sample* proportions, $\hat{p}_c$ denotes the *combined* sample proportion (i.e., the proportion of "successes if the two samples were pooled together as one big sample), and Z represents the familiar standard normal distribution.

The **technical conditions** necessary for the validity of this test procedure are:
1. That the two samples are independently selected simple random samples from the populations of interest,
2. That $n_1\hat{p}_c \geq 5$ and $n_1(1 - \hat{p}_c) \geq 5$ and $n_2\hat{p}_c \geq 5$ and $n_2(1 - \hat{p}_c) \geq 5$.

Activity 24-2: Pregnancy, AZT, and HIV (*cont.*)

Recall from Activity 13-5 on page 287 that medical experimenters randomly assigned 164 pregnant, HIV-positive women to receive the drug AZT during pregnancy, while another 160 such women were randomly assigned to a control group that received a placebo. Of those in the AZT group, 13 had babies who tested HIV-positive, compared to 40 HIV-positive babies in the placebo group.

(a) Let θ_{AZT} denote the proportion of *all* potential AZT-takers who would have HIV-positive babies and θ_{plac} denote the proportion of *all* potential placebo-takers who would have HIV-positive babies. Write (using these symbols) the appropriate null and alternative hypotheses for testing the researchers' conjecture that AZT would prove beneficial for these patients.

(b) Calculate (by hand) the sample proportions of HIV-positive babies in the two experimental groups. Record these along with the symbols used to represent them. Also construct a segmented bar graph to display the distributions between the two groups.

(c) A total of how many women were subjects in this study? How many had HIV-positive babies? Use this information to determine $\hat{p}_c$, the combined sample proportion who had HIV-positive babies.

(d) Use your answers to (b) and (c) to calculate the test statistic for testing the hypotheses of (a).

(e) Based on this test statistic, use the table of normal probabilities (Table II) to calculate the p-value of the test.

(f) If AZT and the placebo were equally effective in preventing babies from having HIV, what would be the probability of chance alone assigning 40 HIV-positive babies to the placebo group and only 13 to the AZT group? Is this outcome very unlikely to occur by chance alone?

(g) Based on the p-value, write a one-sentence conclusion about the extent to which the sample data support the researchers' conjecture.

While this test of significance addresses the question of the extent of evidence that the population proportions differ, it does not provide information about the magnitude of the difference. A confidence interval addresses this issue.

Confidence interval for $\theta_1 - \theta_2$:

$$(\hat{p}_1 - \hat{p}_2) \pm z^* \sqrt{\frac{\hat{p}_1(1 - \hat{p}_1)}{n_1} + \frac{\hat{p}_2(1 - \hat{p}_2)}{n_2}},$$

where z^* is the appropriate critical value from the normal distribution.

> The **technical conditions** necessary for the validity of this interval are:
>
> 1. That the two samples are independently selected simple random samples from the populations of interest,
>
> 2. That there are at least five "successes" and five "failures" in each group.

(h) Find a 95% confidence interval by hand for estimating the difference in population proportions $\theta_{AZT} - \theta_{plac}$.

(i) Check your answer to (h) above by selecting option 2-PropZInt from the STAT TESTS menu. Enter the sample counts, sample sizes, and confidence level.

(j) Write a one-sentence description of what this interval says about the difference in population proportions. In particular, comment on the meaning of whether the interval includes the value zero, contains only negative values, or contains only positive values. [Please be sure that your comments relate to this context involving HIV-positive babies.]

24

Activity 24-3: Perceptions of Self-Attractiveness

A survey conducted by the Gallup organization in 1999 asked American adults whether they are satisfied with their physical attractiveness or wish they could be more attractive. 71% of the women and 81% of the men said that they were satisfied with their appearance.

(a) What additional information is necessary to determine whether a significantly higher proportion of men than of women say that they are satisfied with their appearance?

(b) Describe a circumstance in which these sample proportions would not convince you at all that males are more satisfied with their appearance.

(c) Describe a circumstance in which these sample proportions would strongly convince you that males are more satisfied with their appearance.

(d) Use your calculator to determine the test statistic and p-value of the appropriate significance test for the different sample sizes in the table below. Use the 2-PropZ-Test feature of your calculator (located in the STAT TESTS menu). As with earlier tests for a single proportion, you need only fill in the counts, sample sizes, and form of the alternative hypothesis to complete each test. Also record whether or not the difference in sample proportions is statistically significant at the .10, .05, and .01 levels.

sample size	satisfied women	satisfied men	test stat	p-value	α = .10?	α = .05?	α = .01?
100	71	81					
200	142	162					
500	355	405					

(e) Write a sentence or two commenting on the role of sample size in determining whether a difference between two sample proportions is statistically significant.

This activity should reinforce what you have learned earlier about the effect of sample size on inference procedures. An observed difference in sample proportions is more statistically significant (unlikely to have occurred by chance) with larger sample sizes. Even a small difference can be statistically significant with a large sample size.

Activity 24-4: Graduate Admissions Discrimination (*cont.*)

Recall from Activity 7-20 on page 155 that the University of California at Berkeley was alleged to have committed discrimination against women in its graduate admissions practices. When you analyzed the data from the six largest graduate programs, you found that 1195 of the 2681 male applicants had been accepted and that 559 of 1835 female applicants gained acceptance.

(a) Use your calculator to conduct the appropriate test of significance for assessing whether this difference in sample proportions is statistically significant. Report the null and alternative hypotheses (in words and in symbols), the test statistic, and the p-value of the test. Are these sample results statistically significant at commonly used levels of significance?

(b) Remembering the analysis that you performed in Activity 7-20, do you regard this statistically significant difference as evidence of discrimination? Explain.

24

These inference procedures for comparing two proportions can be used with data from observational studies as well as from controlled experiments. No matter how statistically significant a difference might be, however, one can *not* draw conclusions about *causation* from an observational study.

WRAP-UP

You have investigated tests of significance for comparing proportions between two groups in this topic. You have used physical (card shuffling) and computer simulations to study the reasoning behind the test procedure and have also explored the effects of sample sizes on the test. You have also been reminded that the conclusions one can draw from such a test depend on whether the design of the data collection was an observational study or a controlled experiment. A confidence interval procedure for estimating the magnitude of the difference was also introduced.

In the next topic you will turn again from categorical variables to quantitative ones, and you will study inferential techniques for comparing two means.

HOMEWORK ACTIVITIES

Activity 24-5: Literature for Parolees (*cont.*)

Consider the study described in Activity 13-21 on page 297 concerning the program to enroll convicts in a literature course in the hope that they would be less likely to commit a crime after their release. Of the 32 convicts given the course, only six committed a crime within thirty months of release. Of 40 similar parolees not given the literature course, 18 committed a crime in that period.

To assess how likely these sample results would have occurred by chance if the literature course had no effect, consider the histogram below. These data are the result of 1000 repetitions of the simulated random assignment of convicts to groups, assuming that the 24 (i.e., 6 + 18) who committed a crime would have done so regardless of the group to which they were assigned. The following histogram displays the number of repeat offenders assigned by chance to the group that took the literature course.

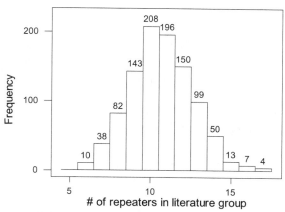

(a) In how many of these 1000 simulated random assignments were the results as favorable to the literature program as the actual results were. Explain what "as favorable" means in this case. What proportion of the 1000 repetitions were as favorable to the program as the actual results?

(b) Conduct the appropriate test of significance to assess whether the proportion of crime-committers is statistically significantly smaller among those who take the course than among those who do not. Report the null and alternative hypotheses, and show the details of your calculations of the test statistic and p-value.

(c) Is the p-value close to the proportion that you found from the simulation in (a)?

(d) Summarize in a paragraph your findings about the effectiveness of this literature course. Be sure to mention some of the experimental design considerations first raised in Activity 13-21.

Activity 24-6: Perceptions of Self-Attractiveness (*cont.*)

Reconsider the survey result concerning perceptions of self-attractiveness that you investigated in Activity 24-3 on page 511.

(a) For each of the three sample sizes listed there, find a 95% confidence interval for the difference in proportions of men and of women who are satisfied with their appearance.

(b) Report the half-width of each of these intervals. Does the half-width increase or decrease as the sample size increases? Is this consistent with what happens with confidence intervals for a single proportion?

(c) With which sample sizes does the 95% confidence interval include the value zero and with which does it not? Explain how this is consistent with your test results in Activity 24-3.

24

Activity 24-7: Perceptions of Self-Attractiveness (*cont.*)

Reconsider again the survey result that you investigated in Activity 24-3. Suppose that the sample proportions who indicated satisfaction with their appearance had been 76% for men and 75% for women. Further suppose that the sample size had been the same for each group.

(a) Find a sample size for which this difference would be statistically significant at the $\alpha = .01$ level. Report the test statistic and p-value of the test.

(b) Construct a 99% confidence interval for the difference in the population proportions using the sample size you found in (a).

(c) Repeat (a) and (b) if the sample proportions had been 75.6% and 75.5%, respectively.

(d) Would you regard the difference as practically significant in the situation of (c)? Explain.

Activity 24-8: Campus Alcohol Habits

Some researchers wanted to investigate whether the proportion of college students who drink alcohol decreased between 1982 and 1991. They analyzed data from two national studies. In a national study conducted in 1982, 4324 of a sample of 5252 college students said that they drank alcohol. In a similar national study conducted in 1991, 3820 of a sample of 4845 college students said that they drank alcohol.

(a) What proportion of the 1982 sample drank alcohol? What proportion of the 1991 sample drank alcohol?

(b) Create a segmented bar graph to display the conditional distributions of alcohol drinking between the 1982 and 1991 samples.

(c) State the null and alternative hypotheses for testing the researchers' conjecture in both symbols and words.

(d) Calculate the appropriate test statistic and the p-value of the test. Is the decrease in sample proportions statistically significant at the .01 level?

(e) Produce a 95% confidence interval for the difference in proportions. Does the interval include the value 0? What does the interval reveal about the question of whether the proportion of alcohol consumers dropped between 1982 and 1991?

(f) Is this study a controlled experiment or an observational study?

(g) Does the design of the study allow you to supply a causal explanation for the decrease in proportions of alcohol drinkers? Explain.

Activity 24-9: Campus Alcohol Habits (*cont.*)

Refer to the study described in Activity 24-8 concerning alcohol use on campus. The national studies also asked students more specific questions about students' drinking habits. Students were asked whether they have gotten into fights after drinking and whether they have had trouble with the law after drinking. Of the 4324 drinkers surveyed in 1982, 502 reported getting into a fight after drinking, while 190 reported getting into trouble with the law due to drinking. Of the 3820 drinkers surveyed in 1991, these numbers were 657 and 290, respectively.

Use your calculator to analyze these data for any differences between 1982 and 1991 with regard to getting into fights or trouble with the law due to drinking. Report the results of significance tests as well as confidence intervals. Also compare your findings with those from Activity 24-8. Include visual displays in your analysis. Write a paragraph or two summarizing your conclusions.

Activity 24-10: Baldness and Heart Disease (*cont.*)

Reconsider the data presented in Activity 7-12 on page 152 concerning baldness and heart disease, the table for which is reproduced here:

	none	little	some	much	extreme
heart disease	251	165	195	50	2
control	331	221	185	34	1

(a) Calculate the sample proportion of the heart disease patients who had some or more baldness (i.e., some, much, or extreme baldness). Calculate the same for the control group.

(b) Conduct (by hand or with your calculator) a test of significance to determine whether these proportions differ significantly. State the null and alternative hypotheses in words as well as in symbols, and report the test statistic and p-value. Also indicate whether the difference in sample proportions is statistically significant at the .05 level.

(c) Construct a 98% confidence interval for the difference in population proportions. Write a sentence or two describing what the interval reveals.

(d) What conclusion can you draw about a possible association between baldness and heart disease? Be sure to keep in mind the design of the study as you address this question.

24

Activity 24-11: Suitability for Politics (*cont.*)

Recall from Activity 7-7 on page 150 that the General Social Survey asked American adults whether they agree with the proposition that men are better suited emotionally for politics than women are. The sample results are summarized in the following table:

	men	women
agree	169	236
disagree	565	777

(a) State the hypotheses for testing whether men tend to agree with this proposition more often than women do.

(b) Calculate the test statistic for conducting this test.

(c) Find the p-value of the test.

(d) Interpret the p-value in the context of this issue.

(e) Construct a 95% confidence interval for the difference in population proportions of agreement between men and women.

(f) Write a sentence or two summarizing your findings.

Activity 24-12: Baseball "Big Bang" (*cont.*)

Recall from Activity 21-2 on page 452 and Activity 21-8 on page 458 the definition of a "big bang" in a baseball game. For the sample of 190 games played between July 26 and August 8, 1999, a big bang occurred in 46 of 87 American League games and in 52 of 103 National League games. For the 99 games won by the home team, 62 contained a big bang, compared to 36 games with a big bang among the 91 won by the visiting team.

Analyze these sample data for evidence of a difference in the proportion of "big bang" games between the leagues and between games won by the home or visiting team. Write a paragraph describing your findings. Be sure to include graphical displays and numerical summaries as well as inferential results. Also discuss whether the technical conditions that underlie the validity of the inferential procedure are satisfied.

Activity 24-13: Solitaire (*cont.*)

Recall from Activity 14-14 on page 322 and Activity 14-15 that author A won 25 times in 217 games of solitaire and that author B won 74 times in 444 games.

(a) Calculate the proportion of games won by each author and construct a segmented bar graph to display the distributions.

(b) Do these data suggest that the winning probabilities for these two people differ at the .10 level? At the .05 level? Show the details of your test procedure.

(c) Find a 90% confidence interval to estimate the difference in the probabilities.

(d) Are the technical conditions satisfied for the inference procedures in (b) and (c)? Explain.

(e) How would the interval differ if you found a confidence interval for $\theta_A - \theta_B$ rather than for $\theta_B - \theta_A$.

(f) Explain how the test result reveals whether a 95% confidence interval for the difference in probabilities would include the value zero.

(g) Explain how the p-value would have differed if you had been testing that author B's probability is higher than author A's.

Activity 24-14: Magazine Advertisements (*cont.*)

Recall from Activity 19-7 on page 416 that the September 13, 1999 issue of *Sports Illustrated* had 54 pages with advertisements among its 116 pages, while the September 14, 1999 issue of *Soap Opera Digest* had 28 pages with advertisements among its 130 pages.

(a) Calculate the sample proportions of pages with ads in these magazines.

(b) Construct a segmented bar graph to compare the proportions of pages with ads between the two magazines.

(c) Conduct the appropriate test of significance to assess whether the magazines' proportions of pages with ads differ significantly at the $\alpha = .01$ level. Report the hypotheses (in symbols and in words), the test statistic, and the p-value.

(d) Write a sentence or two summarizing your conclusion and how it follows from this test result.

(e) What assumption must you make in order for this test procedure to be valid?

Activity 24-15: Gender and Lung Cancer (*cont.*)

Recall from Activity 13-13 on page 295 the study of gender and incidence of lung cancer among smokers. Researchers screened 1000 people who were smokers of age 60 or higher. They found 459 women and 541 men, with 19 of the women and 10 of the men suffering from lung cancer. Test whether these data provide evidence that women smokers contract lung cancer at a higher rate than male smokers. Write a paragraph describing your findings and explaining how they follow from the test result. Include a discussion of the technical conditions of the inference procedure and of the design of the study.

24

Activity 24-16: President's Popularity (*cont.*)

Reconsider Activity 12-12 on page 264, which described a Gallup poll conducted August 17, 1998, following a speech by President Clinton. Respondents were asked "Now thinking about Bill Clinton as a person, do you have a favorable or unfavorable opinion of him?" Forty percent of respondents replied "favorable," and 48% replied "unfavorable." The next day, Gallup asked another random sample the following question: "Now I'd like to get your opinion about some people in the news. As I read the name, please say if you have a favorable or unfavorable opinion of this person." In this poll, 55 percent of these responses about President Clinton were favorable and 42 percent unfavorable.

(a) Explain in your own words what a test of significance would allow us to determine for these data.

(b) Assuming that Gallup used the same sample size for both polls, find the smallest sample size that would lead us to decide that the difference in the percentage of favorable responses could not be attributed entirely to random variability at the .05 level of significance? [*Hint*: You can use trial and error or do some algebra with the test statistic formula and Table II.]

(c) Do you believe that Gallup used a sample size this large? Explain what conclusion this allows you to make about the difference in poll results.

Activity 24-17: BAP Study

Researchers investigating the disease Bacillary Angiomatosis and Peliosis (BAP) conducted a study of 48 BAP patients and a control group of 94 subjects. The following table lists the numbers of people in each group who had the indicated characteristics.

	case patients (*n*=48)	control group (*n*=94)
# male	42	84
# white	43	89
# non-Hispanic	38	75
# with AIDS	24	44
# who own a cat	32	37
# scratched by a cat	30	29
# bitten by cat	21	14

(a) Is this study a controlled experiment or an observational study? Explain.

(b) Use your calculator to perform tests of significance on each of the variables listed in the table. On which variables do the two groups differ significantly at the .05 level?

(c) In light of the type of study involved, can you conclude that any of these variables from (b) cause the disease? Explain.

Activity 24-18: Wording of Surveys

Much research goes into identifying factors that can unduly influence people's responses to survey questions. In a 1974 study researchers conjectured that people are prone to acquiesce, to agree with attitude statements presented to them. They investigated their claim by asking subjects whether they agree or disagree with the following statement. Some subjects were presented with form A and others with form B:

- Form A: "Individuals are more to blame than social conditions for crime and lawlessness in this country."
- Form B: "Social conditions are more to blame than individuals for crime and lawlessness in this country."

The responses are summarized in the table:

	blame individuals	blame social conditions
form A	282	191
form B	204	268

(a) Let θ_A represent the population proportion of all potential form A subjects who would contend that individuals are more to blame, and let θ_B denote the population proportion of all potential form B subjects who would contend that individuals are more to blame. If the researchers' claim about acquiescence is valid, should θ_A be greater than θ_B or vice versa?

(b) Calculate the sample proportion of form A subjects who contended that individuals are more to blame. Then calculate the sample proportion of form B subjects who contended that individuals are more to blame.

(c) Conduct (either by hand or with your calculator) the appropriate test of significance to assess the researchers' conjecture. Report the null and alternative hypotheses, the test statistic, and p-value. Also write a one-sentence conclusion.

Activity 24-19: Wording of Surveys (*cont.*)

Researchers have conjectured that the use of the words "forbid" and "allow" can affect people's responses to survey questions. In a 1976 study one group of subjects were asked, "Do you think the United States should forbid public speeches in favor of communism?" while another group were asked, "Do you think the United States should allow public speeches in favor of communism?" Of the 409 subjects asked the "forbid" version, 161 favored the forbidding of communist speeches. Of the 432 subjects asked the "allow" version, 189 favored allowing the speeches.

(a) Calculate the sample proportion of "forbid" subjects who oppose communist speeches (i.e., favor forbidding them) and the sample proportion of "allow" subjects who oppose communist speeches (i.e., do *not* favor allowing them).

(b) Conduct (either by hand or with your calculator) the appropriate two-sided test of significance to test the researchers' conjecture. Report the null and alternative hypotheses (explaining any symbols that you use) as well as the test statistic and p-value. Indicate whether the difference in sample proportions is statistically significant at the .10, .05, and .01 significance levels.

(c) A 1977 study asked 547 people, "Do you think the government should forbid the showing of X-rated movies?" A total of 224 answered in the affirmative (i.e., to forbid). At the same time a group of 576 people were asked, "Do you think the government should allow the showing of X-rated movies?" A total of 309 answered in the affirmative (i.e., to allow). Repeat (a) and (b) for the results of this experiment. (Be very careful to calculate and compare relevant sample proportions. Do not just calculate proportions of "affirmative" responses.)

(d) When a 1979 study asked 607 people, "Do you think the government should forbid cigarette advertisements on television?" A total of 307 answered in the affirmative (i.e., to forbid). At the same time a group of 576 people were asked, "Do you think the government should allow cigarette advertisements on television?" A total of 134 answered in the affirmative (i.e., to allow). Repeat (a) and (b) for the results of this experiment; heed the same caution as in (c).

(e) Write a paragraph summarizing your findings about the impact of forbid/allow distinctions on survey questions.

Activity 24-20: Questioning Smoking Policies

An undergraduate researcher at Dickinson College examined the role of social fibbing (the tendency of subjects to give responses that they think the interviewer wants to hear) with the following experiment. Students were asked, "Would you favor a policy to eliminate smoking from all buildings on campus?" For half of the subjects, the interviewer was smoking a cigarette when asking the question; the other half were interviewed by a nonsmoker. Prior to conducting the experiment, the researcher suspected that students interviewed by a smoker would be less inclined to indicate that they favored the ban. It turned out that 43 of the 100 students interviewed by a smoker favored the ban, compared to 79 of the 100 interviewed by a nonsmoker. Carry out the appropriate test of significance to assess the researcher's hypothesis. Write a brief conclusion as if to the researcher, addressing in particular the question of how likely her experimental results would have occurred by chance alone.

Activity 24-21: Employment Discrimination

In the legal case of *Teal vs. Connecticut* (1982), a company was charged with discrimination in that blacks passed its employment eligibility exam in smaller proportions than did whites. Of the 48 black applicants to take the test during the year in question, 26 passed; of the 259 white applicants to take the test, 206 passed.

(a) What is the sample proportion of black applicants who passed the test? What is the sample proportion of white applicants who passed?

(b) Perform the significance test of whether the data provide strong evidence that the proportion of blacks who pass the test is statistically significantly less than that of whites. Show your work, indicate whether the difference is significant at the .05 and/or .01 levels, and write a one- or two-sentence conclusion (as if reporting to the jurors in the case).

Activity 24-22: Teen Smoking

A newspaper account of a medical study claimed that the daughters of women who smoked during pregnancy are more likely to smoke themselves. The study surveyed teenagers, asking them whether they had smoked in the last year and then asking the mother whether she had smoked during pregnancy. Only 4% of the daughters of mothers who did not smoke during pregnancy had smoked in the past year, compared to 26% of girls whose mothers had smoked during pregnancy.

(a) What further information do you need to determine whether this difference in sample proportions is statistically significant?

(b) Suppose that there had been 50 girls in each group. Use your calculator to conduct a two-sided significance test. Report the p-value and whether the difference in sample proportions is statistically significant at the .05 level.

(c) Repeat (b) supposing that there had been 50 girls whose mothers had smoked and 200 whose mothers had not.

(d) Repeat (b) supposing that there had been 200 girls in each group.

(e) Construct a segmented bar graph to compare the smoking habits of girls whose mothers smoked and girls whose mothers did not smoke during pregnancy. Does the appearance of this graph change as the sample size increases? Explain.

(f) Is this study a controlled experiment or an observational study? Explain.

(g) Even if the difference in sample proportions is statistically significant, does the study establish that the pregnant mother's smoking caused the daughter's tendency to smoke? Explain.

Activity 24-23: Teen Smoking (*cont.*)

Refer back to the description of the study concerning kids' smoking in Activity 24-22. The researchers also studied sons and found that 15% of the sons of mothers who had not smoked during pregnancy had smoked in the past year, compared to 20% of the sons of mothers who had smoked.

(a) Suppose that there had been 60 boys in each group. Use your calculator to conduct a two-sided significance test. Report the p-value and whether the difference in sample proportions is statistically significant at the .05 level.

(b) Repeat (a) supposing that there had been 200 boys in each group.

(c) Repeat (a) supposing that there had been 500 boys in each group.

(d) Suppose that the two groups had the same number of boys. Try to find the smallest number for this sample size that would make the difference in sample proportions statistically significant at the .05 level. You may either use trial and error with your calculator or solve the problem analytically by hand.

Activity 24-24: Comparing Proportions of Personal Interest

Write a paragraph detailing a real situation in which you would be interested in performing a test of significance to compare two *proportions*. Describe precisely the context involved and explain whether the study would be a controlled experiment or an observational study. Also identify very carefully the observational units and variables involved. Finally, suggest how you might go about collecting the sample data.

Topic 25:

COMPARING TWO MEANS

In the previous topic you studied the application of inference techniques to the comparison of two proportions. With this topic you will examine the case of comparing two sample means where the samples have been collected *independently* (as opposed to the matched pairs design that you studied previously). The inference procedures will again be based on the *t*-distribution; the reasoning behind and interpretation of the procedures remain the same as always. Also as always, you will see the importance of an initial examination of the data, visual and numerical, prior to applying formal inference procedures.

- To learn a significance testing procedure for determining whether two sample means differ "significantly" and to understand the reasoning behind and limitations of the procedure.
- To learn a confidence interval procedure for estimating the magnitude of the difference between two population means.
- To explore the effects of the various factors that affect inference procedures concerning a difference in population means.
- To acquire proficiency using graphical, descriptive, and inferential methods to analyze data concerned with questions of whether a difference exists between two groups or two treatments.

25

PRELIMINARIES

1. Would you expect to find longer sentences on the average in a novel or in a text-book?

2. How many hours per week do you think a typical third or fourth grader spends watching television?

3. If one commuting route has a sample mean travel time of 32 minutes and another has a sample mean commuting time of 28 minutes, when might you be fairly convinced that the first route is slower than the second?

4. Do you suspect that second-grade instructors spend more individualized instructional time with boys or with girls when teaching *mathematics*, or do you not expect to see a gender difference there?

5. Do you suspect that second-grade instructors spend more individualized instructional time with boys or with girls when teaching *reading*, or do you not expect to see a gender difference there?

IN-CLASS ACTIVITIES

Activity 25-1: Sentence Lengths (*cont.*)

Recall from Activity 20-8 on page 437 that a sample of sentences from John Grisham's novel *The Testament* revealed the following sentence lengths, measured as number of words per sentence:

17	21	8	32	13	16	17	37	27	20	30	15	64	34
18	26	23	17	5	10	29	9	22	18	7	16	13	10

A sample of words from David Moore's *The Basic Practice of Statistics* contains the following numbers of words per sentence:

20	3	12	3	30	13	16	18	22	21	14	14	9	11	13
22	13	15	5	12	20	23	7	15	17	18	31	21	17	13
14	23	15	20	12	16	14	20	22	4					

Summary statistics for these distributions follow:

	size	mean	std. dev.	min	Q_1	median	Q_3	max
Grisham	28	20.50	12.00	5	13	17.5	26.75	64
Moore	40	15.70	6.42	3	12.25	15	20	31

(a) Create boxplots to compare the numbers of words per sentence of the two authors. [*Hint:* Use the five-number summaries provided. Make them modified boxplots, with Grisham's 64-word sentence being the only outlier.]

25

(b) Comment on any differences you find in these two authors' word per sentence distributions.

(c) Would it be possible to obtain sample means this far apart even if the population means were equal?

Because of sampling variability, one cannot conclude that because these sample means differ, the means of the respective populations must differ as well. As always, one can use a test of significance to establish whether a sample result (in this case, the observed difference in sample mean sentence lengths) is "significant" in the sense of being unlikely to have occurred by chance alone. Also as always, one can use a confidence interval to estimate the magnitude of the difference in population means.

Inference procedures for comparing the population means of two different groups are similar to those for comparing population proportions in that they take into account sample information from both groups; they are similar to inference procedures for a single population mean in that the sample sizes, sample means, and sample standard deviations are the relevant summary statistics. These statistics are denoted by the following notation:

	first group	second group
sample size	n_1	n_2
sample mean	$\bar{x}_1$	$\bar{x}_2$
sample std dev	s_1	s_2

The forms for confidence intervals and significance tests concerning the difference between two population means, which will be denoted by $\mu_1 - \mu_2$, are presented below. (The use of the subscripts will indicate the population from which the measurements come.) An important point to remember is that the reasoning behind and interpretation of these procedures is always the same.

Confidence interval for $\mu_1 - \mu_2$:

$$(\bar{x}_1 - \bar{x}_2) \pm t_k^* \sqrt{\frac{s_1^2}{n_1} + \frac{s_2^2}{n_2}},$$

where t_k^* is the appropriate critical value from the t-distribution with degrees of freedom k equal to the smaller of $n_1 - 1$ and $n_2 - 1$.

Significance test of equality of μ_1 and μ_2:

$H_o: \mu_1 = \mu_2$

$H_a: \mu_1 < \mu_2$ or $H_a: \mu_1 > \mu_2$ or $H_a: \mu_1 \neq \mu_2$

test statistic: $\quad t = \dfrac{\bar{x}_1 - \bar{x}_2}{\sqrt{\dfrac{s_1^2}{n_1} + \dfrac{s_2^2}{n_2}}}$

p-value $= \Pr(T_k \leq t)$ or $\Pr(T_k \geq t)$ or $2\Pr(T_k \geq |t|)$,

where T_k represents a t-distribution with number of degrees of freedom equal to the smaller of $n_1 - 1$ and $n_2 - 1$.

Technical conditions:

1. That the two samples are independently selected simple random samples from the populations of interest,

2. *Either* that both sample sizes are large ($n_1 \geq 30$ and $n_2 \geq 30$ as a rule of thumb) *or* that both populations are normally distributed.

(d) Let μ_G denote the mean sentence length in Grisham's book and μ_M denote the mean sentence length in Moore's book. Write out, in these symbols and also in words, the hypotheses for testing whether these authors have different population means.

25

(e) Use the summary statistics provided above to calculate the test statistic.

(f) Use the *t*-table to find (as accurately as possible) the p-value of the test.

(g) Which of the following is a correct interpretation of the p-value?
- The p-value is the probability that Grisham and Moore have the same mean number of words per sentence in these samples.
- The p-value is the probability that Grisham and Moore have the same mean number of words per sentence in these two books.
- The p-value is the probability that Grisham has a higher mean number of words per sentence than Moore does.
- The p-value is the probability of getting sample data so extreme if in fact Grisham and Moore have the same mean number of words per sentence in these books.

(h) Is this p-value small enough to reject the hypothesis that these population means are equal at the $\alpha = .05$ level?

(i) Is the observed difference in sample means statistically significant at the $\alpha = .01$ level?

(j) State the technical conditions necessary for this procedure to be valid. Do the sample data provide any reason to doubt the validity?

Activity 25-2: Hypothetical Commuting Times

Suppose that a commuter Alex wants to determine which of two possible driving routes gets him to work more quickly. Suppose that he drives route 1 for ten randomly chosen days and route 2 for ten other randomly selected days, recording the commuting times (in minutes) and displaying them as follows:

route 1	19.3	20.5	23.0	25.8	28.0	28.8	30.6	32.1	33.5	38.4
route 2	23.7	24.5	27.7	30.0	31.9	32.5	32.6	35.5	38.7	42.9

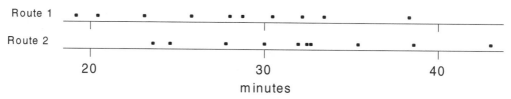

(a) For this sample of days, does one route *always* get Alex to work more quickly than the other?

(b) For this sample of days, does one route *tend* to get Alex to work more quickly than the other? If so, identify the route that seems to be quicker.

(c) Download HYPOCOMM.83g into your calculator. The data for Alex's commuting times are stored in ALEX1 and ALEX2. Use your calculator to determine the sample means and sample standard deviations of these commuting times for each route; record them in the table below:

	sample size	sample mean	sample std. dev.	p-value
Alex route 1				
Alex route 2				

25

(d) Use your calculator to conduct a signficiance test of whether Alex's sample commuting times provide evidence that the mean commuting times with these two routes differ:

- Select 2-SampTTest from the STAT TESTS menu.
- You can enter either the two lists or the summary statistics you found in (c). Leave the "Pooled" option set to No. Scroll down to chose either the "Calculate" or "Draw" option.

Record the p-value of the test in the table above.

(e) Are Alex's sample results statistically significant at any of the commonly used significance levels? Can Alex reasonably conclude that one route is faster than the other for getting to work? Explain.

(f) Use the 2-SampTInt feature of your calculator (in the STAT TESTS menu) to determine the 90% confidence interval for the difference in Alex's mean commuting times between route 1 and route 2. Record the confidence interval below.

(g) Does this interval include the value zero? Explain the importance of this.

Now consider three other commuters who conduct similar experiments to compare travel times for two different driving routes. For the sake of comparison, Alex's results are also reproduced below:

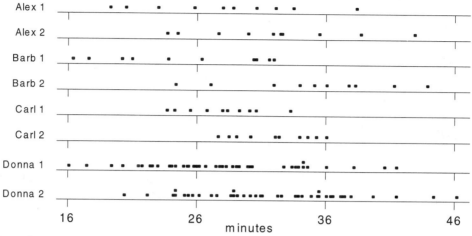

(**h**) Based on your visual analysis of these pairs of dotplots, what strikes you as the most important difference between Alex's and Barb's results?

(**i**) Based on your visual analysis of these pairs of dotplots, what strikes you as the most important difference between Alex's and Carl's results?

(**j**) Based on your visual analysis of these pairs of dotplots, what strikes you as the most important difference between Alex's and Donna's results?

(**k**) For each commuter (data are in lists named BARB1, BARB2, CARL1, CARL2, and DONNA1, DONNA2), use your calculator to conduct a significance test of whether the difference in his or her sample mean commuting times differ significantly. Record the p-values of these tests below, along with the appropriate sample statistics.

	sample size	sample mean	sample std. dev.	p-value
Barb route 1				
Barb route 2				
Carl route 1				
Carl route 2				
Donna route 1				
Donna route 2				

(1) For each of Barb, Carl, and Donna, explain why his or her sample results differ significantly, while Alex's do not.

Barb:

Carl:

Donna:

These comparisons should help you to see the roles of sample sizes, means, and standard deviations in the two-sample *t*-test. All else being the same, the test result becomes more statistically significant as:
- the difference in sample means increases,
- the sample sizes increase,
- the sample standard deviations decrease.

Activity 25-3: Children's Television Viewing (*cont.*)

Recall from Activity 13-11 on page 294 the study in which one group of third and fourth graders was randomly assigned to receive an 18-lesson, 6-month classroom curriculum intended to reduce their use of television, videotapes, and video games. The other group of children received their usual elementary school curriculum. All children were asked to report how many hours per week they spent on these activities, both before the curriculum intervention and afterward.

(a) If the randomization achieved its goal, would there be a significant difference between the two groups prior to the curriculum intervention? Explain.

The following summary statistics pertain to the reports of television watching, in hours per week, prior to the intervention:

Baseline	sample size	sample mean	sample std. dev.
Control group	103	15.46	15.02
Intervention group	95	15.35	13.17

(b) State in words and in symbols the null and alternative hypotheses for testing whether the two groups are indeed similar in television viewing habits prior to the intervention.

(c) Use the summarized sample information to calculate the test statistic.

(d) Use the *t*-table to determine the p-value of the test as accurately as possible.

(e) Would you reject the null hypothesis that the population means are equal prior to the intervention at the .05 level?

The following summary statistics pertain to the reports of television watching at the conclusion of the study:

Follow-up	sample size	sample mean	sample std. dev.
Control group	103	14.46	13.82
Intervention group	95	8.80	10.41

25

(f) Conduct a two-sample *t*-test of whether the mean number of hours of television viewing per week is higher in the control group than in the intervention group at the conclusion of the study. State the hypotheses, and report the test statistic and p-value. Also indicate whether you would reject the null hypothesis (of no difference) at the .05 level.

(g) Even though you do not have the raw data, explain how the summary statistics reveal that the distributions of television viewing hours must be nonnormal.

(h) Explain why the nonnormality of these distributions does not hinder the validity of using this test procedure.

(i) Write a paragraph or two summarizing your findings about the comparison of television viewing habits between these two groups before and after the study. Would you conclude that the curriculum intervention succeeded in reducing television viewing? Also refer to the role of randomization in addressing this question.

This activity should remind you of the important role of randomization in designing experiments. Randomization serves to make the groups similar with regard to variables such as television watching prior to the imposition of the treatment, so that differences observed after the experiment can reasonably be attributed to that treatment. Had this been an observational study, the difference between the groups could not have been attributed solely to the explanatory variable.

WRAP-UP

This topic has extended your knowledge of inference techniques to include the goal of comparing two means. You have discovered the role that the sample sizes and sample standard deviations play in the inference procedures, and you have once again seen that the structure and reasoning of inference procedures are the same in all situations. You have also encountered again the important differences in the conclusions one can draw from controlled experiments as opposed to observational studies.

We have noted previously that comparisons between two groups can be thought of as relationships between two variables, considering the group designation as a categorical variable. The next topic will shift your attention from inference for comparisons to inference for relationships between variables. In particular, you will examine relationships when both variables are categorical. You will review the graphical and descriptive analysis of two-way tables and then encounter the chi-square test of association in two-way tables.

25

HOMEWORK ACTIVITIES

Activity 25-4: Sentence Lengths (*cont.*)

Consider again the sentence lengths that you analyzed in Activity 25-1 on page 527. Describe the effect (decrease, increase, or remain the same) on the p-value of the significance test if everything remained the same except:
(a) all of Grisham's sentences were one word longer than in the sample.
(b) all of Moore's sentences were one word longer than in the sample.
(c) both sample standard deviations were larger.
(d) both sample sizes were larger.
Explain your answer in each case.

Activity 25-5: Sentence Lengths (*cont.*)

Refer again to your analysis in Activity 25-1 comparing sentence lengths in a sample from a John Grisham novel and from a David Moore textbook (SENTENCES.83g).
(a) Remove Grisham's outlier and repeat your analysis of whether the sample data provide evidence that the population mean sentence lengths differ between these two authors. Report the hypotheses (in symbols and in words), test statistic, and p-value.
(b) Write a few sentences describing how the results of this analysis differ from the analysis in Activity 25-1. Also explain why removing the outlier has the effect that it does.

Activity 25-6: Hypothetical Commuting Times (*cont.*)

Reconsider the commuting time experiments presented Activity 25-2 on page 531. For each of the following people (Earl, Fred, Grace, Harry, and Ida), determine whether or not his or her sample results are statistically significant. Do *not* perform any calculations; simply compare the sample results to those of Alex, Barb, Carl, or Donna. In each case indicate whose results you use for the comparison and explain your answer.

	sample size	sample mean	sample std. dev.
Earl 1	10	29.0	6.0
Earl 2	10	31.0	6.0
Fred 1	75	28.0	6.0
Fred 2	75	32.0	6.0

Grace 1	10	20.0	6.0
Grace 2	10	40.0	6.0
Harry 1	10	28.0	9.0
Harry 2	10	32.0	9.0
Ida 1	10	28.0	1.5
Ida 2	10	32.0	1.5

Activity 25-7: Hypothetical Commuting Times (*cont.*)

Consider four more hypothetical commuters who are trying out two different routes for driving to work (Jerry, Karla, Larry, Morrie). The following boxplots present distributions of their sample driving times. Suppose that each person does a two-sample *t*-test of whether one route tends to be faster than the other. Arrange these four people in order from the one who will have the highest p-value to the one who will have the smallest p-value. Explain your choices. [*Hint*: Note the different sample sizes indicated with the boxplots.]

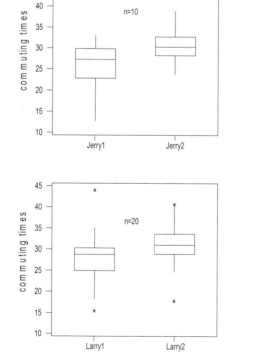
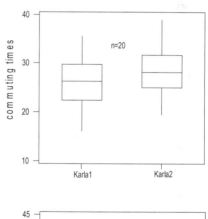
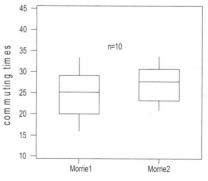

Activity 25-8: Children's Television Viewing (*cont.*)

In addition to asking the children in the obesity study about their television viewing habits, researchers also asked them to report how many hours they spend per week watching videotapes and playing video games. Summary statistics for the baseline comparisons are reported below:

Baseline: videotapes	sample size	sample mean	sample std. dev.
Control group	103	5.52	10.44
Intervention group	95	4.74	6.57

Baseline: video games	sample size	sample mean	sample std. dev.
Control group	103	3.85	9.17
Intervention group	95	2.57	5.10

(a) Conduct appropriate tests of whether the control and intervention group means differ significantly on either of these variables. Report the test statistics and p-values, and write a few sentences detailing your findings.

Summary statistics for the follow-up comparisons on these variables are reported below:

Follow-up: videotapes	sample size	sample mean	sample std. dev.
Control group	103	5.21	8.41
Intervention group	95	3.46	4.86

Follow-up: video games	sample size	sample mean	sample std. dev.
Control group	103	4.24	10.00
Intervention group	95	1.32	2.72

(b) Repeat (a) for these follow-up data.

Activity 25-9: Classroom Attention

Researchers in a 1979 study recorded the lengths of individual instructional time (in seconds) that second-grade instructors spent with their students. They compared these times between girls and boys in the subjects of reading and mathematics. Numerical summaries of their results appear below:

Reading	sample size	sample mean	sample std. dev.
boys	372	35.90	18.46
girls	354	37.81	18.64

Mathematics	sample size	sample mean	sample std. dev.
boys	372	38.77	18.93
girls	354	29.55	16.59

(a) Conduct appropriate tests of significance to determine whether the sample mean instructional times differ significantly between boys and girls in either subject.
(b) Also produce confidence intervals for the difference in population means between boys and girls.
(c) Comment on whether the technical conditions necessary for the validity of these procedures seem to be satisfied.
(d) Write a paragraph or two summarizing your conclusions.

Activity 25-10: UFO Sighters' Personalities (*cont.*)

Reconsider the data you analyzed in Activity 22-8 on page 474. The group of 25 people who claimed to have had an intense experience with a UFO had a mean IQ of 101.6, and the standard deviation of these IQs was 8.9. A control group of 53 community members who had not reported UFO experiences had a mean IQ of 100.6 with a standard deviation of 12.3.
(a) Is this an observational study or a controlled experiment? Explain.
(b) Identify the explanatory and response variables. Also identify each as quantitative or categorical.
(c) Use this sample information together with that presented in Activity 22-8 to test whether the mean IQs of community members and UFO sighters differ significantly. Show the details of the test and write a one-sentence conclusion.
(d) Find a confidence interval for the difference in population means and explain what the interval reveals.
(e) Given the type of study involved here, even if the sample data had revealed a significantly higher mean IQ for the control group, would you be able to draw any causal conclusion that seeing a UFO affects intelligence? Explain.

25

Activity 25-11: Parents' Ages (*cont.*)

Refer to the data on ages at which samples of 35 mothers (Activity 13-11 on page 294) and of 35 fathers (Activity 6-11 on page 130) first had a child (AGECHILD.83g).

(a) Are these data from a matched pairs design? Explain.

(b) Consider testing the hypotheses H_o: $\mu_{mother} = \mu_{father}$ vs. H_a: $\mu_{mother} < \mu_{father}$ Explain what the symbols μ_{mother} and μ_{father} represent, and restate these hypotheses in words.

(c) Conduct the test of the hypotheses listed in (b). Would you reject H_0 at the $\alpha = .05$ level?

(d) Find a 95% confidence interval for $\mu_{mother} - \mu_{father}$

(e) Do the sample data suggest that all mothers in the population have their first child at an earlier age than do all fathers? Explain.

(f) Are there any fathers in the sample who had their first child at an earlier age than any of the mothers in the sample? Explain.

(g) Summarize your findings with regard to whether mothers and fathers tend to have their first child at different ages.

Activity 25-12: Memory Experiment (*cont.*)

Consider the data that you collected in Topic 13, concerning how many letters you could memorize in 20 seconds depending on the form in which the letters were presented. The research hypothesis prior to collecting the data was that those who saw the familiar three-letter packets would memorize more letters on average than those who saw haphazardly organized letters.

(a) Conduct a preliminary examination of the data with regard to this issue. Construct graphical displays and calculate summary statistics. Write a paragraph comparing the distributions of scores (consecutive letters correctly memorized) between the two groups.

(b) Carry out a significance test of whether the experimental data support the research hypothesis. Report the hypotheses (in words and in symbols) as well as the test statistic and p-value. Write a one-sentence conclusion.

(c) Identify the smallest significance level α at which the difference between the mean scores of the two groups is statistically significant. [*Hint:* Do not limit yourself to considering only common α values.]

(d) If there were no effect of the different presentations of letter sequences, what is the probability that the sample mean scores would be as far apart as these are? [*Hint:* You have already done this calculation.]

(e) Are there any reasons to doubt that the technical conditions of this test are satisfied? Explain.

Activity 25-13: Female President (*cont.*)

Consider again the data on students' predictions for the year in which the United States will first have a female President, collected in Topic 6. Conduct a full analysis of whether men and women tend to respond differently to this issue. Include a graphical, numerical, and inferential component to your analysis. Be sure to address the question of whether the technical conditions for the inference procedures appear to be satisfied. Write a paragraph or two reporting on your findings.

Activity 25-14: Lengths and Scores of Baseball Games

Recall that Activity 19-9 on page 417 and Activity 19-10 referred to a sample of 190 Major League Baseball games played between July 26 and August 8, 1999. In addition to margin of victory, other variables recorded included total runs scored, playing time in minutes, and league (National or American). The data are stored in BASE-BALL.83g (MARGN, RUNS, TIME, LEAG).

(a) For each of the four variables mentioned in the preceding sentence, classify it as categorical or quantitative.

(b) Compare the game lengths in the National League vs. the American League by producing visual displays and calculating numerical summaries. Write a paragraph comparing key features of these distributions.

(c) Construct a 95% confidence interval for the difference in mean game lengths between the two leagues. Write a sentence or two interpreting the interval.

(d) Based solely on the confidence interval, indicate whether a significance test of whether the mean game lengths differ between the two leagues would be significant at the .05 level. Explain how you can answer this question based on the confidence interval alone.

(e) Repeat (b), (c), and (d) for the total runs scored in a game.

(f) Repeat (b), (c), and (d) for a game's margin of victory.

(g) Write a paragraph summarizing your findings on the comparison of these three variables between the two leagues.

Activity 25-15: Lengths and Scores of Baseball Games (*cont.*)

It seems reasonable to conjecture that baseball games containing a "big bang" (see Activity 21-2 on page 452 for a definition) would tend to last longer, involve more total runs, and have bigger margins of victory than games that do not contain a "big bang."

(a) Use graphical displays and numerical summaries to analyze the sample data from the 190 Major League games played during July 26–August 8, 1999 (BASE-BALL.83g) with regard to this question.

25

(b) Translate the conjecture stated above into null and alternative hypotheses for a series of three significance tests. Are these one-sided or two-sided hypotheses? Explain.

(c) Conduct the two-sample tests of significance on all three variables. Report the test statistics and p-values for each test. Are any of the results significant at the .05 level? How about the .01 level?

(d) Find 95% confidence intervals to estimate the difference in population means for all three variables. Interpret the intervals, being sure to mention the order in which you subtracted.

(e) This sample was not strictly a simple random sample from the population of Major League Baseball games played in 1999. Do you suspect that the sample is likely to be biased in one direction or the other with respect to any of these three variables? Explain.

(f) If the variable were game-time temperature, do you think the sample would be biased? Explain.

(g) Are the other technical conditions for the validity of the *t*-procedures satisfied here? Explain.

Activity 25-16: Top American Films (*cont.*)

Refer to the data collected in Topic 6 concerning the ranks and ages of films in the American Film Institute's "top 100" list. Recall that you indicated which of the films you have seen and which you have not.

(a) Conduct a significance test to assess whether the mean age of the films you have seen is significantly smaller than the mean age of films you have not seen. Would you reject the null hypothesis at the α = .10 level? [As always, report all of the details of the test procedure.]

(b) If this test leads to rejecting the null hypothesis at the α = .10 level, calculate a 90% confidence interval for estimating the difference in group means. Write a sentence explaining what the interval reveals.

Activity 25-17: Top American Films (*cont.*)

Consider again the data on the American Film Institute's "top 100" list. The following summary statistics pertain to the ages of the films seen and not seen by author A and author B (as of December 21, 1999):

	Sample size	Mean age	Std. dev. age
Author A seen	46	31.9	16.4
Author A not seen	54	44.2	18.5
Author B seen	53	31.1	17.6

Author B not seen	47	46.9	15.9

For each author, test whether the data provide evidence that the mean age of films seen is significantly less than that of films not seen. Show the details of your analyses, and write a paragraph or two summarizing your findings.

Activity 25-18: Birth and Death Rates (*cont.*)

Consider the birth rates of the fifty states presented in Activity 6-1 on page 118 (BIRTHDEATH.83g). Explain why it would be inappropriate to conduct a test of significance of whether states east of the Mississippi River have a different mean birth rate than do states west of the Mississippi. [*Hint:* Try to think about what the populations and parameters would be in this situation.]

Activity 25-19: Lifetimes of Notables (*cont.*)

Refer to Activity 6-16 on page 133, where you analyzed distributions of lifetimes of "noted personalities" (LIFETIMES.83g). Recall that the categories of notables are scientists (SCI), writers (WRI), politicians (POL), military leaders (MIL), artists (ART), philosophers (PHIL), social reformers(SREF), historians (HIST), and business leaders (BUS).

(a) Pick out any *two* of these groups of notables that are of interest to you. List your choices and designate one as group 1 and the other as group 2.

(b) Use your calculator to look at boxplots of the distributions of lifetimes in these two groups. Write a few sentences comparing and contrasting their key features, particularly with regard to the question of whether one group seems to live longer on the average than the other group.

(c) Use your calculator to calculate the sample sizes, sample means, and sample standard deviations of lifetimes for your two groups.

(d) Use your calculator to perform the significance test to assess whether the data provide evidence that the means of the *populations* truly differ. Be sure to state the null and alternative hypotheses (identifying whatever symbols you introduce) and to report the test statistic and p-value. Also state exactly what the p-value says in this context, and write a one-sentence conclusion.

(e) Use your calculator to find a 95% confidence interval for the difference in population means.

(f) Write a sentence or two interpreting the interval that you have found in (e). In particular, comment on whether the interval contains only negative values, only positive values, or some of both; also indicate what this says about the mean lifetimes of the populations represented by your two groups.

25

(g) Indicate specifically how your results in (d) and (e) would have differed if you had labeled your two groups in the opposite manner (i.e., if the group you labeled as group 1 had been group 2, and vice versa).

(h) List the technical conditions that underlie the validity of these procedures and comment on whether they seem to be satisfied. In particular, do you need to assume anything about the distributions of individual lifetimes?

(i) Use your calculator to determine a 95% confidence interval for the population mean lifetime of the first group that you chose. Then do the same for your second group of notables.

(j) Determine and record the half-widths of the three confidence intervals that you have found (i.e., for the difference in population means, for the population mean of group 1, and for the population mean of group 2).

(k) Compare these half-widths. Specifically, is the half-width of the interval estimating the difference in population means larger than either of the individual group half-widths? Is the half-width of the interval estimating the difference in population means as large as the sum of the two individual group half-widths?

Activity 25-20: Word Lengths (*cont.*)

Consider the word lengths that you analyzed in Activity 6-20 on page 136. The lengths of our 26 words are reproduced below:

10	2	3	7	2	9	4	4	2	1	7	5	2
5	4	5	2	2	9	4	2	5	2	3	4	4

(a) Use your calculator to create visual displays comparing your distribution of word lengths with ours. Comment on similarities and differences that you observe in these distributions.

(b) Use your calculator to calculate summary statistics for these two distributions of word lengths. Record the sample sizes, sample means, and sample standard deviations.

(c) Use your calculator to conduct a test of significance to assess whether these sample data provide evidence that your mean number of letters per word differs from ours. Report the null and alternative hypotheses (identifying whatever symbols you introduce), and show the calculations of the test statistic and p-value. Finally, write a one-sentence conclusion about the question of interest, whether the data provide evidence that your mean number of letters per word differs significantly from ours.

(d) If you find evidence at the $\alpha = .10$ level that the population means differ, estimate the magnitude of that difference with a 90% confidence interval.

(e) Comment on whether the technical conditions that underlie the validity of these inference procedures are satisfied by these data.

Activity 25-21: Tennis Simulations (*cont.*)

Refer again to the data of Activity 3-16 on page 62, Activity 4-9 on page 82, and Activity 6-10 on page 129 concerning three different scoring systems for tennis (TENNSIM.83g).

(a) Use your calculator to analyze whether the sample data suggest that the mean game length with conventional scoring differs significantly from the mean game length with no-ad scoring and, if so, by about how much. Write a paragraph describing your findings.

(b) Repeat (a), comparing no-ad scoring with handicap scoring.

(c) Do the inference techniques that you employed in (a) and (b) say anything about whether the variability of games' lengths differs among the scoring methods? Do they say anything about whether the shapes of the distributions differ?

Activity 25-22: Marriage Ages (*cont.*)

Reconsider the data on marriage ages of 100 couples from Activity 22-11 on page 476. Since these data are *paired*, you were asked in that activity to perform the correct analysis on the *differences* in marriage ages. Suppose now (incorrectly) that the data had been independent samples, one sample of 100 husbands' ages and another sample of 100 wives' ages. The sample statistics are as follows:

	sample size	sample mean	sample std. dev.
husband	100	33.08	12.31
wife	100	31.16	11.00

(a) Use this information to conduct (by hand) a two-sample test (not a paired test) of whether the sample mean age for husbands exceeds that of wives by a statistically significant margin. Report the details of the test procedure and write a one-sentence conclusion.

(b) Find (by hand) a 95% confidence interval for the difference in mean ages between husbands and wives in the population. (Continue to assume that these are two independent samples.)

(c) Comment on how these results differ from those of Activity 22-11 on page 476, where you performed the appropriate matched pairs analysis of these data. Also comment on how these differences highlight the usefulness of the matched pairs design.

25

Activity 25-23: Hypothetical ATM Withdrawals (*cont.*)

Reconsider the hypothetical ATM withdrawals first presented in Activity 5-24 on page 110 (HYPOATM.83g). In Activity 23-6 on page 489 you calculated sample sizes, sample means, and sample standard deviations of the withdrawal amounts for each machine.

(a) Choose any pair of machines and calculate a 90% confidence interval for estimating the difference in population mean withdrawal amounts between the two machines.

(b) Does this interval include the value 0?

(c) Would this interval be any different if you had chosen a different pair of machines? Explain.

(d) Are these three machines identical in their distributions of withdrawal amounts? Explain.

Activity 25-24: Hypothetical Bowlers' Scores

Suppose that three bowlers Chris, Fran, and Pat bowl 36 games each with the following results (HYPOBOWL.83g):

	132	135	136	140	140	142	142	143	144	147	147	148
Chris	149	149	150	152	153	154	155	155	156	158	158	159
	160	161	161	162	162	163	165	165	167	169	169	170
	72	76	79	88	93	99	104	170	171	172	172	172
Fran	173	173	174	174	175	175	175	176	178	178	179	179
	181	183	183	184	185	186	186	186	187	191	191	194
	170	171	172	172	172	173	173	173	174	174	174	175
Pat	175	175	176	176	178	178	178	179	179	179	181	183
	183	183	184	185	186	186	186	187	188	191	191	194

(a) Use your calculator to produce dotplots of the three bowlers' scores on the same scale. Write a few sentences comparing and contrasting the three distributions of scores.

Consider these scores as random samples from the hypothetically infinite population of scores that these bowlers would achieve in the long run. Let μ_C, μ_F, and μ_P represent the population means of the three respective bowlers.

(b) Use your calculator to perform the appropriate significance test of whether μ_C differs from μ_F. Report the test statistic and p-value of the test. Is the difference in sample means statistically significant at any of the common significance levels?

(c) Now use your calculator to perform the appropriate significance test of whether μ_F differs from μ_P. Report the test statistic and p-value of the test. Is the difference in sample means statistically significant at any of the common significance levels?

(d) Comment on how this example illustrates that one should look at an entire distribution of values rather than just concentrate on the mean. In other words, what would you have missed about the bowlers if you had just performed the significance tests without looking at the data first?

Activity 25-25: Hypothetical Bowlers' Scores (*cont.*)

Suppose that two bowlers Jack and Jill each bowl 1000 games. Suppose further that Jack's scores have a mean of 166.8 and a standard deviation of 14.9, while Jill's scores have a mean of 169.3 and a standard deviation of 14.6.

(a) Use this sample information to determine whether Jack's and Jill's population means differ significantly at the $\alpha = .01$ level. Report the null and alternative hypotheses, and show the calculations of the test statistic and p-value.

(b) Construct a 95% confidence interval for the difference in Jack's and Jill's population means. Show the details of your calculation, and comment on whether or not the interval contains the value zero.

(c) Identify which of the following statements is more accurate:
- The sample data provide extremely strong evidence that Jill has a higher mean bowling score than does Jack.
- The sample data provide strong evidence that Jill has a substantially higher mean bowling score than does Jack.

Explain your reasoning and also the subtle difference between these two statements.

Activity 25-26: Students' Data (*cont.*)

Select any quantitative variable and any binary categorical variable on which student data have been collected together. Before studying the data closely, formulate a research question about comparing the quantitative variable between two groups of the categorical variable. Then employ graphical, numerical, and inferential techniques to investigate that question. Write a paragraph or two, accompanied by appropriate graphs and calculations, to summarize and explain your findings.

25

Topic 26:

INFERENCE FOR TWO-WAY TABLES

OVERVIEW

You have studied applications of inference procedures to a wide variety of statistical questions. In particular, you have conducted inferences concerning a population proportion, concerning a population mean, concerning a difference in population means, and concerning a difference in population proportions. In each of these cases you have studied how to combine the inferential procedures with graphical, numerical, and verbal descriptions of the data as studied earlier in the course. In this topic you will study inference techniques that can be applied to the question of whether a relationship exists between two categorical variables. You will discover a formal inference technique, the **chi-square test**, for analyzing sample data summarized as counts in a two-way table.

OBJECTIVES

- To understand the notion of statistical independence as it applies to categorical variables represented in two-way tables of counts.
- To discover how to apply the **chi-square test** to test for independence of two categorical variables.
- To recall how to calculate appropriate proportions from two-way tables and to discover how to combine such an analysis with the chi-square test procedure.

26

PRELIMINARIES

1. Would you expect to find an association between a person's political leaning (liberal, moderate, conservative) and his or her opinion about whether the federal government spends too much on various programs? If so, in what direction would you expect to find the association?

2. Would you expect a stronger association in question 1 with regard to federal spending on the space program or on the environment?

3. Do you expect that women read a newspaper more than men, vice versa, or about the same?

4. Think of a categorical variable with which you would expect to see an association with gender.

5. Think of a categorical variable containing more than two categories with which you would expect to see no association with gender.

IN-CLASS ACTIVITIES

Activity 26-1: Suitability for Politics (*cont .*)

Recall from Activity 7-7 on page 150 that the 1998 General Social Survey asked a sample of American adults whether they agree or disagree with the statement, "Most men are better suited emotionally for politics than are most women." They were also asked to describe their political views as liberal, moderate, or conservative.

(a) Identify very clearly the observational units in this study, the two variables involved, and whether the variables are categorical or quantitative.

To investigate whether there is any association between political view and agreement that men are better suited for politics than women, the following table summarizes the survey responses:

	liberal	moderate	conservative	total
agree	74	139	169	382
disagree	410	471	422	1303
total	484	610	591	1685

One can calculate the conditional distributions:

<u>liberal</u>: .153 agree   <u>moderate</u>:   .228 agree   <u>conservative</u>:   .245 agree
        .847 disagree   .772 disagree   .755 disagree

A segmented bar graph follows:

26

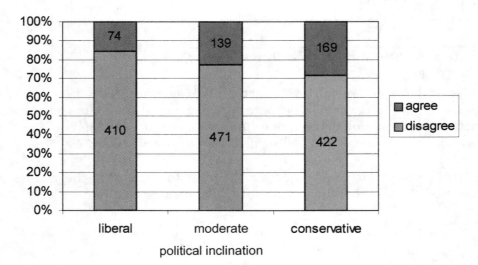

(b) Does there appear to be any association between political view and agreement with the statement? If so, describe the direction and strength of the association, and explain your reasoning.

When asking whether these sample data provide evidence that the proportions agreeing with the statement differ among the three political groups *in the population*, the issue of sampling variability arises again. As we have done with all of the other inference techniques we have studied, we ask how likely it is to have observed the sample data if in fact the agreement proportions are the same for the three groups in the population. We will explore this question by comparing the *observed* counts with the counts that would be *expected* under the hypothesis that the variables "agreement with assertion" and "political leaning" are **independent**. (Recall from Topic 7 that two categorical variables are independent if the conditional distribution of one is the same for all categories of the other.)

(c) Overall, what proportion of the survey respondents agree with the assertion?

(d) If this same percentage (your answer to (c)) of the 484 liberals had responded with agreement, how many people would that have been? (Please record your answer with two decimal places.)

    Liberals:

(e) Repeat (d) for the 610 moderates and for the 591 conservatives.

    Moderates: Conservatives:

You have calculated the **expected counts** under the hypothesis that the two variables are not associated (i.e., that the agreement proportions are the same in the three groups). A more general technique for calculating the expected count of the cell in row i and column j of the table is to take the marginal total for row i times the marginal total for column j divided by the grand total. Our notation for this is

$$E_{ij} = \frac{R_i C_j}{n} = \frac{(\text{row total})(\text{column total})}{\text{grand total}}.$$

(f) Verify that this formula gives the same expected count for the (agree, liberal) cell as your calculation in (d).

(g) Use your answers to (d) and (e) to fill in the expected counts in parentheses in the table below:

	liberal	moderate	conservative	total
agree	74 ( )	139 ( )	169 ( )	382
disagree	410 (374.27)	471 (471.71)	422 (457.02)	1303
total	484	610	591	1685

Now our task is to devise a *test statistic* to measure how far the observed counts fall from the expected counts. Rather than just look at the magnitude of the differences, we look at their squares. Also, since the importance of the difference depends on the size of the counts involved, we divide by the expected count. Finally, we add these quantities over all cells of the table. Our notation for this test statistic is

26

$$X^2 = \sum_{i,j} \frac{(O_{ij} - E_{ij})^2}{E_{ij}} = \sum_{\text{all cells}} \frac{(\text{observed} - \text{expected})^2}{\text{expected}}.$$

(h) Calculate the value of this test statistic for these sample data.

(i) What kind of values (e.g., large or small) of the test statistic constitute evidence against the null hypothesis of no association between the two variables?

Finally, we need to know how to calculate the *p-value* of the test, which will tell us how often we would get such a large test statistic just by sampling variability if in fact the variables were independent. It turns out that under the hypothesis of independence, the test statistic X^2 has what is called a ***chi-square*** distribution with number of degrees of freedom equal to $(r-1)(c-1)$, where r is the number of rows and c is the number of columns. Critical values for the chi-square distribution have been tabulated in Table IV. One reads the chi-square table just as one reads the *t*-table. Since large values of the test statistic provide evidence against the null hypothesis, the p-value for the chi-square test is the probability of exceeding the value of the test statistic. As always, the *smaller* the p-value, the less likely that the observed data would have occurred by sampling variability. Thus, the *smaller* the p-value, the *stronger* the evidence that there *is* an association between the two variables.

(j) Use the chi-square table to determine (as accurately as possible) the p-value for these sample data. Would these sample data have been very unlikely to occur by chance if agreement with the proposition were independent of political leaning?

(k) Summarize and explain your conclusion in a few well-constructed sentences.

Chi-square test of independence in a two-way table:

H_o: There is no association between the row and column variables.

H_a: There is an association between the two variables.

test statistic: $X^2 = \sum\limits_{\text{all cells}} \dfrac{(O_{ij} - E_{ij})^2}{E_{ij}}$

p-value $= \Pr(\chi_k^2 > X^2)$,

where χ_k^2 represents a chi-square distribution with df $= (r-1)(c-1)$ degrees of freedom.

The **technical conditions** necessary for the test to be valid are:
1. that the observations constitute a simple random sample from the population of interest, and

2. that the *expected* counts are at least 5 for each cell of the table.

Activity 26-2: Government Spending

The 1998 General Social Survey asked a sample of adult Americans whether they think the federal government spends too little, the right amount, or too much money on various programs. The responses concerning the space program, categorized by political leaning, are summarized in the table below. Some expected counts appear in parentheses.

	liberal	moderate	conservative	total
too little	50 ( )	47 (46.85)	36 (45.66)	133
just right	164 (179.63)	214 ( )	212 (202.56)	590
too much	162 (155.88)	174 (108.34)	176 ( )	512
total	376	435	424	1235

(a) State the hypotheses for applying a chi-square test to this table of data.

26

(b) Fill in the missing expected counts.

(c) Fill in the missing pieces (which correspond to the missing expected counts) in the calculation of the test statistic:

$$X^2 = \underline{\hspace{1.5cm}} \quad + 0.001 \quad + 2.044$$
$$+ 1.360 \quad + \underline{\hspace{1cm}} \quad + 0.440$$
$$+ 0.240 \quad + 0.223 \quad + \underline{\hspace{0.8cm}} \quad = \underline{\hspace{1.5cm}}$$

(d) Use the chi-square table to determine (as accurately as possible) the p-value of the test.

(e) What does the test reveal about a possible relationship between political leaning and opinion about federal spending on the space program? Do the different political groups seem to have different opinions on this issue? Explain.

Activity 26-3: Newspaper Reading (*cont.*)

Recall from Activity 19-17 on page 421 that the 1998 General Social Survey asked a sample of adult Americans how often they read a newspaper. The results, tabulated by gender, appear in the table below:

	men	women
every day	375	430
a few times per week	182	238
once per week	121	173
less than once per week	133	218

(a) You will need to enter the observed counts and expected counts into your calculator as matrices:

- Select the [MATRIX] menu and then select EDIT.
- Press ENTER to edit Matrix A.
- Change the dimensions to be 4×2 and then enter the values from the above table.

To conduct a chi-square test of whether newspaper reading is independent of gender, choose χ^2 – Test in the STAT TESTS menu. Keep matrix [A] specified as containing the observed counts and matrix [B] as where you want the expected counts stored. Highlight the "Calculate" or "Draw" option and press ENTER. To view matrix B, select the [MATRIX] menu, select B, and press ENTER. Use the right arrow to scroll to the right.

Report the expected counts next to the observed counts. Also report the null and alternative hypotheses, the test statistic, and p-value. Does the test provide evidence (at the .05 level, say) that newspaper reading differs between the two genders?

By itself, the chi-square test determines only whether the data provide evidence of a relationship between the two variables. If the result is significant, one can go on to identify the source of that relationship by finding the cells of the table that contribute the most to the X^2 value (i.e., those cells with the biggest discrepancy between the observed and expected counts) and by noting whether the observed count falls above or below the observed count in those cells.

(b) Identify the cell that contributes the most to the test statistic. Is the observed count higher or lower than the expected count for that cell? Explain what this signifies in the context of the data.

26

(c) Identify the next three cells making the largest contributions to the text statistic. Comment on what they further reveal about the association between gender and newspaper reading.

Activity 26-4: Suitability for Politics (*cont* .)

Reconsider the data presented in Activity 7-7 on page 150 concerning a sample of responses to the assertion that "most men are better suited emotionally for politics than are most women." The table of summarized responses is reproduced below:

	men	women
agree	169	236
disagree	565	777

(a) Use your calculator to conduct a chi-square test of independence. Clearly state the null and alternative hypotheses in context. Record the expected counts in the table above next to the observed ones. Also report the test statistic and p-value. Finally, summarize your conclusion in a sentence or two.

(b) Use your calculator to conduct a two-sample *z*-test of whether the proportion of agreement among men differs significantly from that among women. Again clearly state the null and alternative hypotheses in context. Also report the test statistic and p-value, and summarize your conclusion in a sentence or two.

(c) How do the p-values of the two tests compare? Is the conclusion the same with both tests?

(d) Compare the test statistic z that you found in (b) with the test statistic X^2 that you found in (a). If you square the value of z, is the result equal to (roughly, anyway) the value of X^2?

This activity has shown that for tables with two rows and two columns, the chi-square test of independence is equivalent to the test of whether two population proportions differ. In fact, the square of the z test statistic for comparing two proportions equals the X^2 statistic from the chi-square test.

Activity 26-5: Government Spending (*cont .*)

Reconsider the data of Activity 26-2 on page 557 concerning attitudes toward spending on the space program:

	liberal	moderate	conservative	total
too little	50	47	36	133
just right	164	214	212	590
too much	162	174	176	512
total	376	435	424	1235

This activity asks you to approximate the sampling distribution of the X^2 statistic under the hypothesis of no association between the row and column variables. To do this you will keep the marginal totals fixed and randomly assign, for example, the 133 "too little" responses among the 376 liberals, 435 moderates, and 424 conservatives.

(a) Download CHISIM.83p and run this progam to carry out this random assignment once (enter 1 at the first prompt to specify one random assignment). Record

26

below the table that results (note that the marginal totals should remain the same) and the X^2 statistic for this table:

	liberal	moderate	conservative	total
too little				133
just right				590
too much				512
total	376	435	424	1235

X^2 statistic:

(b) Is this simulated X^2 statistic, based on the hypothesis of no association, larger than the X^2 statistic from the actual sample data?

(c) Again use the CHISIM program to repeat this simulation for a total of 100 repetitions (type 100 at the prompt). The results are stored in a list named CHI. Look at a dotplot of the distribution of your resulting X^2 statistics from these repetitions. Sketch its shape below (labeling both axes).

This distribution approximates the sampling distribution of the X^2 statistic under the assumption that there is no association between the two variables. In other words, it approximates the chi-square distribution with 4 degrees of freedom.

(d) How many of the 100 simulated tables produced a X^2 statistic greater than or equal to the one from the actual sample data? What proportion of the 100 repetitions is this? [*TI Hint:* You may want to sort these data using SortA(found in the STAT menu.]

(e) Is your answer to (e) reasonably close to the p-value of the test that you found in (d) of Activity 26-2? Explain why it should be.

This topic has asked you to reconsider the issue of analyzing two-way tables and exploring whether two categorical variables are related, as you first did in Topic 7. You have revisited the concept of **statistical independence**, and you have learned to apply the **chi-square test** of independence to such data.

The next topic expands your study of inference procedures for relationships between variables by considering quantitative variables. You will learn inference techniques concerning a correlation coefficient and the slope coefficient of a linear regression model.

HOMEWORK ACTIVITIES

Activity 26-6: Government Spending (*cont.*)

The following table reports survey responses to a question about the level of the federal government's spending on the environment, with respondents classified according to their political leaning:

	liberal	moderate	conservative	total
too little	292	271	239	802
just right	85	142	133	360
too much	11	27	64	102
total	388	440	436	1264

(a) Conduct a chi-square test of whether opinions about spending on the environment are related to political leaning. Be sure to report the hypotheses, test statistic, and p-value of the test.

(b) Is the test result significant at the .05 level? At the .01 level? At the .001 level?

(c) If the test result is significant, identify the three cells that make the largest contribution to the test statistic. Also indicate whether the observed count exceeds or falls short of the expected count for each of those cells.

(d) Write a paragraph summarizing your findings. Also compare your findings with regard to spending on the environment with those about spending on the space program from Activity 26-2 on page 557.

26

Activity 26-7: Newspaper Reading (*cont.*)

The table below summarizes the results of the newspaper reading survey by political leaning:

	liberal	moderate	conservative
every day	209	292	184
a few times per week	122	140	143
once per week	94	104	83
less than once per week	93	119	104

(a) Begin with a descriptive analysis: Calculate the conditional distributions for each of the three political groups, construct a segmented bar graph, and comment on whether the data suggest a relationship between political leaning and newspaper reading.

(b) Proceed to an inferential analysis: Calculate the expected counts of each cell under the hypothesis of independence, determine the test statistic and the p-value of the chi-square test, and comment on your findings.

Activity 26-8: Age and Political Interest (*cont.*)

Reconsider the data from Activity 7-2 on page 140 concerning an association between age group and level of interest in politics, which are reproduced below along with some of the expected counts:

	18–35	36–55	56–94	total
not much	146 (115.96)	146 (159.93)	89 ( )	381
somewhat	192 ( )	260 (254.38)	154 (167.19)	606
very much	47 (84.61)	125 ( )	106 (76.70)	278
total	385	531	349	1265

(a) Fill in the remaining expected counts.

(b) Fill in the corresponding missing values in the calculation of the test statistic:

$$X^2 \quad = \quad 7.784 \quad\quad + 1.283 \quad\quad + \underline{\quad\quad}$$
$$+ \underline{\quad\quad} \quad\quad + 0.124 \quad\quad + 1.040$$
$$+ 16.717 \quad\quad + \underline{\quad\quad} \quad\quad + 11.195 \quad\quad = \underline{\quad\quad}$$

(c) Determine (as accurately as possible from the chi-square table) the p-value of the test.

(d) Does the test reveal strong evidence of an association between age and political interest?

(e) Which cells contribute the most to the test statistic? What do they reveal about the direction of the association?

Activity 26-9: Pregnancy, AZT, and HIV (*cont.*)

Recall from Activity 7-3 on page 144 that in an experiment to assess the effectiveness of AZT for reducing the proportion of HIV positive babies born to HIV positive women, 13 of 164 babies of mothers in the AZT group were born HIV positive, compared to 40 of 160 babies from the placebo group.

(a) Identify the explanatory and response variables in this experiment. For each variable, indicate whether the variable is quantitative or categorical.

(b) Construct a two-way table from these sample data, remembering to put the explanatory variable in columns and the response variable in rows.

(c) A two-sample proportion test produces a test statistic of $z = 4.25$. Use only this information to calculate the test statistic for the chi-square test of independence.

(d) Use the chi-square table and the test statistic that you calculated in (c) to determine the p-value of the test.

(e) Summarize what the chi-square test reveals about whether the sample data provide evidence that the baby's HIV status is related to whether the mother received AZT or a placebo.

Activity 26-10: Women Senators (*cont.*)

The following table reveals the composition of the 1999 U.S. Senate classified according to sex and party:

	Republican	Democrat	total
men	52	39	91
women	3	6	9
total	55	45	100

(a) Calculate the conditional distributions of gender within each party. Which party has the higher proportion of women Senators?

(b) Conduct a chi-square test of independence, reporting the hypotheses, test statistic, and p-value.

(c) Is this test a sensible one for these data? Are the technical conditions satisfied? Explain.

26

Activity 26-11: Baldness and Heart Disease (*cont.*)

Recall from Activity 7-12 on page 152 that a study classified samples of males, one group suffering from heart disease and one group not, according to their degree of baldness. The resulting table is reproduced here:

	none	little	some	much	extreme
heart disease	251	165	195	50	2
control	331	221	185	34	1

(a) If the table were to be analyzed as is, would one of the technical conditions for the validity of the chi-square test be violated? Explain.

(b) Combine the "much" and "extreme" categories into a "much or more" category and produce the resulting 2×4 table.

(c) Use your calculator to conduct a chi-square test of independence on this 2×4 table. Report the hypotheses (in context), test statistic, and p-value.

(d) Write a few sentences summarizing your findings about a possible relationship between baldness and heart disease. Be sure to mention the type of study involved here.

Activity 26-12: Student-Generated Data (*cont.*)

Select any pair of categorical variables from the student-generated data collected in class, and conduct a chi-square test of independence. Write a paragraph summarizing your findings about a possible relationship between the two variables.

Activity 26-13: Government Spending (*cont.*)

Return to the data in the table of Activity 26-2 on page 557.

(a) Report the test statistic and p-value of the chi-square test of independence for those sample data concerning opinions on government spending on the space program among three political classifications.

Now suppose that the sample size had been doubled, with every cell containing exactly twice as many responses as in the actual sample.

(b) How would the conditional distributions of opinions about spending among the three political groups change from the actual sample? Explain.

(c) Use your calculator to recompute the test statistic and p-value for the new, larger sample. How do they compare to the original sample?

(d) Repeat (b) and (c) supposing that the sample size had been multiplied by ten, with every cell containing exactly ten times as many responses as in the actual sample.

(e) Explain what this activity reveals about the role of sample size in chi-square tests of independence.

Activity 26-14: Spirituality and Addict Recovery

Betty Jarusiewicz investigated the role of spirituality in recovery of addicts. She asked a series of questions of addicts, including whether or not they personally identified with a particular religion. Of 19 subjects who did identify with a particular religion, 13 suffered a relapse within two years. Of the 21 subjects who did not identify with a particular religion, 7 relapsed.

(a) Is this an observational study or a controlled experiment? Explain.

(b) Identify the explanatory and response variables in this study. For each variable, indicate whether it is quantitative or categorical.

(c) Organize the data in a two-way table, remembering to put the explanatory variable in columns and the response variable in rows.

(d) Calculate the conditional distributions of relapse for each group (whether or not they identified with a religion).

(e) Construct a segmented bar graph to display these two conditional distributions.

(f) Comment on whether there seems to be a relationship between identifying with a religion and suffering a relapse.

(g) Conduct a chi-square test of whether relapse is associated with identification with a particular religion. Report the expected counts, test statistic, and p-value. Write a sentence or two summarizing your conclusion.

(h) If you had conducted a two-sided z-test comparing the proportions of relapse between the two groups, what would the test statistic z equal? What would the p-value be? Explain.

Activity 26-15: Suitability for Politics (*cont.*)

Consider again the data of Activity 26-1 on page 553 concerning the reactions of those with differing political views to the statement that men are better suited emotionally for politics than women are. Show that the expected counts do indeed reflect independence between the two variables. In other words, find the conditional distribution of reaction to the assertion (agree or disagree) for each of the three political groups and note that they are identical. Also produce a segmented bar graph to illustrate this.

Activity 26-16: Credit Cards and Cellular Phones (*cont.*)

Reconsider the data collected in Topic 23 and analyzed in Activity 23-27 on page 499 concerning students' ownership of credit cards and/or cellular phones.

(a) Summarize the data in a two-way table, remembering to put the explanatory variable in columns and the response variable in rows.

26

(b) Create a segmented bar graph to display the association between ownership of a credit card and of a cellular phone. Comment on whether the graph suggests any such association.

(c) Conduct a chi-square test to assess whether the sample data provide evidence of an association between ownership of a credit card and of a cellular phone in the population. Report the hypotheses, test statistic, and p-value along with your conclusion.

(d) Conduct a two-sample z-test comparing the proportions of cell phone owners between the "own a credit card" group and the "do not own a credit card" group. Verify that the square of this test statistic produces the chi-square test statistic from (c) and that the p-values are the same.

(e) Are the technical conditions about sample sizes and expected counts satisfied in (c) and (d)? Explain.

(f) To what population would you suggest that the results of these tests generalize? Explain.

Topic 27:

INFERENCE FOR CORRELATION AND REGRESSION

OVERVIEW

In the previous topic you studied the chi-square test as an inferential procedure for assessing the evidence of association between two categorical variables. In this topic you will study the analogous situation for two quantitative variables. First you will encounter a test procedure concerning a correlation coefficient, and then you will learn inferential techniques for the slope coefficient of a regression equation. You will continue to use graphical and descriptive methods from earlier in the course, but you will study sampling distributions of the relevant sample statistics as an introduction to the inferential methods based again on the *t*-distribution. Once again you will see that the structure, reasoning, and interpretation of confidence intervals and significance tests remain unchanged.

OBJECTIVES

- To develop a sense for the sampling distributions of correlation coefficients, slope coefficients, and regression lines.
- To learn how to apply a test procedure involving the correlation coefficient.
- To acquire the ability to apply inferential techniques concerning the slope coefficient in a regression equation.
- To understand the use of residual plots for checking technical conditions related to regression inference.
- To continue to apply graphical and descriptive methods learned earlier in conjunction with these inferential techniques.

PRELIMINARIES

1. Do you think that a professional baseball team's total player payroll is associated with its winning percentage? If so, in what direction?

2. Make a guess for the correlation between a baseball team's payroll and winning percentage.

3. Do you think that there is a positive association between the amount of time a student spends studying per week and his or her grade point average?

4. Make a guess as to the correlation between students' studying times and grade point averages.

IN-CLASS ACTIVITIES

Activity 27-1: Baseball Payrolls

The data below are the player payrolls (in millions of dollars) and won/lost records of the sixteen National League baseball teams in 1999. A scatterplot is also provided:

	team	payroll	wins	losses	win pct		team	payroll	wins	losses	win pct
1	Arizona	70.4	100	62	61.7%	9	Milwaukee	42.9	74	87	46.0%
2	Atlanta	75.1	103	59	63.6%	10	Montreal	16.4	68	94	42.0%
3	Chicago	55.4	67	95	41.4%	11	New York	71.3	97	66	59.5%
4	Cincinnati	42.1	96	67	58.9%	12	Philadelphia	30.5	77	85	47.5%
5	Colorado	54.4	72	90	44.4%	13	Pittsburgh	24.2	78	83	48.4%
6	Florida	15.2	64	98	39.5%	14	St. Louis	46.2	75	86	46.6%
7	Houston	55.6	97	65	59.9%	15	San Diego	45.9	74	88	45.7%
8	Los Angeles	71.1	77	85	47.5%	16	San Francisco	46.0	86	76	53.1%

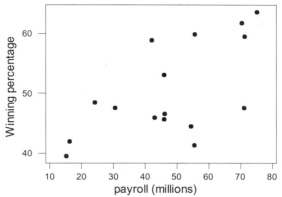

(a) Does the scatterplot reveal an association between payroll and winning percentage? If so, is the association positive or negative? Explain.

(b) The four teams that made the play-offs in the National League were Arizona, Atlanta, Houston, and New York. Circle and label their points on the scatterplot. Do these teams seem to have had large payrolls as well as high winning percentages?

(c) Download BASESAL.83g. The list names are PAY, WINS, LOSS, and WINP. Based on the scatterplot, make an educated guess as to the value of the correlation coefficient between winning percentage and size of payroll. Then use your calculator to calculate this correlation.

   Guess: Correlation (actual):

The question of statistical significance once again asks how likely such an extreme sample would arise by chance alone even when there is in fact no association between the two variables (zero correlation). To investigate this question through simulation, we will randomly distribute these sixteen winning percentages among the sixteen teams and their payrolls. If it turns out that a correlation as large as you found in (c) rarely occurs through this chance mechanism, we will have grounds for concluding that the observed correlation value differs significantly from zero.

27

(d) Implement this simulation by writing these sixteen winning percentages on sixteen index cards or pieces of paper. Shuffle them well, and then deal them out one at a time, the first being randomly assigned to Arizona with its $70.4 million payroll, the second to Atlanta with its $75.1 million payroll, and so on alphabetically through the sixteen teams. Then enter these winning percentages into your calculator along with the team payrolls to which they were randomly assigned. Calculate the correlation between payroll and winning percentage for your "scrambled" sample.

(e) Is this correlation as large as the one in the actual sample?

(f) Combine the correlation from your "scrambled" sample with those of your classmates, and display the results in a dotplot below:

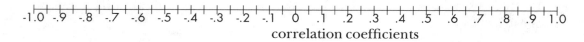

correlation coefficients

(g) Write a few sentences describing this distribution.

(h) For how many and what proportion of these scrambled samples does the correlation exceed the correlation in the actual sample? Does this indicate that a correlation as large as the one actually found is unlikely to occur by chance if in fact payroll and winning percentage are not associated? Explain.

This activity has reminded you of the concept of statistical significance as applied to investigating whether a sample correlation is large enough to suggest an association between two variables. The details of a formal test of significance for whether a population correlation coefficient, denoted by the Greek letter ρ ("rho"), differs from zero follow:

Test of significance for population correlation ρ:

$H_o: \rho = 0$

$H_a: \rho < 0$ or $H_a: \rho > 0$ or $H_a: \rho \neq 0$

test statistic: $t = \dfrac{r\sqrt{n-2}}{\sqrt{1-r^2}}$

p-value: $\Pr(T_{n-2} \leq t)$ or $\Pr(T_{n-2} \geq t)$ or $2\Pr(T_{n-2} \geq |t|)$

The **technical conditions** required for the validity of this test procedure are that:
1. the data comprise a simple random sample from the population, and
2. both variables are normally distributed.

Note that the structure, reasoning, and interpretation of this test are no different from the others you have studied. In particular, the test statistic is equal to the sample statistic minus the hypothesized value (zero), divided by the standard error. (The above formula has been rearranged for easier calculation.) Also, as always, the p-value measures how unlikely such an extreme sample (or more extreme) would be if the null hypothesis were true.

(i) Apply this test to the baseball payroll data by calculating the test statistic from the sample correlation coefficient you found in (c) and then using the *t*-table (Table III) to find the p-value. Would you reject the null hypothesis at the $\alpha = .05$ level? Write a one-sentence conclusion about whether the data strongly suggest an association between payroll and winning percentage.

Activity 27-2: Studying and Grades

Inference about relationships is not limited to the correlation coefficient. One can make inferences about the parameters of a regression equation for predicting one variable from another. Of particular interest are inferences about the slope coefficient, since a slope coefficient of zero can be interpreted as no linear association between the variables.

Students conducting a project at the University of the Pacific investigated whether an association exists between hours spent studying per week and grade point average. They surveyed eighty of their fellow students, with responses appearing in the table below:

ID	hours	GPA	ID	hours	GPA	ID	hours	GPA	ID	hours	GPA
1	0.5	1.90	21	2.0	2.94	41	4.0	3.30	61	2.0	3.60
2	1.0	2.30	22	5.5	2.94	42	5.0	3.30	62	4.0	3.60
3	4.0	2.30	23	3.0	3.00	43	7.0	3.40	63	2.0	3.67
4	5.0	2.40	24	5.0	3.00	44	6.0	3.40	64	2.0	3.70
5	1.5	2.50	25	2.0	3.00	45	4.5	3.40	65	5.0	3.70
6	4.0	2.50	26	5.0	3.00	46	7.0	3.40	66	3.0	3.71
7	3.0	2.50	27	2.0	3.00	47	5.0	3.40	67	4.5	3.74
8	6.0	2.50	28	4.5	3.00	48	7.0	3.44	68	1.5	3.75
9	2.0	2.50	29	3.0	3.00	49	4.0	3.45	69	3.0	3.75
10	1.5	2.70	30	3.5	3.00	50	3.5	3.48	70	8.0	3.83
11	2.0	2.70	31	7.0	3.00	51	8.0	3.49	71	7.0	3.87
12	5.0	2.77	32	4.0	3.00	52	6.0	3.50	72	8.0	3.90
13	2.0	2.78	33	4.0	3.05	53	5.0	3.50	73	2.0	3.90
14	4.0	2.80	34	3.5	3.10	54	3.0	3.50	74	4.0	3.92
15	2.0	2.80	35	1.0	3.20	55	3.0	3.50	75	5.0	3.92
16	4.0	2.80	36	2.0	3.20	56	4.0	3.50	76	4.0	3.94
17	4.0	2.80	37	3.5	3.20	57	3.0	3.50	77	4.0	3.96
18	3.0	2.80	38	4.0	3.20	58	2.0	3.50	78	6.0	3.98
19	0.25	2.83	39	3.0	3.25	59	4.0	3.55	79	6.0	3.98
20	5.0	2.90	40	3.5	3.30	60	8.0	3.58	80	3.5	4.00

(a) Identify the observational units and the variables measured. Indicate whether the variables are categorical or quantitative.

(b) Download UOPGPA.83g. The list names are HOURS and GPA. Use your calculator to construct a scatterplot with GPA on the vertical axis and hours studied per week on the horizontal. Comment on whether the scatterplot reveals an association between the two.

(c) Use your calculator to compute the regression equation for predicting grade point average from hours of study per week. Record the equation, being sure to write it in the context of these two variables. Also report the correlation coefficient.

(d) By how many grade points does this equation predict the GPA to rise or fall for each additional hour of study? Explain where this number appears in the regression equation.

(e) If the student researchers were to take another sample of eighty students from the population, would they be likely to get exactly the same regression equation from the new sample? Explain.

To investigate how often a sample regression line this extreme would occur by chance if in fact there were no association between hours studied and GPA in the population, you will generate a random sample from a hypothetical population in which there is no association between hours studied and GPA, as our null hypothesis asserts. You will do this by assuming that the population of hours studied and GPAs follow normal distributions (Note: This assumption is not one of the technical conditions you will see later, but it will improve the speed of your simulation.)

(f) Generate 80 values for hours studied according to a normal distribution with mean 4 and standard deviation 1.5 and generate 80 values of GPA according to a normal distribution with mean 3.2 and standard deviation .25. [*TI Hint:* Use the randNorm feature located in your MATH PRB menu. The parameters for this function are mean, standard deviation, and sample size. For example, to get the 80 values of hours you would use randNorm(4,1.5,80). This function will return a list with 80 entries, so do not forget to store this into a list named HOUR2. Likewise, store the data for the GPA into a list named GPA2.]

(g) Determine the regression line for predicting GPA from hours studied in your simulated sample. Record the value of the slope coefficient and *y*-intercept below. Then pool your results with those of your classmates until you have recorded a total of twenty slope coefficients in the table:

Sample #	1	2	3	4	5	6	7	8	9	10
Sample slope										
y-intercept										
Sample #	11	12	13	14	15	16	17	18	19	20
Sample slope										
y-intercept										

(h) Enter the data above into lists named SLOPE and INTER. Construct a dotplot of these slope coefficients. [If you use your calculator, reproduce the dotplot below.]

(i) Where does the slope coefficient from the UOP students' actual sample data fall in relation to the sample slope coefficients taken from the population with no association between GPA and study time? Does your simulation indicate that a slope coefficient at least as extreme as the sample one obtained by the UOP students would rarely happen by chance if there were no association in the population? Explain.

(j) Use your calculator to compute the mean and standard deviation of these 20 slope coefficients. Is the mean reasonably close to the population's slope value of zero?

Mean of slopes: Standard deviation of slopes:

Close to 0?

If you were to repeat this process many more times, you would obtain a close approximation to the sampling distribution of the sample slope coefficient b. As you have seen with many other sampling distributions, the shape is approximately normal, and it is centered at the value of the population parameter. This sampling distribution forms the basis for inference (significance test, confidence interval) procedures concerning the population slope coefficient, which will be denoted by β ("beta").

The details of the procedures for conducting a test of whether the population slope β differs from zero and for a confidence interval to estimate the population slope β are as follows:

Inference procedures for the population slope β:

Confidence interval: $b \pm (t^*_{n-2}) S.E.(b)$

$H_o: \beta = 0$

$H_a: \beta < 0$ or $H_a: \beta > 0$ or $H_a: \beta \neq 0$

test statistic: $t = \dfrac{b-0}{S.E.(b)}$,

p-value: $Pr(T_{n-2} \leq t)$ or $Pr(T_{n-2} \geq t)$ or $2Pr(T_{n-2} \geq |t|)$

where $S.E.(b)$ denotes the standard error of the sample slope coefficient, b, which will be calculated using the STDER program for your calculator.

(k) Download STDER.83p. For the original students' sample data at UOP, find the standard error of the slope coefficient in your regression output from (c) using the program STDER and specifying HOURS as the X-LIST and GPA as the Y-List. Is this reasonably close to the standard deviation of your 20 simulated slope coefficients?

(l) Use this standard error from (k) to calculate (by hand) the test statistic for testing whether the population slope differs from zero.

(m) Use the *t*-table and your test statistic in (l) to find the p-value of the test.

(n) Is this p-value consistent with your simulation results? Explain.

(o) Does the p-value suggest that the association between GPA and hours of study found in the sample of 80 students would be very unlikely if in fact no association existed in the population? Explain.

(p) Use the LinRegTTest feature of your calculator (found in the STAT TEST menu) to check your work above. [*TI Hint:* Place Y_1 (located in the VARS Y-VARS FUNCTION menu) at the RegEQ prompt and make sure that the frequency is set to 1.] Choose the correct form of the alternative, highlight "Calculate," and press ENTER .

(q) Find a 95% confidence interval for the population slope coefficient.

(r) Interpret this interval, remembering what a slope coefficient represents and remembering to relate your interpretation to the context.

(s) Download LINES.83p and use this program to plot the twenty regression lines from these simulated samples. This program will also plot the students' actual sample regression line using a thicker line to distinguish it from the rest. [This gives you a sense for the sampling variability in these lines.] Comment on how the actual sample line compares to the simulated sample lines from the population with no association.

(t) Write a few sentences summarizing your conclusion about the evidence found in the students' sample concerning an association between GPA and hours of study.

Activity 27-3: Studying and Grades (*cont.*)

The **technical conditions** required for the validity of these inference procedures about the slope coefficient are:

1. that the data are a simple random sample from the population,

2. that the two variables are linearly related,

3. that for a given x-value, the distribution of the y-values in the population is normal,

4. that the standard deviations of those normal distributions are the same at all x-values.

27

One checks these last three conditions by examining the *residuals* of the regression analysis. If the conditions are met, the residuals will reveal no pattern when plotted against the *x*-values, will have about the same variability at all *x*-values (equal spread across the graph), and will roughly follow a normal-shaped distribution.

(a) Use your calculator to compute the residuals (remember that the residuals are stored in the list named RESID every time you use your calculator to compute the least squares line) from the regression analysis of the GPA study. Produce a histogram, dotplot, and boxplot of the residuals. Describe the distributions of the residuals. Do these plots reveal any marked features suggesting nonnormality?

(b) Produce a scatterplot of the residuals vs. hours per week of study, the explanatory variable. Describe the plot. Does it reveal a strong pattern (such as curvature) suggesting that a linear model is not appropriate? Does the plot suggest that the variability of the residuals differs substantially at various *x*-values (hours per week)?

WRAP-UP

This topic has deepened your knowledge of methods related to least squares regression. You have studied the sampling distributions of correlation coefficients, slope coefficients, and regression lines as an introduction to the inferential methods for regression. You have continued to learn the important role of graphical and descriptive methods, particularly with regard to checking technical conditions of procedures. You have also seen similarities between these test and interval procedures and all of the others that you have studied previously.

HOMEWORK ACTIVITIES

Activity 27-4: Baseball Payrolls (*cont.*)

Refer again to the data on baseball teams' player payrolls and winning percentages from Activity 27-1 on page 570 (BASESAL.83g).
(a) Use your calculator to find the regression equation for predicting a team's winning percentage from its payroll.
(b) Conduct a test of whether the slope coefficient differs from zero at the $\alpha = .05$ level. Report the hypotheses, test statistic, and p-value, along with a one-sentence conclusion.
(c) Verify that the test statistic and p-value match those concerning the correlation coefficient that you calculated in (i) of Activity 27-1.
(d) Examine residual plots and comment on whether the technical conditions for this inference procedure appear to be satisfied.

Activity 27-5: Studying and Grades (*cont.*)

Recall that the study analyzed in Activity 27-2 on page 574 found a sample correlation of .343 between grade point average and study hours per week. Let ρ denote the correlation between these two variables in the *population* of all students at the school.
(a) Conduct a test of significance of whether these sample data provide evidence at the .05 level that ρ is not zero. State the hypotheses (in words and in symbols), and show the details of your calculation of the test statistic and p-value. [Recall that the sample size in the study was $n = 80$.]
(b) Verify that the test statistic and p-value from (a) based on the correlation coefficient match those found in (l) and (m) of Activity 27-2 based on the regression slope.
(c) Suppose that a sample correlation of .343 were found in a study that involved a sample size of $n = 20$. Would you expect the p-value of this test to be larger or smaller than it was when $n = 80$? Explain.
(d) Calculate the test statistic and p-value for the test described in (c). Is your expectation confirmed or refuted?
(e) Repeat (c) and (d) for a sample size of $n = 200$.
(f) Suppose that a study involves $n = 5000$ subjects. Determine how small the sample correlation r could be and still be statistically significantly different from zero at the $\alpha = .05$ level.

Activity 27-6: Airfares (*cont.*)

Consider the data that you examined in Activity 10-1 on page 206 concerning the airfare and distances to various destinations (AIRFARE.83g).

(a) Produce a scatterplot of airfare vs. distance, and comment on the association revealed by the graph.

(b) Determine the regression equation for predicting airfare from distance, and sketch the regression line on the scatterplot.

(c) Conduct a test of whether the sample data provide strong evidence that the population slope coefficient is positive. Report the hypotheses (in symbols and in words), test statistic, and p-value. Also write a brief conclusion.

(d) Construct a 99% confidence interval for the population slope coefficient, and interpret the interval in context.

(e) Does this confidence interval suggest that it is plausible that the price of flying to a destination increases by a quarter for each additional mile? How about a dime? A nickel?

(f) Examine residual plots, and comment on what they reveal concerning the technical conditions for this procedure.

Activity 27-7: Climatic Conditions (*cont.*)

Refer to the data presented in Activity 9-17 on page 202 concerning some climatic variables for a sample of 25 American cities (CLIMATE.83g).

(a) Consider the four variables January high, January low, July high, and July low temperatures. Use your calculator to compute the correlation coefficient between all pairs of these variables. [There should be six pairs.] Then arrange them from highest to lowest.

(b) For which of these six pairs does the correlation differ significantly from zero at the $\alpha = .10$ level? Report the test statistic and p-value for each pair.

(c) Select one pair for which the correlation is significant. Designate one variable to be the explanatory variable and the other to be the response variable. Use your calculator to produce a scatterplot and then to determine the regression equation.

(d) Construct a 90% confidence interval for the population slope coefficient of your regression line in (c). Interpret this interval in context.

(e) Explain why the result of your test in (b) indicates that this interval should not include the value 0.

(f) Examine residual plots from the regression, and comment on whether they provide reason to doubt that the procedure's technical conditions are satisfied.

Activity 27-8: Marriage Ages (*cont.*)

Consider again the data on marriage ages for a sample of 24 couples that appear in Activity 3-13 on page 60. Suppose that you are interested in predicting the wife's age based on the husband's age (MARRIAGE.83g).

(a) Determine the appropriate regression equation, and sketch it on a scatterplot.

(b) Test whether the slope coefficient differs significantly from zero. Write a paragraph showing the details and summarizing your conclusion.

(c) Find a 95% confidence interval for the slope coefficient in the population of all married couples in this county.

(d) Does this confidence interval include the value 1? Comment on the importance of this value in this context.

(e) Conduct a significance test of whether the population slope coefficient differs from the value 1. [*Hints*: Use the significance test procedure given for testing whether the slope is zero, but adjust the hypotheses accordingly and subtract 1 rather than 0 in the numerator of the test statistic.] Write a paragraph showing the details and summarizing your conclusion.

(f) Which of the 24 husbands has the most potential *influence* over the regression line by virtue of his fairly extreme age? Also explain what "influence" means in this setting.

(g) Suppose that the 71-year-old husband had actually married a 21-year-old wife rather than a 73-year-old wife. Repeat (a)–(d) in this case, and also comment on how much this one change altered your findings.

Activity 27-9: Cars' Data (*cont.*)

The following table reports the weight (in pounds) and time to travel one-quarter mile (in seconds) for ten cars classified in the "upscale" category of *Consumer Reports 1999 New Car Buying Guide*:

Car	weight	1/4 mile	car	weight	1/4 mile
Acura TL	3460	16.4	Infiniti I30	3195	16.8
Audi A4	3345	18.4	Lexus ES300	3390	16.6
Audi A6	3785	17.5	Mercedes/Benz C/Class	3320	16.7
BMW 3-series	3265	16.6	Saab 9/5	3475	17.4
Cadillac Catera	3805	17	Volvo S70-V70	3305	15.9

(a) Construct a scatterplot of time to cover 1/4 mile vs. weight, and comment on whether the plot reveals any association between the two.

(b) Calculate the value of the correlation coefficient between these two variables.

27

(c) Conduct a simulation to investigate whether such an extreme correlation would occur rarely by chance if the variables were not associated. Do this by mixing up the 1/4 mile times and randomly assigning them to the cars and their weights. [*TI Hint*: Assign integers to the cars and use randInt located in the MATH PRB menu to do the random assignment.] Calculate the correlation for the "scrambled" sample. Then repeat this at least twenty times. Create a dotplot of the correlations, and identify the ones that are as extreme (in absolute value) as the actual correlation for these data.

(d) Write a few sentences summarizing your conclusions from this simulation and explaining why it follows from your results.

Activity 27-10: College Football Players (*cont.*)

Consider again the data presented in Activity 12-16 on page 265 and analyzed in Activity 10-16 on page 225 (CPFOOTBALL.83g). Test whether the data suggest an association between a player's weight and his jersey number. Show the details of the procedure that you use, and explain your conclusion. Also indicate whether you can conclude that the relationship is a causal one.

Activity 27-11: Comparison Shopping (*cont.*)

Consider again the comparison shopping data from Activity 8-19 on page 182 (SHOPPING.83g).

(a) Use your calculator to produce a scatterplot of one store's prices vs. the other's.

(b) Determine the regression equation for predicting one store's prices from the other's. Be sure to identify the explanatory and response variables clearly.

(c) Find a 90% confidence interval for the slope coefficient in the population of all items that could have been purchased in the two stores. Interpret this interval in context, and comment on whether it contains the value 1.

(d) Repeat (a)–(c) upon removing the most unusual item (in terms of its pair of prices). Comment on how much this one item affects your analysis and conclusions.

(e) Examine residual plots and discuss whether the technical conditions of this procedure seem to be satisfied.

(f) Conduct a significance test of whether the population slope coefficient differs from the value one. [*Hint*: See the hints in (e) of Activity 27-8 on page 583.] Perform this test both including and excluding the unusual item. Write a paragraph showing the details of your tests and summarizing your conclusions.

Activity 27-12: College Tuitions (*cont.*)

Consider again the data on founding dates and tuition charges of Pennsylvania colleges described in Activity 6-15 on page 133 (COLLEGES.83g). The correlation between founding date and tuition charge is –.505 for the 67 private four-year colleges and is .507 for the 25 public four-year colleges.

(a) Before you do any calculations, indicate which of these groups you expect to have a smaller p-value when you test the significance of the association. Explain.
(b) Use the given correlation values and sample sizes to test whether the correlations differ significantly from zero at the $\alpha = .05$ level.
(c) Examine scatterplots of tuition vs. founding date for both groups of colleges, and comment on which seems to satisfy the technical conditions of the inference procedure and which does not. Explain your answer.

Activity 27-13: Residual Plots

For each of the following, use your calculator to determine the regression equation indicated. Then use your calculator to create residual plots. Finally, describe the residual plots and comment on what they reveal about the technical conditions required for the regression inference procedures. In particular, specify any conditions that appear to be violated and explain your conclusion.

(a) Predicting life expectancy from people per television set (Activity 9-4 on page 193, TVLIFE.83g)
(b) Predicting gross receipts from attendance for Broadway shows (Activity 2-8 on page 36, BROADWAY.83g)
(c) Predicting an animal's gestation period from its longevity (Activity 11-1 on page 228, ANIMALS.83g)
(d) Predicting the rent from the price of a Monopoly property (Activity 9-2 on page 189, MONOPOLY.83g)
(e) Predicting the second weekend's box office revenue from the first weekend's for Hollywood blockbusters (Activity 2-9 on page 37, MOVIES.83g)

Activity 27-14: Signature Measurements (*cont.*)

Recall the data on signature lengths that you collected in Topic 1. Answer the following for each of the three regression lines for predicting a signature's length from its number of letters described in Activity 10-11 on page 223.

(a) Test whether the slope coefficient differs significantly from zero at the $\alpha = .05$ significance level.
(b) Construct a 95% confidence interval for the population slope.
(c) Examine residual plots to check for potential violations of the technical conditions required for these procedures to be valid.

27

(d) Write a paragraph summarizing your findings.

Activity 27-15: Height and Foot Length (*cont.*)

Refer to the data on heights and foot lengths gathered in Topic 8. Use these sample data to examine whether foot length is of use in predicting height. Conduct a complete analysis involving graphical, descriptive, and inferential components. Write a paragraph or two reporting your findings.

Table I:

RANDOM NUMBER TABLE

Line										
1	17139	27838	19139	82031	46143	93922	32001	05378	42457	94248
2	20875	29387	32582	86235	35805	66529	00886	25875	40156	92636
3	34568	95648	79767	15307	71133	15714	44142	44293	19195	30569
4	11169	41277	01417	34656	80207	33362	71878	31767	04056	52582
5	15529	30766	70264	86253	07179	24757	57502	51033	16551	66731
6	33241	87844	41420	10084	55529	68560	50069	50652	76104	42086
7	83594	48720	96632	39724	50318	91370	68016	06222	26806	86726
8	01727	52832	80950	27135	14110	92292	17049	60257	01638	04460
9	86595	21694	79570	74409	95087	75424	57042	27349	16229	06930
10	65723	85441	37191	75134	12845	67868	51500	97761	35448	56096
11	82322	37910	35485	19640	07689	31027	40657	14875	07695	92569
12	06062	40703	69318	95070	01541	52249	56515	59058	34509	35791
13	54400	22150	56558	75286	07303	40560	57856	22009	67712	19435
14	80649	90250	62962	66253	93288	01838	68388	55481	00336	19271
15	70749	78066	09117	62350	58972	80778	46458	83677	16125	89106
16	50395	30219	03068	54030	49295	48985	16247	28818	83101	18172
17	48993	89450	04987	02781	37935	76222	93595	20942	90911	57643
18	77447	34009	20728	88785	81212	08214	93926	66687	58252	18674
19	24862	18501	22362	37319	33201	88294	55814	67443	77285	36229
20	87445	26886	66782	89931	29751	08485	44910	83844	56013	26596
21	14779	15506	62210	44517	14721	99774	19102	44921	80165	03984
22	79801	42412	09555	05280	00534	10592	36738	03573	77510	66222
23	05297	26648	86929	47906	09699	36563	66286	23137	79434	51560
24	10384	40923	29328	82914	16875	15622	39567	68096	14770	89189
25	31058	85234	87674	21263	42429	99728	16261	16519	65635	59197
26	33946	72507	42425	33159	77435	31639	41819	06220	52297	04770
27	33443	10708	05353	97028	24889	93727	55007	40152	09817	02700
28	13371	02271	79272	26697	63927	01803	74220	93901	07809	69131
29	64066	18676	20977	69271	31705	38898	30547	39412	93606	89007
30	13054	79473	09691	66466	33599	89378	38296	20518	00716	56617

31	34282	50764	73730	49650	13380	49874	81896	57540	66339	76256
32	47982	45080	98353	50960	90679	61396	87926	65972	12968	90622
33	43609	07061	43180	70246	59960	36413	02223	57282	79836	45778
34	85025	13341	31367	83655	04989	88434	96865	46110	60777	54349
35	54749	23767	77894	97636	61682	77530	47755	16204	80388	66027
36	44494	67165	26455	35332	42455	00522	74952	08265	53974	26516
37	24228	49167	38008	24678	74906	90101	03858	64494	01655	82336
38	28260	54215	25038	12030	55249	21816	85465	91448	06029	30113
39	80761	07480	46048	97445	80484	42483	63428	80705	19955	31948
40	35969	42174	57469	65149	75981	50368	78941	76123	94717	36337
41	56033	79072	43090	73725	35831	46077	30078	98594	09732	49898
42	06185	53730	84649	40715	25446	90966	40349	81553	36166	97485
43	96249	21285	85912	53370	13489	20214	81912	19510	03250	44944
44	60504	57017	38892	76889	05930	44207	60717	70562	74812	15672
45	27651	90466	80992	03306	75644	45917	30889	36652	75794	34054
46	64410	74350	60581	08972	26284	71779	84450	62976	36142	01946
47	95202	01751	86883	28174	38999	94490	49858	02425	40593	83850
48	78892	99777	18049	66117	78028	70955	75476	98203	01512	23591
49	12709	43956	61118	90635	09299	60528	80161	53155	90775	11483
50	62053	67152	49996	39482	68025	94097	83991	05527	48252	47713
51	19060	58444	64059	97800	16489	87323	05381	50739	06795	67614
52	93142	09643	63074	51502	27243	17788	80518	31654	37937	28985
53	53647	72844	72831	04314	34827	85166	06378	78712	57505	94858
54	74440	32656	13285	82425	29910	75178	48183	52694	82881	30463
55	00875	39143	30131	20830	55687	45897	51699	16687	14893	62227
56	92403	73577	01847	44960	01413	80761	94012	53683	01197	90235
57	73889	36517	94526	09074	79816	79249	68788	04445	12270	69892
58	73717	50576	62638	93857	05737	80588	02481	15325	64990	37749
59	42176	41558	40318	03808	77255	76219	80048	77046	06268	84857
60	03102	55200	52076	68954	36760	59891	34203	52293	14493	39429
61	43468	85530	17692	86887	43360	89660	09245	98723	49664	77925
62	27948	48317	12879	33738	08930	35252	88590	70673	53033	66886
63	56256	00738	97390	28189	50935	07846	65929	45204	96515	36730
64	73057	32866	18205	78341	16474	11094	45487	59284	89071	94574
65	04941	79671	06593	80638	10957	18261	91490	08417	39933	50915
66	33311	58411	97188	03098	06639	30810	65231	46429	95749	68249
67	96459	68595	13745	30662	02707	02666	01937	40702	75257	44362
68	29868	71432	07187	65108	35046	30404	69061	30546	87289	09953
69	27335	84350	60204	13434	07806	80533	07612	29126	58743	68703
70	13524	26818	44665	01502	56917	01550	60298	65433	08407	65654
71	30149	51222	45593	75282	11560	66997	30922	77893	21988	96037
72	67244	94022	05995	37383	42375	87856	69908	44703	92659	09319
73	96190	84676	29074	98926	60127	76094	26079	20548	22746	08079
74	64413	69462	40328	95740	80855	96865	94793	47329	73979	29327
75	90438	55215	07235	50602	69957	61746	05753	97097	44279	97163

Table II:

STANDARD NORMAL PROBABILITIES

The table reports the area to the left of the value z under the standard normal curve.

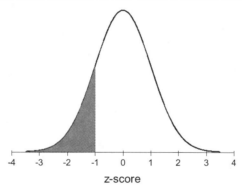

z-score

z	_0	_1	_2	_3	_4	_5	_6	_7	_8	_9
-3.4	.0003	.0003	.0003	.0003	.0003	.0003	.0003	.0003	.0003	.0002
-3.3	.0005	.0005	.0005	.0004	.0004	.0004	.0004	.0004	.0004	.0004
-3.2	.0007	.0007	.0006	.0006	.0006	.0006	.0006	.0005	.0005	.0005
-3.1	.0010	.0009	.0009	.0009	.0008	.0008	.0008	.0008	.0007	.0007
-3.0	.0014	.0013	.0013	.0012	.0012	.0011	.0011	.0011	.0010	.0010
-2.9	.0019	.0018	.0018	.0017	.0016	.0016	.0015	.0015	.0014	.0014
-2.8	.0026	.0025	.0024	.0023	.0023	.0022	.0021	.0021	.0020	.0019
-2.7	.0035	.0034	.0033	.0032	.0031	.0030	.0029	.0028	.0027	.0026
-2.6	.0047	.0045	.0044	.0043	.0041	.0040	.0039	.0038	.0037	.0036
-2.5	.0062	.0060	.0059	.0057	.0055	.0054	.0052	.0051	.0049	.0048
-2.4	.0082	.0080	.0078	.0075	.0073	.0071	.0069	.0068	.0066	.0064
-2.3	.0107	.0104	.0102	.0099	.0096	.0094	.0091	.0089	.0087	.0084
-2.2	.0139	.0136	.0132	.0129	.0125	.0122	.0119	.0116	.0113	.0110
-2.1	.0179	.0174	.0170	.0166	.0162	.0158	.0154	.0150	.0146	.0143
-2.0	.0228	.0222	.0217	.0212	.0207	.0202	.0197	.0192	.0188	.0183
-1.9	.0287	.0281	.0274	.0268	.0262	.0256	.0250	.0244	.0239	.0233
-1.8	.0359	.0351	.0344	.0336	.0329	.0322	.0314	.0307	.0301	.0294

z	_0	_1	_2	_3	_4	_5	_6	_7	_8	_9
-1.7	.0446	.0436	.0427	.0418	.0409	.0401	.0392	.0384	.0375	.0367
-1.6	.0548	.0537	.0526	.0516	.0505	.0495	.0485	.0475	.0465	.0455
-1.5	.0668	.0655	.0643	.0630	.0618	.0606	.0594	.0582	.0571	.0559
-1.4	.0808	.0793	.0778	.0764	.0749	.0735	.0721	.0708	.0694	.0681
-1.3	.0968	.0951	.0934	.0918	.0901	.0885	.0869	.0853	.0838	.0823
-1.2	.1151	.1131	.1112	.1093	.1075	.1057	.1038	.1020	.1003	.0985
-1.1	.1357	.1335	.1314	.1292	.1271	.1251	.1230	.1210	.1190	.1170
-1.0	.1587	.1562	.1539	.1515	.1492	.1469	.1446	.1423	.1401	.1379
-0.9	.1841	.1814	.1788	.1762	.1736	.1711	.1685	.1660	.1635	.1611
-0.8	.2119	.2090	.2061	.2033	.2005	.1977	.1949	.1922	.1894	.1867
-0.7	.2420	.2389	.2358	.2327	.2297	.2266	.2236	.2207	.2177	.2148
-0.6	.2743	.2709	.2676	.2643	.2611	.2578	.2546	.2514	.2483	.2451
-0.5	.3085	.3050	.3015	.2981	.2946	.2912	.2877	.2843	.2810	.2776
-0.4	.3446	.3409	.3372	.3336	.3300	.3264	.3228	.3192	.3156	.3121
-0.3	.3821	.3783	.3745	.3707	.3669	.3632	.3594	.3557	.3520	.3483
-0.2	.4207	.4168	.4129	.4090	.4052	.4013	.3974	.3936	.3897	.3859
-0.1	.4602	.4562	.4522	.4483	.4443	.4404	.4364	.4325	.4286	.4247
-0.0	.5000	.4960	.4920	.4880	.4840	.4801	.4761	.4721	.4681	.4641
0.0	.5000	.5040	.5080	.5120	.5160	.5199	.5239	.5279	.5319	.5359
0.1	.5398	.5438	.5478	.5517	.5557	.5596	.5636	.5675	.5714	.5753
0.2	.5793	.5832	.5871	.5910	.5948	.5987	.6026	.6064	.6103	.6141
0.3	.6179	.6217	.6255	.6293	.6331	.6368	.6406	.6443	.6480	.6517
0.4	.6554	.6591	.6628	.6664	.6700	.6736	.6772	.6808	.6844	.6879
0.5	.6915	.6950	.6985	.7019	.7054	.7088	.7123	.7157	.7190	.7224
0.6	.7257	.7291	.7324	.7357	.7389	.7422	.7454	.7486	.7517	.7549
0.7	.7580	.7611	.7642	.7673	.7704	.7734	.7764	.7794	.7823	.7852
0.8	.7881	.7910	.7939	.7967	.7995	.8023	.8051	.8079	.8106	.8133
0.9	.8159	.8186	.8212	.8238	.8264	.8289	.8315	.8340	.8365	.8389
1.0	.8413	.8438	.8461	.8485	.8508	.8531	.8554	.8577	.8599	.8621
1.1	.8643	.8665	.8686	.8708	.8729	.8749	.8770	.8790	.8810	.8830
1.2	.8849	.8869	.8888	.8907	.8925	.8944	.8962	.8980	.8997	.9015
1.3	.9032	.9049	.9066	.9082	.9099	.9115	.9131	.9147	.9162	.9177
1.4	.9192	.9207	.9222	.9236	.9251	.9265	.9279	.9292	.9306	.9319
1.5	.9332	.9345	.9357	.9370	.9382	.9394	.9406	.9418	.9429	.9441
1.6	.9452	.9463	.9474	.9484	.9495	.9505	.9515	.9525	.9535	.9545
1.7	.9554	.9564	.9573	.9582	.9591	.9599	.9608	.9616	.9625	.9633
1.8	.9641	.9649	.9656	.9664	.9671	.9678	.9686	.9693	.9699	.9706
1.9	.9713	.9719	.9726	.9732	.9738	.9744	.9750	.9756	.9761	.9767
2.0	.9773	.9778	.9783	.9788	.9793	.9798	.9803	.9808	.9812	.9817
2.1	.9821	.9826	.9830	.9834	.9838	.9842	.9846	.9850	.9854	.9857
2.2	.9861	.9864	.9868	.9871	.9875	.9878	.9881	.9884	.9887	.9890
2.3	.9893	.9896	.9898	.9901	.9904	.9906	.9909	.9911	.9913	.9916
2.4	.9918	.9920	.9922	.9925	.9927	.9929	.9931	.9932	.9934	.9936
2.5	.9938	.9940	.9941	.9943	.9945	.9946	.9948	.9949	.9951	.9952
2.6	.9953	.9955	.9956	.9957	.9959	.9960	.9961	.9962	.9963	.9964
2.7	.9965	.9966	.9967	.9968	.9969	.9970	.9971	.9972	.9973	.9974
2.8	.9974	.9975	.9976	.9977	.9977	.9978	.9979	.9979	.9980	.9981

z	_0	_1	_2	_3	_4	_5	_6	_7	_8	_9
2.9	.9981	.9982	.9983	.9983	.9984	.9984	.9985	.9985	.9986	.9986
3.0	.9987	.9987	.9987	.9988	.9988	.9989	.9989	.9989	.9990	.9990
3.1	.9990	.9991	.9991	.9991	.9992	.9992	.9992	.9992	.9993	.9993
3.2	.9993	.9993	.9994	.9994	.9994	.9994	.9994	.9995	.9995	.9995
3.3	.9995	.9995	.9996	.9996	.9996	.9996	.9996	.9996	.9996	.9997
3.4	.9997	.9997	.9997	.9997	.9997	.9997	.9997	.9997	.9997	.9998

Table III:

t-DISTRIBUTION CRITICAL VALUES

The table reports the critical value for which the area to the right is as indicated.

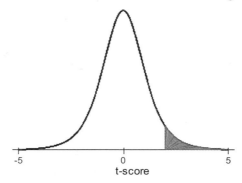

Area to right	0.2	0.1	0.05	0.025	0.01	0.005	0.001	0.0005
Conf. level	60%	80%	90%	95%	98%	99%	99.80%	99.90%
d.f.								
1	1.376	3.078	6.314	12.706	31.821	63.657	318.317	636.607
2	1.061	1.886	2.920	4.303	6.965	9.925	22.327	31.598
3	0.978	1.638	2.353	3.182	4.541	5.841	10.215	12.924
4	0.941	1.533	2.132	2.776	3.747	4.604	7.173	8.610
5	0.920	1.476	2.015	2.571	3.365	4.032	5.893	6.869
6	0.906	1.440	1.943	2.447	3.143	3.708	5.208	5.959
7	0.896	1.415	1.895	2.365	2.998	3.500	4.785	5.408
8	0.889	1.397	1.860	2.306	2.897	3.355	4.501	5.041
9	0.883	1.383	1.833	2.262	2.821	3.250	4.297	4.781
10	0.879	1.372	1.812	2.228	2.764	3.169	4.144	4.587
11	0.876	1.363	1.796	2.201	2.718	3.106	4.025	4.437
12	0.873	1.356	1.782	2.179	2.681	3.055	3.930	4.318
13	0.870	1.350	1.771	2.160	2.650	3.012	3.852	4.221
14	0.868	1.345	1.761	2.145	2.625	2.977	3.787	4.140

Area to right	0.2	0.1	0.05	0.025	0.01	0.005	0.001	0.0005
Conf. level	60%	80%	90%	95%	98%	99%	99.80%	99.90%
d.f.								
15	0.866	1.341	1.753	2.131	2.602	2.947	3.733	4.073
16	0.865	1.337	1.746	2.120	2.583	2.921	3.686	4.015
17	0.863	1.333	1.740	2.110	2.567	2.898	3.646	3.965
18	0.862	1.330	1.734	2.101	2.552	2.878	3.611	3.922
19	0.861	1.328	1.729	2.093	2.539	2.861	3.579	3.883
20	0.860	1.325	1.725	2.086	2.528	2.845	3.552	3.850
21	0.859	1.323	1.721	2.080	2.518	2.831	3.527	3.819
22	0.858	1.321	1.717	2.074	2.508	2.819	3.505	3.792
23	0.858	1.319	1.714	2.069	2.500	2.807	3.485	3.768
24	0.857	1.318	1.711	2.064	2.492	2.797	3.467	3.745
25	0.856	1.316	1.708	2.060	2.485	2.787	3.450	3.725
26	0.856	1.315	1.706	2.056	2.479	2.779	3.435	3.707
27	0.855	1.314	1.703	2.052	2.473	2.771	3.421	3.690
28	0.855	1.313	1.701	2.048	2.467	2.763	3.408	3.674
29	0.854	1.311	1.699	2.045	2.462	2.756	3.396	3.659
30	0.854	1.310	1.697	2.042	2.457	2.750	3.385	3.646
31	0.853	1.309	1.696	2.040	2.453	2.744	3.375	3.633
32	0.853	1.309	1.694	2.037	2.449	2.738	3.365	3.622
33	0.853	1.308	1.692	2.035	2.445	2.733	3.356	3.611
34	0.852	1.307	1.691	2.032	2.441	2.728	3.348	3.601
35	0.852	1.306	1.690	2.030	2.438	2.724	3.340	3.591
36	0.852	1.306	1.688	2.028	2.434	2.719	3.333	3.582
37	0.851	1.305	1.687	2.026	2.431	2.715	3.326	3.574
38	0.851	1.304	1.686	2.024	2.429	2.712	3.319	3.566
39	0.851	1.304	1.685	2.023	2.426	2.708	3.313	3.558
40	0.851	1.303	1.684	2.021	2.423	2.704	3.307	3.551
50	0.849	1.299	1.676	2.009	2.403	2.678	3.261	3.496
60	0.848	1.296	1.671	2.000	2.390	2.660	3.232	3.460
80	0.846	1.292	1.664	1.990	2.374	2.639	3.195	3.416
100	0.845	1.290	1.660	1.984	2.364	2.626	3.174	3.391
500	0.842	1.283	1.648	1.965	2.334	2.586	3.107	3.310
Infinity	0.842	1.282	1.645	1.960	2.326	2.576	3.090	3.291

Table IV:

CHI-SQUARE DISTRIBUTION CRITICAL VALUES

The table reports the critical value for which the area to the right is as indicated.

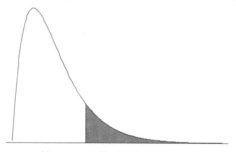

chi-square critical values

Area to right	0.2	0.1	0.05	0.025	.02	0.01	0.005	0.001	0.0005
d.f.									
1	1.64	2.71	3.84	5.02	5.41	6.63	7.88	10.83	12.12
2	3.22	4.61	5.99	7.38	7.82	9.21	10.60	13.82	15.20
3	4.64	6.25	7.81	9.35	9.84	11.34	12.84	16.27	17.73
4	5.99	7.78	9.49	11.14	11.67	13.28	14.86	18.47	20.00
5	7.29	9.24	11.07	12.83	13.39	15.09	16.75	20.52	22.11
6	8.56	10.64	12.59	14.45	15.03	16.81	18.55	22.46	24.10
7	9.80	12.02	14.07	16.01	16.62	18.48	20.28	24.32	26.02
8	11.03	13.36	15.51	17.53	18.17	20.09	21.96	26.12	27.87
9	12.24	14.68	16.92	19.02	19.68	21.67	23.59	27.88	29.67
10	13.44	15.99	18.31	20.48	21.16	23.21	25.19	29.59	31.42
11	14.63	17.28	19.68	21.92	22.62	24.73	26.76	31.26	33.14
12	15.81	18.55	21.03	23.34	24.05	26.22	28.30	32.91	34.82
13	16.98	19.81	22.36	24.74	25.47	27.69	29.82	34.53	36.48
14	18.15	21.06	23.68	26.12	26.87	29.14	31.32	36.12	38.11
15	19.31	22.31	25.00	27.49	28.26	30.58	32.80	37.70	39.72
16	20.47	23.54	26.30	28.85	29.63	32.00	34.27	39.25	41.31

Area to right	0.2	0.1	0.05	0.025	.02	0.01	0.005	0.001	0.0005
17	21.61	24.77	27.59	30.19	31.00	33.41	35.72	40.79	42.88
18	22.76	25.99	28.87	31.53	32.35	34.81	37.16	42.31	44.43
19	23.90	27.20	30.14	32.85	33.69	36.19	38.58	43.82	45.97
20	25.04	28.41	31.41	34.17	35.02	37.57	40.00	45.31	47.50
21	26.17	29.62	32.67	35.48	36.34	38.93	41.40	46.80	49.01
22	27.30	30.81	33.92	36.78	37.66	40.29	42.80	48.27	50.51
23	28.43	32.01	35.17	38.08	38.97	41.64	44.18	49.73	52.00
24	29.55	33.20	36.42	39.36	40.27	42.98	45.56	51.18	53.48
25	30.68	34.38	37.65	40.65	41.57	44.31	46.93	52.62	54.95
26	31.79	35.56	38.89	41.92	42.86	45.64	48.29	54.05	56.41
27	32.91	36.74	40.11	43.19	44.14	46.96	49.64	55.48	57.86
28	34.03	37.92	41.34	44.46	45.42	48.28	50.99	56.89	59.30
29	35.14	39.09	42.56	45.72	46.69	49.59	52.34	58.30	60.73
30	36.25	40.26	43.77	46.98	47.96	50.89	53.67	59.70	62.16
40	47.27	51.81	55.76	59.34	60.44	63.69	66.77	73.40	76.09
50	58.16	63.17	67.50	71.42	72.61	76.15	79.49	86.66	89.56
60	68.97	74.40	79.08	83.30	84.58	88.38	91.95	99.61	102.70
80	90.41	96.58	101.88	106.63	108.07	112.33	116.32	124.84	128.26
100	111.67	118.50	124.34	129.56	131.14	135.81	140.17	149.45	153.17

Appendix A:

STUDENT GLOSSARY

This "do-it-yourself" glossary is meant to provide you with an opportunity to organize definitions of key terms in one location. The following key terms are all marked in the text with **_bold italics_**. You should feel free to write definitions or properties of these terms in the glossary as you encounter them in the text. You can also record several topic numbers or activity numbers or page numbers for each key word, as each term may appear in more than one activity throughout the text.

Key Words	Reference	Notes
45° line		
alternative hypothesis		
anecdote		
association		
bar graph		
biased sampling		
binary		
binomial distribution		
blindness		
boxplot		
categorical		
causation		
center		
Central Limit Theorem		
comparison		
conditional distribution		

Key Words	Reference	Notes
confidence		
confidence interval		
confidence level		
confounding variable		
control		
control group		
controlled experiment		
correlation coefficient		
critical value		
data		
degrees of freedom		
density curve		
dependent variable		
distribution		
dotplot		
double blindness		
duality		
equally likely		
empirical rule		
expected value		
experiment		
explanatory variable		
extrapolation		
fitted value		
five-number summary		
granularity		
histogram		
independence		
independent variable		
influential observation		
intercept coefficient		
interquartile range		
labeled scatterplot		
least squares line		
lurking variable		

Key Words	Reference	Notes
margin of error		
marginal distribution		
matched pairs		
mean		
median		
mode		
modified boxplots		
normal distributions		
null hypothesis		
observational study		
observational unit		
outlier (regression)		
outlier test		
outliers		
p-value		
paired *t*-test		
parameter		
peaks/clusters		
placebo effect		
population		
practical significance		
prediction		
probability		
proportion of variability explained		
randomization		
range		
rate		
regression line		
relative frequency		
relative risk		
residual		
residual plot		
resistant		
response variable		

Key Words	Reference	Notes
robust		
sample		
sample size		
sample size determination		
sample space		
sampling		
sampling distribution		
sampling frame		
sampling variability		
scatterplot		
segmented bar graph		
shape		
side-by-side stemplot		
significance level		
simple random sampling		
Simpson's paradox		
skewed left		
skewed right		
slope coefficient		
standard deviation		
standard error		
standard normal distribution		
standard normal table		
standardization		
standardized score		
statistic		
statistical significance		
statistical tendency		
stemplot		
stratified sample		
symmetric		
systematic sample		
t distribution		
t-interval		

Key Words	Reference	Notes
t-test		
table of random digits		
tally		
test decision		
test of significance		
test statistic		
transformation		
two sample *t*-interval		
two sample *t*-test		
two sample *z*-interval		
two sample *z*-test		
two-sided test		
two-way table		
unbiased		
variability		
variable		
z-score		

Appendix B:

SOURCES FOR DATA SETS

Topic 1: Data and Variables

- *Activity 1-6:* The data on ages at which parents had their children are from the 1998 General Social Survey conducted by the National Opinion Research Center (www.norc.uchicago.edu).
- *Activity 1-12:* The study on natural light and achievement is from the December 21, 1999 issue of *The Tribune*, San Luis Obispo County.
- *Activity 1-13:* The study on children's television viewing is from "Reducing Children's Television Viewing to Prevent Obesity: A Randomized Controlled Trial," by Thomas N. Robinson, *Journal of the American Medical Association*, 282, 1999, pp. 1561–1567.

Topic 2: Data, Variables, and Calculators

- *Activity 2-1:* The names of famous statisticians are taken from *The Basic Practice of Statistics*, by David S. Moore, W.H. Freeman, 1996. Point values are from Scrabble brand crossword game, by Selchow and Righter.
- *Activity 2-2:* The data on gender of physicians are from *The 1999 World Almanac and Book of Facts*, p. 881.
- *Activity 2-3:* The fan cost index data are from Team Marketing Research.
- *Activity 2-4:* The data on SAT averages are from the South Carolina Information Highway (www.sciway.net/statistics/satstates96-98.html).
- *Activity 2-7:* The data on hazardousness of sports are from the publication *Injury Facts*, National Safety Council, 1999.
- *Activity 2-8:* The data on Broadway shows are from www.buybroadway.com, provided by the League of American Theatres and Producers.
- *Activity 2-9:* The data on movies are from www.showbizdata.com.

- *Activity 2-11:* The data on uninsured Americans are from *The 1999 World Almanac and Book of Facts*, p. 883.
- *Activity 2-13:* The data on drivers' ages and fatalities are from the publication *Traffic Safety Facts*, 1997, U.S. Department of Transportation National Highway Traffic Safety Administration.

Topic 3: Displaying and Describing Distributions

- *Activity 3-2:* The data on the 1999 Cal Poly football team are from the September 18, 1999 Official Game Program. The data on snowfall amounts are from *The 1992 Statistical Abstract of the United States*. The data on Major League Baseball games were compiled from www.espn.com by Jason Herr on July 26–August 8, 1999. The Monopoly data are from the Monopoly real estate trading game, by Parker Brothers. The data on mothers' ages are from the 1998 General Social Survey conducted by the National Opinion Research Center. The data on cars are from *Consumer Reports 1999 New Car Buying Guide*. The data on exam scores were compiled by one of the authors.
- *Activity 3-3:* The data on rowers' weights are from the NBC Olympic Web Site www.olympics.nbc.com/sports/rowing/events.html.
- *Activity 3-4:* The data on British rulers' reigns are from *The 1995 World Almanac and Book of Facts*, pp. 534–535.
- *Activity 3-5:* The data on geyser eruptions were originally reported in "A Look at Some Data on the Old Faithful Geyser," by A. Azzalini and A.W. Bowman, *Journal of the Royal Statistical Society, Series C*, 39, 1990, pp. 357–366.
- *Activity 3-6:* The car data are from *Consumer Reports 1999 New Car Buying Guide*.
- *Activity 3-8:* The placement exam scores are from a 1992 exam administered by the Dickinson College Department of Mathematics and Computer Science.
- *Activity 3-10:* The tuition data are from *The 1999 World Almanac and Book of Facts*, pp. 247–271.
- *Activity 3-13:* The marriage age data were compiled by Matthew Parks at the Cumberland County (PA) courthouse during June–July, 1993.
- *Activity 3-14:* The data on Hitchcock films were gathered at a Blockbuster Video store in Carlisle, PA, in January 1995.
- *Activity 3-15:* The data on dinosaur heights are from *Jurassic Park*, by Michael Crichton, Ballantine Books, 1990, p. 165.
- *Activity 3-16:* The data on tennis simulations are a random sample from the analysis in "Computer Simulation of a Handicap Scoring System for Tennis," by Allan Rossman and Matthew Parks, *Stats: The Magazine for Students of Statistics*, 10, 1993, pp. 14–18.

Topic 4: Measures of Center

- *Activity 4-1:* The data on Supreme Court justices are from *The 1999 World Almanac and Book of Facts*, p. 91.
- *Activity 4-4:* The data on cancer pamphlets and patients are from "Readability of Educational Materials for Cancer Patients," by Thomas H. Short, Helene Moriarty, and Mary Cooley, *Journal of Statistics Education*, 3, 1995.
- *Activity 4-5:* The data on planetary measurements are from *The 1993 World Almanac and Book of Facts*, p. 251.

Topic 5: Measures of Spread

- *Activity 5-1:* The data on city temperatures are from *The 1999 World Almanac and Book of Facts*, p. 220.
- *Activity 5-10:* The climate data are from Tables 368–375 of *The 1992 Statistical Abstract of the United States*.

Topic 6: Comparing Distributions I: Quantitative Variables

- *Activity 6-1:* The data on birth and death rates are from *The 1999 World Almanac and Book of Facts*, p. 874.
- *Activity 6-2:* The data on golfers' winnings are from www.espn.go.com/golfonline.
- *Activity 6-3:* The data on college degrees, marital status, and Mexican descent are from the 1990 U.S Census. The data on teen motherhood are from *The 1996 Statistical Abstract of the United States*.
- *Activity 6-4:* The data on the American Film Institute's list of top 100 American films are from their web site www.afionline.org.
- *Activity 6-12:* The data on female percentages are from the 1990 U.S. Census.
- *Activity 6-13:* The data on vehicle theft rates are from Table 209 of *The 1992 Statistical Abstract of the United States*.
- *Activity 6-14:* The video rental data were collected at a Blockbuster Video store in July of 1999 by participants in the SEQuaL workshop at Manheim Township (PA) School District.
- *Activity 6-16:* The data on lifetimes are from *The 1991 World Almanac and Book of Facts*, pp. 336–374.
- *Activity 6-18:* The voting data are from *The 1993 World Almanac and Book of Facts*, p. 73.

Topic 7: Comparing Distributions II: Categorical Variables

- *Activity 7-2:* The data on political interest are from the 1998 American National Election Study conducted by the Center for Political Studies of the Institute for Social Research (www.umich.edu/~nes/).
- *Activity 7-3:* The data on AZT and HIV are reported in the March 7, 1994 issue of *Newsweek.*
- *Activity 7-5:* The data on U.S. Senators are from *The 1999 World Almanac and Book of Facts*, pp. 80–81.
- *Activity 7-6:* The data concerning attitudes about suitability for politics are from the 1998 General Social Survey.
- *Activity 7-8:* The data on toy advertising were supplied by Dr. Pamela Rosenberg.
- *Activity 7-10:* The data on physicians' gender are from *The 1995 World Almanac and Book of Facts*, p. 966.
- *Activity 7-11:* The data on children's living arrangements are from *The 1995 World Almanac and Book of Facts*, p. 961.
- *Activity 7-12:* The data on baldness and heart disease are from "A Case-Control Study of Baldness in Relation to Myocardial Infarction in Men," by Samuel M. Lasko et al., *Journal of the American Medical Association*, 269, 1993, pp. 998–1003.
- *Activity 7-13:* The data on numbers of drivers in the age groups are from the publication *Traffic Safety Facts*, 1997, U.S. Department of Transportation National Highway Traffic Safety Administration.
- *Activity 7-14:* The data on gender and lung cancer were reported in the November 1, 1999 issue of *USA Today.*
- *Activity 7-20:* The data on graduate admissions are from "Is There Sex Bias in Graduate Admission?" by P.J. Bickel and J.W. O'Connell, *Science*, 187, pp. 398–404.

Topic 8: Graphical Displays of Association

- *Activity 8-4:* The space shuttle data are from "Lessons Learned from *Challenger*: A Statistical Perspective," by Siddhartha R. Dalal, Edward B. Folkes, and Bruce Hoadley, *Stats: The Magazine for Students of Statistics*, 2, 1989, pp. 14–18.
- *Activity 8-5:* The fast food data are from Arby's "Comprehensive Guide of Quality Ingredients" obtained in November 1999.
- *Activity 8-15:* The data on alumni donations to Harvey Mudd College are from Harvey Mudd College Alumni Office promotional material, September 1999.
- *Activity 8-16:* The data on peanut butter are from the September 1990 issue of *Consumer Reports.*
- *Activity 8-18:* The data on governor salaries and per capita income are from *The 1999 World Almanac and Book of Facts*, pp. 88, 635–660.

- *Activity 8-19:* The data on grocery prices were collected by Cal Poly students in San Luis Obispo (CA) in November 1999.

Topic 9: Correlation Coefficient

- *Activity 9-2:* The Monopoly data are from the Parker Brothers board game.
- *Activity 9-4:* The data on televisions and life expectancy are from *The 1993 World Almanac and Book of Facts*, pp. 727–817.
- *Activity 9-14:* The data on solitaire were collected by the authors in the summer of 1999.

Topic 10: Least Squares Regression I

- *Activity 10-1:* The data on airfares are from the January 8, 1995 issue of *The Harrisburg Sunday Patriot-News*. The data on distances are from the Delta Air Lines worldwide timetable guide effective December 15, 1994.
- *Activity 10-10:* The data on electricity bills were collected by one of the authors.

Topic 11: Least Squares Regression II

- *Activity 11-1:* The data on gestation and longevity are from *The 1993 World Almanac and Book of Facts*, p. 676.
- *Activity 11-12:* The data on college enrollments are a sample from *The 1991 World Almanac and Book of Facts*, pp. 214–239.

Topic 12: Sampling

- *Activity 12-1:* The data on the Elvis poll are reported in the August 18, 1989 issue of *The Harrisburg Patriot-News*. The data on the *Literary Digest* poll are from "Why the *Literary Digest* Poll Failed," by Peverill Squire, *Public Opinion Quarterly,* 52, 1988, pp. 125–133.
- *Activity 12-3:* The data on U.S. Senators are from *The 1999 World Almanac and Book of Facts*, pp. 80–81.
- *Activity 12-10:* The web addiction study was reported in the August 23, 1999 issue of the *Tampa Tribune.*
- *Activity 12-11:* The data on polls regarding emotional support are reported in *The Superpollsters,* by David W. Moore, Four Walls Eight Windows Publishers, p. 19.
- *Activity 12-12:* The Gallup poll results were reported in the August 20, 1998 issue of *The New York Times.*

- *Activity 12-13:* The data on alternative medicine are from a March 1994 issue of *Self* magazine.
- *Activity 12-14:* The data on courtroom cameras were reported in the October 4, 1994 issue of *The Harrisburg Evening-News.*
- *Activity 12-17:* The data from the phone book were gathered by the authors from the 1998–99 San Luis Obispo County (CA) telephone book in September of 1999.
- *Activity 12-20:* The self-reported election data are from the 1998 General Social Survey.

Topic 13: Designing Studies

- *Activity 13-4:* The Parkinson's disease study was reported in the February 4, 1999 issue of *The Boston Globe.* The placebo study was reported in the February 22, 1999 issue of *Time.*
- *Activity 13-9:* The heart attack study was reported in the October 12, 1999 issue of *USA Today.*
- *Activity 13-10:* The description of the psychology study and the data are from "The Trouble with Friendly Faces: Skilled Performance with a Supportive Audience," by Jennifer L. Butler and Roy F. Baumeister, *Journal of Personality and Social Psychology,* 75, 1998, pp. 1213–1230.
- *Activity 13-14:* The zinc nasal spray study was reported in the November 1, 1999 issue of *USA Today.*
- *Activity 13-15:* The study of religion and longevity was reported in the August 9, 1999 issue of *USA Today.*
- *Activity 13-21:* The parolee study and data are from the October 6, 1993 issue of *The New York Times,* p. B10.
- *Activity 13-22:* The parking meter study was reported in the September 8, 1998 issue of *The Los Angeles Times.* The study is elaborated on by Laurie Snell in edition 7.09 of the *Chance News* newsletter, available at www.dartmouth.edu/ ~chance/.
- *Activity 13-23:* The therapeutic touch study is from "A Close Look at Standards for Therapeutic Touch," by Linda Rosa et al., *Journal of the American Medical Association,* 279, 1998, pp. 1005–1010.

Topic 14: Probability

- *Activity 14-4:* The data on male/female births are from the *National Vital Statistics Report,* 47, 1999, available at www.cdc.gov/nchs/data/nvs47_18.pdf.

Topic 15: Normal Distributions

- *Activity 15-2:* The data on birth weights are from Table 43 of the *National Vital Statistics Report*, 47, 1999.
- *Activity 15-5:* The data on pregnancy durations are from Table 43 of the *National Vital Statistics Report*, 47, 1999.
- *Activity 15-11:* The data on heights are from Table 240 of *The 1998 Statistical Abstract of the United States*.
- *Activity 15-12:* The data on weights are from Table 241 of *The 1998 Statistical Abstract of the United States*.

Topic 16: Sampling Distributions I: Proportions

- *Activity 16-6:* The data on 1996 Presidential election voting percentages are from *The 1999 World Almanac and Book of Facts*, p. 462.
- *Activity 16-11:* The claim about calling heads is from *Statistics You Can't Trust*, by Steve Campbell, Think Twice Publishing, p. 158.
- *Activity 16-13:* The survey data about Halloween practices are from a Gallup Poll conducted on October 21–24, 1999, and released on October 29, 1999, at www.gallup.com.
- *Activity 16-14:* The survey data about Halloween beliefs are from a Gallup Poll conducted on October 21–24, 1999, and released on October 29, 1999, at www.gallup.com.

Topic 17: Sampling Distributions II: Means

- *Activity 17-3:* The survey data about Christmas shopping are from a Gallup Poll conducted on November 18–21, 1999, and released on November 26, 1999, at www.gallup.com.

Topic 18: Central Limit Theorem

- *Activity 18-1:* The smoking data are from the November 19, 1999 issue of *USA Today*.
- *Activity 18-14:* The data on non-English speakers are from *The 1995 World Almanac and Book of Facts*, p. 600.

Topic 19: Confidence Intervals I: Proportions

- *Activity 19-7:* The data on magazine advertisements were gathered by the authors in September 1999.

- *Activity 19-9*: The data on Major League Baseball games were compiled from www.espn.com by Jason Herr on July 26–August 8, 1999.
- *Activity 19-10*: The data on Major League Baseball games were compiled from www.espn.com by Jason Herr on July 26–August 8, 1999.
- *Activity 19-12*: The data on television characters were reported in the June 28, 1994 issue of *USA Today*.
- *Activity 19-13*: The data on cat households are from Table 431 of *The 1998 Statistical Abstract of the United States*.
- *Activity 19-14*: The data on charitable contributions are from Tables 639–640 of *The 1998 Statistical Abstract of the United States*.
- *Activity 19-16*: The data on newspaper reading are from the 1998 General Social Survey.
- *Activity 19-18*: The data on marital problems are from "Why Does Military Combat Experience Adversely Affect Marital Relations?" by Cynthia Gimbel and Alan Booth, *Journal of Marriage and the Family*, 56, 1994, pp. 691–703.

Topic 20: Confidence Intervals II: Means

- *Activity 20-8*: The data on sentence lengths were compiled by one of the authors from the opening sentences of chapter 3 of *The Testament*, by John Grisham, Doubleday, 1999.

Topic 21: Tests of Significance I: Proportions

- *Activity 21-2*: The "Ask Marilyn" column appeared in the May 3, 1998 issue of *Parade* magazine. The data were compiled from www.espn.com by Jason Herr on July 26–August 8, 1999.
- *Activity 21-3*: The campus legend is recounted in an "Ask Marilyn" column that appeared in the March 3, 1996 issue of *Parade* magazine. The story is elaborated upon by Laurie Snell in edition 5.04 of the *Chance News* newsletter, available at www.dartmouth.edu/~chance/.
- *Activity 21-8*: The baseball data were reported in edition 7.05 of the *Chance News* newsletter, available at www.dartmouth.edu/~chance/.
- *Activity 21-12*: The data on volunteer work are from Table 638 of *The 1998 Statistical Abstract of the United States*.
- *Activity 21-13*: The data on teacher hiring are reported in *Statistics for Lawyers*, by Michael O. Finkelstein and Bruce Levin, Springer-Verlag, 1990, pp. 161–162.

Topic 22: Tests of Significance II: Means

- *Activity 22-1:* The data on basketball scoring were compiled for NBA games played on December 10–12, 1999, by one of the authors from www.espn.com.

- *Activity 22-8:* The data on UFO sightings are from "Close Encounters: An Examination of UFO Experiences," by Nicholas P. Spanos et al., *Journal of Abnormal Psychology*, 102, 1993, pp. 624–632.
- *Activity 22-10:* The data on exam scores were compiled by one of the authors.

Topic 24: Comparing Two Proportions

- *Activity 24-3:* The survey data about personal attractiveness are from a Gallup Poll conducted on July 22–25, 1999 and released on September 15, 1999 at www.gallup.com.
- *Activity 24-8:* The data on campus alcohol habits are from "Boozing and Brawling on Campus: A National Study of Violent Problems Associated with Drinking over the Past Decade," by Ruth C. Enge and David J. Hanson, *Journal of Criminal Justice*, 22, 1994, pp. 171–180.
- *Activity 24-17:* The data on BAP are from "The Epidemiology of Bacillary Angiomatosis and Bacillary Peliosis," by Jordan W. Tappero et al., *Journal of the American Medical Association*, 269, 1993, pp. 770–775.
- *Activity 24-18:* The data on wording of questions are from Table 8.1 of *Questions and Answers in Attitude Surveys*, by Howard Schuman and Stanley Presser, Academic Press, 1981.
- *Activity 24-19:* The data on wording of questions are from Tables 11.2, 11.3, and 11.4 of *Questions and Answers in Attitude Surveys*, by Howard Schuman and Stanley Presser, Academic Press, 1981.
- *Activity 24-20:* The data on smoking policies were supplied by Janet Meyer.
- *Activity 24-21:* The data on employment discrimination are reported in *Statistics for Lawyers*, by Michael O. Finkelstein and Bruce Levin, Springer-Verlag, 1990, p. 123.
- *Activity 24-22:* The data on teenagers' smoking were reported in the October 4, 1994 issue of *The Harrisburg Patriot-News*.

Topic 25: Comparing Two Means

- *Activity 25-1:* The data on sentence lengths in Moore's writing are from the beginning of part 1 of *The Basic Practice of Statistics*, by David S. Moore, W.H. Freeman, 1996.
- *Activity 25-9:* The data on classroom attention are from "Learning What's Taught: Sex Differences in Instruction," by Gaea Leinhardt, Andrea Mar Seewald, and Mary Engel, *Journal of Educational Psychology*, 71, 1979, pp. 432–439.

Topic 26: Inference for Two-Way Tables

- *Activity 26-2:* The data on government spending are from the 1998 General Social Survey.
- *Activity 26-14:* The data on addict recovery were supplied by Betty Jarusiewicz.

Topic 27: Inference for Correlation and Regression

- *Activity 27-1:* The data on baseball payrolls are from the November 19, 1999 issue of *USA Today.*
- *Activity 27-2:* The data on studying and grades were gathered by a group of students at the University of the Pacific.

Index

Distributed by Winterset Water
800-617-3676•Hebron, OH 43025
Source: Aristocrat Spring, Saegertown, PA Permit #200

Lutheran Northwest

Home of the Crusaders!

Thank You For Supporting Lutheran Northwest Students!

NATURAL SPRING WATER

www.lhnw.lhsa.com